Student Solutions Guide
and
Technology Guide

for

FUNCTIONS AND CHANGE:
A Modeling Approach
to College Algebra

Third Edition

Bruce Crauder   Benny Evans   Alan Noell

Oklahoma State University

Produced with the aid of the National Science Foundation

HOUGHTON MIFFLIN COMPANY   BOSTON   NEW YORK

Senior Sponsoring Editor: Leonid Tunik
Associate Editor: Melissa Parkin
Manufacturing Coordinator: Brian Pieragostino
Senior Marketing Manager: Jennifer Jones

Printed in the U.S.A.

ISBN 13: 978-0-618-64303-5
ISBN 10: 0-618-64303-6

3 4 5 6 7 8 9-POO-10 09 08 07

Contents

Solution Guide for Prologue: Calculator Arithmetic

CALCULATOR ARITHMETIC

E-1. **Order of operations when no parentheses are present**: We use the four rules for the order of operations.

(a) According to rule 2, to evaluate $2 + 3 \times 4 + 5$ we should perform the multiplication before we evaluate the sums:

$$2 + \mathbf{3 \times 4} + 5 = 2 + \mathbf{12} + 5 = 19.$$

(b) According to rule 2, to evaluate $6 \div 3 \times 3 \wedge 2$ we should calculate the exponential first. This gives

$$6 \div 3 \times \mathbf{3 \wedge 2} = 6 \div 3 \times \mathbf{9}.$$

Then we should perform the multiplication and division, working from left to right. This gives

$$\mathbf{6 \div 3} \times 9 = \mathbf{2} \times 9 = 18.$$

(c) According to rule 2, to evaluate $3 \times 16 \div 2 \wedge 2$ we should calculate the exponential first. This gives

$$3 \times 16 \div \mathbf{2 \wedge 2} = 3 \times 16 \div \mathbf{4}.$$

Then we should perform the multiplication and division, working from left to right. This gives

$$\mathbf{3 \times 16} \div 4 = \mathbf{48} \div 4 = 12.$$

(d) According to rule 2, to evaluate $5 - 4 \times 3 + 8 \div 2 \times 3$ we should perform the multiplications and the division before performing the addition and the subtraction. We work from left to right. First we have

$$5 - \mathbf{4 \times 3} + 8 \div 2 \times 3 = 5 - \mathbf{12} + 8 \div 2 \times 3.$$

Then we have

$$5 - 12 + \mathbf{8 \div 2} \times 3 = 5 - 12 + \mathbf{4} \times 3.$$

Finally we have

$$5 - 12 + \mathbf{4 \times 3} = 5 - 12 + \mathbf{12} = 5.$$

S-1. **Basic calculations**: In typewriter notation, $\dfrac{2.6 \times 5.9}{6.3}$ is $(2.6 \times 5.9) \div 6.3$, which equals 2.434... and so is rounded to two decimal places as 2.43.

S-3. **Basic calculations**: In typewriter notation, $\dfrac{e}{\sqrt{\pi}}$ is $e \div (\sqrt{}(\pi))$, which equals 1.533... and so is rounded to two decimal places as 1.53.

S-5. **Parentheses and grouping**: When we add parentheses, $\dfrac{7.3 - 6.8}{2.5 + 1.8}$ becomes $\dfrac{(7.3 - 6.8)}{(2.5 + 1.8)}$, which, in typewriter notation, becomes $(7.3 - 6.8) \div (2.5 + 1.8)$. This equals 0.116... and so is rounded to two decimal places as 0.12.

S-7. **Parentheses and grouping**: When we add parentheses, $\dfrac{\sqrt{6 + e} + 1}{3}$ becomes $\dfrac{(\sqrt{}(6 + e) + 1)}{3}$, which, in typewriter notation, becomes $(\sqrt{}(6 + e) + 1) \div 3$. This equals 1.317... and so is rounded to two decimal places as 1.32.

S-9. **Subtraction versus sign**: Noting which are negative signs and which are subtraction signs, we see that $\dfrac{-3}{4 - 9}$ means $\dfrac{negative\ 3}{4\ subtract\ 9}$. Adding parentheses and putting it into typewriter notation yields *negative* $3 \div (4\ subtract\ 9)$, which equals 0.6.

S-11. **Subtraction versus sign**: Noting which are negative signs and which are subtraction signs, we see that $-\sqrt{8.6 - 3.9}$ means *negative* $\sqrt{8.6\ subtract\ 3.9}$. In typewriter notation this is

$$negative\ \sqrt{}(8.6\ subtract\ 3.9),$$

which equals $-2.167...$ and so is rounded to two decimal places as -2.17.

S-13. **Chain calculations**:

(a) To do this as a chain calculation, we first calculate $\dfrac{3}{7.2 + 5.9}$ and then complete the calculation by adding the second fraction to this first answer. In typewriter notation $\dfrac{3}{7.2 + 5.9}$ is $3 \div (7.2 + 5.9)$, which is calculated as 0.2290076336; this is used as *Ans* in the next part of the calculation. Turning to the full expression, we calculate it as *Ans* $+ \dfrac{7}{6.4 \times 2.8}$ which is, in typewriter notation, *Ans* $+ 7 \div (6.4 \times 2.8)$. This is 0.619..., which rounds to 0.62.

(b) To do this as a chain calculation, we first calculate the exponent, $1 - \dfrac{1}{36}$, and then the full expression becomes

$$\left(1 + \dfrac{1}{36}\right)^{Ans}.$$

In typewriter notation, the first calculation is $1 - 1 \div 36$, and the second is $(1 + 1 \div 36) \wedge Ans$. This equals 1.026... and so is rounded to two decimal places as 1.03.

S-15. **Evaluate expression**: In typewriter notation, $e^{-3} - \pi^2$ is $e \wedge (negative\ 3) - \pi \wedge 2$, which equals $-9.819\ldots$ and so is rounded to two decimal places as -9.82.

1. **Arithmetic**: We will show each calculation in typewriter notation and give the numerical answer rounded to two decimal places. We will use $-$ to denote subtraction and *negative* to denote a minus sign.

   (a) $(4.3 + 8.6)(8.4 - 3.5) = 63.21$

   (b) $(2 \wedge 3.2 - 1) \div (\sqrt{(3)} + 4) = 1.43$

   (c) $\sqrt{(2 \wedge \textit{negative } 3 + e)} = 1.69$

   (d) $(2 \wedge \textit{negative } 3 + \sqrt{(7)} + \pi)(e \wedge 2 + 7.6 \div 6.7) = 50.39$

   (e) $(17 \times 3.6) \div (13 + 12 \div 3.2) = 3.65$

3. **A bad investment**: The total value of your investment today is:

$$\text{Original investment} - 7\% \text{ loss} = 720 - 0.07 \times 720 = \$669.60.$$

5. **Pay raise**: The percent pay raise is obtained from

$$\frac{\text{Amount of raise}}{\text{Original hourly pay}}.$$

The raise was $7.50 - 7.25 = 0.25$ dollar while the original hourly pay is \$7.25, so the fraction is $\frac{0.25}{7.25} = 0.0345$. Thus we have received a raise of 3.45%.

7. **Trade discount**:

   (a) The cost price is $9.99 - 40\% \times 9.99 = 5.99$ dollars.

   (b) The difference between the suggested retail price and the cost price is $65.00 - 37.00 = 28.00$ dollars. We want to determine what percentage of \$65 this difference represents. We find the percentage by division: $\frac{28.00}{65.00} = 0.4308$ or 43.08%. This is the trade discount used.

9. **Present value**: We are given that the future value is \$5000 and that $r = 0.12$. Thus the present value is

$$\frac{\text{Future value}}{1 + r} = \frac{5000}{1 + 0.12} = 4464.29 \text{ dollars.}$$

11. **The Rule of 72:**

   (a) The Rule of 72 says our investment should double in

   $$\frac{72}{\% \text{ interest rate}} = \frac{72}{13} = 5.54 \text{ years.}$$

   (b) Using Part (a), the future value interest factor is

   $$(1 + \text{ interest rate})^{\text{years}} = (1 + 0.13)^{5.54} = 1.97.$$

   This is less than the doubling future value interest factor of 2.

   (c) Using our value from Part (c), the future value of a $5000 investment is

   $$\text{Original investment} \times \text{ future value interest factor } = 5000 \times 1.97 = \$9850.$$

   So our investment did not exactly double using the Rule of 72.

13. **The size of the Earth**:

   (a) The equator is a circle with a radius of approximately 4000 miles. The distance around the equator is its circumference, which is

   $$2\pi \times \text{ radius } = 2\pi \times 4000 = 25{,}132.74 \text{ miles,}$$

   or approximately 25,000 miles.

   (b) The volume of the Earth is

   $$\frac{4}{3}\pi \times \text{ radius }^3 = \frac{4}{3}\pi \times 4000^3 = 268{,}082{,}573{,}100 \text{ cubic miles.}$$

   Note that the calculator gives 2.680825731E11, which is the way the calculator writes numbers in scientific notation. It means $2.680825731 \times 10^{11}$ and should be written as such. That is about 268 billion cubic miles or 2.68×10^{11} cubic miles.

   (c) The surface area of the Earth is about

   $$4\pi \times \text{ radius }^2 = 4\pi \times 4000^2 = 201{,}061{,}929.8 \text{ square miles,}$$

   or approximately 201,000,000 square miles.

15. **The length of the Earth's orbit:**

(a) If the orbit is a circle then its circumference is the distance traveled. That circumference is

$$2\pi \times \text{ radius } = 2\pi \times 93 = 584.34 \text{ million miles},$$

or about 584 million miles. This can also be calculated as

$$2\pi \times \text{ radius } = 2\pi \times 93,000,000 = 584,336,233.6 \text{ miles}.$$

(b) Velocity is distance traveled divided by time elapsed. The velocity is given by

$$\frac{\text{Distance traveled}}{\text{Time elapsed}} = \frac{584.34 \text{ million miles}}{1 \text{ year}} = 584.34 \text{ million miles per year},$$

or about 584 million miles per year. This can also be calculated as

$$\frac{584,336,233.6 \text{ miles}}{1 \text{ year}} = 584,336,233.6 \text{ miles per year}.$$

(c) There are 24 hours per day and 365 days per year. So there are $24 \times 365 = 8760$ hours per year.

(d) The velocity in miles per hour is

$$\frac{\text{Miles traveled}}{\text{Hours elapsed}} = \frac{584.34}{8760} = 0.0667 \text{ million miles per hour}.$$

This is approximately 67,000 miles per hour. This can also be calculated as

$$\frac{\text{Miles traveled}}{\text{Hours elapsed}} = \frac{584,336,233.6}{8760} = 66,705.05 \text{ miles per hour}.$$

17. **Newton's second law of motion:** A man with a mass of 75 kilograms weighs $75 \times 9.8 = 735$ newtons. In pounds this is 735×0.225, or about 165.38.

19. **Frequency of musical notes:** The frequency of the next higher note than middle C is $261.63 \times 2^{1/12}$, or about 277.19 cycles per second. The D note is one note higher, so its frequency in cycles per second is

$$(261.63 \times 2^{1/12}) \times 2^{1/12},$$

or about 293.67.

21. **Lean body weight in females:** The lean body weight of a young adult female who weighs 132 pounds and has wrist diameter of 2 inches, abdominal circumference of 27 inches, hip circumference of 37 inches, and forearm circumference of 7 inches is

$$19.81 + 0.73 \times 132 + 21.2 \times 2 - 0.88 \times 27 - 1.39 \times 37 + 2.43 \times 7 = 100.39 \text{ pounds}.$$

It follows that her body fat weighs $132 - 100.39 = 31.61$ pounds. To compute the body fat percent we calculate $\frac{31.61}{132}$ and find 23.95%.

Prologue Review Exercises

1. **Parentheses and grouping**: In typewriter notation, $\dfrac{5.7 + 8.3}{5.2 - 9.4}$ is $(5.7 + 8.3) \div (5.2 - 9.4)$, which equals $-3.333\ldots$ and so is rounded to two decimal places as -3.33.

2. **Evaluate expression**: In typewriter notation, $\dfrac{8.4}{3.5 + e^{-6.2}}$ is $8.4 \div (3.5 + e\wedge(\text{ negative } 6.2))$, which equals $2.398\ldots$ and so is rounded to two decimal places as 2.40.

3. **Evaluate expression**: In typewriter notation, $\left(7 + \dfrac{1}{e}\right)^{\left(\frac{5}{2+\pi}\right)}$ is $(7 + 1 \div e) \wedge (5 \div (2 + \pi))$, which equals $6.973\ldots$ and so is rounded to two decimal places as 6.97. This can also be done as a chain calculation.

4. **Gas mileage**: The number of gallons required to travel 27 miles is

$$g = \frac{27}{15} = 1.8 \text{ gallons.}$$

The number of gallons required to travel 250 miles is

$$g = \frac{250}{15} = 16.67 \text{ gallons.}$$

5. **Kepler's third law**: The mean distance from Pluto to the sun is

$$D = 93 \times 249^{2/3} = 3680.86 \text{ million miles,}$$

or about 3681 million miles. For Earth we have $P = 1$ year, and the mean distance is

$$D = 93 \times 1^{2/3} = 93 \text{ million miles.}$$

6. **Traffic signal**: If the approach speed is 80 feet per second then the length of the yellow light should be

$$n = 1 + \frac{80}{30} + \frac{100}{80} = 4.92 \text{ seconds.}$$

Solution Guide for Chapter 1: Functions

1.1 FUNCTIONS GIVEN BY FORMULAS

E-1. **Determining when a correspondence is a function:**

(a) The correspondence assigns to each element of the set D exactly one element of the set R, so it defines a function f with domain D and range R.

(b) The correspondence assigns to each element of the set D exactly one element of the set R, so it defines a function f with domain D and range R.

(c) The correspondence assigns to the element 1 both 8 and 5, so it does not define a function.

(d) The correspondence assigns to each element of the set D exactly one element of the set R, so it defines a function f with domain D and range R.

E-3. **Functions on other sets:**

(a) This defines a function because each president has exactly one last name.

(b) This does not define a function because there are some last names (such as Johnson) shared by different presidents. Another reason that this fails to be a function is that, at the time this text was written, not all last names were represented by presidents.

(c) This does not define a function because there are some automobiles for which more than one color appears on the body.

E-5. **Finding the range:**

(a) The smallest range is the set of all numbers of the form $x + 2$ for some real number x. Because every real number is of this form (just subtract 2 to find x), the smallest range is the set of all real numbers.

(b) The smallest range is the set of all numbers of the form x^2 for some real number x. A number of that form is nonnegative, and every nonnegative number can be written as the square of either of its square roots. Thus the smallest range is the set of all nonnegative real numbers.

(c) The smallest range is the set of all numbers of the form x^3 for some real number x. Every number can be written as the cube of its cube root, so the smallest range is the set of all real numbers.

(d) The smallest range is the set of all numbers of the form $x^8 + 7$ for some real number x. A number of that form is greater than or equal to 7. Furthermore, every number greater than or equal to 7 can be written in this form: Simply subtract 7 and take an eighth root to find x. Thus the smallest range is the set of all real numbers greater than or equal to 7.

E-7. **Onto functions**:

(a) This function is onto: Given a number y in the range, y is a positive even integer, so $\frac{y}{2}$ is a positive integer and $f\left(\frac{y}{2}\right) = 2\frac{y}{2} = y$.

(b) This function is not onto: Because x^8 is nonnegative when x is a real number, -1 is not a function value. (In fact, no negative number is a function value.)

(c) This function is not onto: Because $\dfrac{1}{x^2 + 1}$ is nonnegative when x is a real number, -1 is not a function value. (In fact, all function values are greater than 0 and less than or equal to 1.)

(d) This function is onto because every element of the range is a function value.

(e) This function is not onto because the element 2 of the range is not a function value.

E-9. **Bijections**:

(a) First, this function is one-to-one: Assume that $f(x) = f(y)$. Then $2x = 2y$, and thus $x = y$ (divide both sides by 2). This shows that if $x \neq y$ then $f(x) \neq f(y)$. Second, we saw from Part (a) of Exercise E-7 that this function is onto. Thus it is a bijection.

(b) This function is not a bijection because (by Part (b) of Exercise E-7) it is not onto.

(c) This function is not a bijection because (by Part (c) of Exercise E-7) it is not onto.

(d) First, this function is one-to-one because no two distinct points in the domain are assigned to the same element of the range. Second, we saw from Part (d) of Exercise E-7 that this function is onto. Thus it is a bijection.

(e) This function is not a bijection because (by Part (e) of Exercise E-7) it is not onto. Another reason that this fails to be a bijection is that it is not one-to-one (because $f(1) = f(3)$).

S-1. **Evaluating formulas**: To evaluate $f(x) = \dfrac{\sqrt{x+1}}{x^2+1}$ at $x = 2$, simply substitute 2 for x. Thus the value of f at 2 is $\dfrac{\sqrt{2+1}}{2^2+1}$, which equals 0.346... and so is rounded to 0.35.

S-3. **Evaluating formulas**: To evaluate $g(x,y) = \dfrac{x^3+y^3}{x^2+y^2}$ at $x = 4.1$, $y = 2.6$, simply substitute 4.1 for x and 2.6 for y. Thus the value of g when $x = 4.1$ and $y = 2.6$ is $\dfrac{4.1^3+2.6^3}{4.1^2+2.6^2}$, which equals 3.669... and so is rounded to 3.67.

S-5. **Getting function values**: To get the function value $f(6.1)$, substitute 6.1 for s in the formula $f(s) = \dfrac{s^2+1}{s^2-1}$. Thus $f(6.1) = \dfrac{6.1^2+1}{6.1^2-1}$, which equals 1.055... and so is rounded to 1.06.

S-7. **Evaluating functions of several variables**: To get the function value $h(3, 2.2, 9.7)$, substitute 3 for x, 2.2 for y, and 9.7 for z in the formula $h(x,y,z) = \dfrac{x^y}{z}$. Thus $h(3, 2.2, 9.7) = \dfrac{3^{2.2}}{9.7}$, which equals 1.155... and so is rounded to 1.16.

S-9. **Using formulas**: To express the cost of buying 2 bags of potato chips, 3 sodas, and 5 hot dogs, note that these values correspond to $p = 2$, $s = 3$, and $h = 5$, and so the cost is expressed by $c(2, 3, 5)$.

S-11. **Practicing calculations with formulas**: In each case, we want to find the value of $f(3)$, given the formula for $f(x)$, so we simply substitute 3 for x:

(a) $f(3) = 3 \times 3 + \dfrac{1}{3}$, which equals 9.333... and so is rounded to 9.33.

(b) $f(3) = 3^{-3} - \dfrac{3^2}{3+1}$, which equals $-2.212...$ and so is rounded to -2.21.

(c) $f(3) = \sqrt{2 \times 3 + 5}$, which equals 3.316... and so is rounded to 3.32.

S-13. **Evaluating functions of several variables**: To evaluate $M = P(e^r - 1)/(1 - e^{-rt})$ at $r = 0.1$, $P = 8300$, and $t = 24$, substitute these values in the formula. The value is $8300(e^{0.1} - 1)/(1 - e^{-0.1 \times 24})$, which equals 960.01.

1. **Tax owed**:

(a) In functional notation the tax owed on a taxable income of $13,000 is $T(13,000)$. The value is

$$T(13,000) = 0.11 \times 13,000 - 500 = 930 \text{ dollars}.$$

(b) The tax owed on a taxable income of $14,000 is

$$T(14,000) = 0.11 \times 14,000 - 500 = 1040 \text{ dollars}.$$

Using the answer to Part (a), we see that the tax increases by $1040 - 930 = 110$ dollars.

(c) The tax owed on a taxable income of $15,000 is

$$T(15,000) = 0.11 \times 15,000 - 500 = 1150 \text{ dollars.}$$

Thus the tax increases by $1150 - 1040 = 110$ dollars again.

3. **Flying ball**:

(a) In functional notation the velocity 1 second after the ball is thrown is $V(1)$. The value is

$$V(1) = 40 - 32 \times 1 = 8 \text{ feet per second.}$$

Because the upward velocity is positive, the ball is rising.

(b) The velocity 2 seconds after the ball is thrown is

$$V(2) = 40 - 32 \times 2 = -24 \text{ feet per second.}$$

Because the upward velocity is negative, the ball is falling.

(c) The velocity 1.25 seconds after the ball is thrown is

$$V(1.25) = 40 - 32 \times 1.25 = 0 \text{ feet per second.}$$

Because the velocity is 0, we surmise from Parts (a) and (b) that the ball is at the peak of its flight.

(d) Using the answers to Parts (a) and (b), we see that from 1 second to 2 seconds the velocity changes by

$$V(2) - V(1) = -24 - 8 = -32 \text{ feet per second.}$$

Because

$$V(3) = 40 - 32 \times 3 = -56 \text{ feet per second,}$$

from 2 seconds to 3 seconds the velocity changes by

$$V(3) - V(2) = -56 - (-24) = -32 \text{ feet per second.}$$

Because

$$V(4) = 40 - 32 \times 4 = -88 \text{ feet per second,}$$

from 3 seconds to 4 seconds the velocity changes by

$$V(4) - V(3) = -88 - (-56) = -32 \text{ feet per second.}$$

Over each of these 1-second intervals the velocity changes by -32 feet per second. In practical terms, this means that the velocity decreases by 32 feet per second for each second that passes. This indicates that the *downward* acceleration of the ball is constant at 32 feet per second per second, which makes sense because the acceleration due to gravity is constant near the surface of Earth.

5. **A population of deer:**

(a) Now $N(0)$ represents the number of deer initially on the reserve and

$$N(0) = \frac{12.36}{0.03 + 0.55^0} = 12 \text{ deer.}$$

So there were 12 deer in the initial herd.

(b) We calculate using

$$N(10) = \frac{12.36}{0.03 + 0.55^{10}} = 379.92 \text{ deer.}$$

This says that after 10 years there should be about 380 deer in the reserve.

(c) The number of deer in the herd after 15 years is represented by $N(15)$, and this value is

$$N(15) = \frac{0.36}{0.03 + 0.55^{15}} = 410.26 \text{ deer.}$$

This says that there should be about 410 deer in the reserve after 15 years.

(d) The difference in the deer population from the tenth to the fifteenth year is given by $N(15) - N(10) = 410.26 - 379.92 = 30.34$. Thus the population increased by about 30 deer from the tenth to the fifteenth year.

7. **Radioactive substances:**

(a) The amount of carbon 14 left after 800 years is expressed in functional notation as $C(800)$. This is calculated as

$$C(800) = 5 \times 0.5^{800/5730} = 4.54 \text{ grams.}$$

(b) There are many ways to do this part of the exercise. The simplest is to note that half the amount is left when the exponent of 0.5 is 1 since then the 5 is multiplied by $0.5 = \frac{1}{2}$. The exponent in the formula is 1 when $t = 5730$ years.

Another way to do this part is to experiment with various values for t, increasing the value when the answer is less than 2.5 and decreasing it when the answer comes out more than 2.5. Students are in fact discovering and executing a crude version of the bisection method.

9. **What if interest is compounded more often than monthly?**

    (a) We would expect our monthly payment to be higher if the interest is compounded daily since additional interest is charged on interest which has been compounded.

    (b) Continuous compounding should result in a larger monthly payment since the interest is compounded at an even faster rate than with daily compounding.

    (c) We are given that $P = 7800$ and $t = 48$. Because the APR is 8.04% or 0.0804, we compute that

    $$r = \frac{\text{APR}}{12} = \frac{0.0804}{12} = 0.0067.$$

    Thus the monthly payment is

    $$M(7800, 0.0067, 48) = \frac{7800(e^{0.0067} - 1)}{1 - e^{-0.0067 \times 48}} = 190.67 \text{ dollars.}$$

    Our monthly payment here is 10 cents higher than if interest is compounded monthly as in Example 1.2 (where the payment was $190.57).

11. **How much can I borrow?**

    (a) Since we will be paying $350 per month for 4 years, then we will be making 48 payments, or $t = 48$. Also, r is the monthly interest rate of 0.75%, or 0.0075 as a decimal. The amount of money we can afford to borrow in this case is given in functional notation by $P(350, 0.0075, 48)$. It is calculated as

    $$P(350, 0.0075, 48) = 350 \times \frac{1}{0.0075} \times \left(1 - \frac{1}{(1 + 0.0075)^{48}}\right) = \$14{,}064.67.$$

    (b) If the monthly interest rate is 0.25% then we can afford to borrow

    $$P(350, 0.0025, 48) = 350 \times \frac{1}{0.0025} \times \left(1 - \frac{1}{(1 + 0.0025)^{48}}\right) = \$15{,}812.54.$$

    (c) If we make monthly payments over 5 years then we will make 60 payments in all. So now we can afford to borrow

    $$P(350, 0.0025, 60) = 350 \times \frac{1}{0.0025} \times \left(1 - \frac{1}{(1 + 0.0025)^{60}}\right) = \$19{,}478.33.$$

13. **Brightness of stars**: Here we have $m_1 = -1.45$ and $m_2 = 2.04$. Thus

    $$t = 2.512^{m_2 - m_1} = 2.512^{2.04 - (-1.45)} = 2.512^{3.49} = 24.89.$$

    Hence Sirius appears 24.89 times brighter than Polaris.

15. **Parallax**: We are given that $p = 0.751$. Thus the distance from Alpha Centauri to the sun is about

    $$d(0.751) = \frac{3.26}{0.751} = 4.34 \text{ light-years.}$$

17. **Mitscherlich's equation**:

(a) We are given that $b = 1$. Thus the percentage (as a decimal) of maximum yield is

$$Y(1) = 1 - 0.5^1 = 0.5.$$

Hence 50% of maximum yield is produced if 1 baule is applied.

(b) In functional notation the percentage (as a decimal) of maximum yield produced by 3 baules is $Y(3)$. The value is

$$Y(3) = 1 - 0.5^3 = 0.875,$$

or about 0.88. This is 88% of maximum yield.

(c) Now 500 pounds of nitrogen per acre corresponds to $\dfrac{500}{223}$ baules, so the percentage (as a decimal) of maximum yield is $1 - 0.5^{500/223}$, or about 0.79. This is 79% of maximum yield.

19. **Thermal conductivity**: We are given that $k = 0.85$ for glass and that $t_1 = 24$, $t_2 = 5$.

(a) Because $d = 0.007$, the heat flow is

$$Q = \frac{0.85(24 - 5)}{0.007} = 2307.14 \text{ watts per square meter.}$$

(b) The total heat loss is

$$\text{Heat flow} \times \text{Area of window} = 2307.14 \times 2.5,$$

or about 5767.85 watts.

21. **Fault rupture length**: Here we have $M = 6.5$, so the expected length is

$$L(6.5) = 0.0000017 \times 10.47^{6.5} = 7.25 \text{ kilometers.}$$

23. **Research project**: Answers will vary.

1.2 FUNCTIONS GIVEN BY TABLES

E-1. **Average rates of change using numbers:**

(a) The average rate of change is

$$\frac{f(3) - f(2)}{3 - 2} = \frac{27 - 8}{1} = 19.$$

(b) The average rate of change is

$$\frac{f(5) - f(2)}{5 - 2} = \frac{19 - 10}{3} = 3.$$

(c) The average rate of change is

$$\frac{f(4) - f(2)}{4 - 2} = \frac{\frac{1}{4} - \frac{1}{2}}{2} = -\frac{1}{8}.$$

(d) The average rate of change is

$$\frac{f(3) - f(1)}{3 - 1} = \frac{12 - 2}{2} = 5.$$

E-3. **Linear functions:** If $f(x) = mx + b$ then the average rate of change is

$$\frac{f(q) - f(p)}{q - p} = \frac{(mq + b) - (mp + b)}{q - p} = \frac{mq - mp}{q - p} = \frac{m(q - p)}{q - p} = m.$$

Thus the average rate of change is m, and this does not depend on either p or q.

E-5. **Sketching a graph:** Because the average rate of change from $x = 0$ to $x = 4$ is 2, the change in f over this interval is $2 \times (4 - 0) = 8$. Thus $f(4) - f(0) = 8$. Because $f(0) = 1$, we have $f(4) = 8 + 1 = 9$. Also, because the average rate of change from $x = 4$ to $x = 6$ is -1, the change in f over this interval is $-1 \times (6 - 4) = -2$. Thus $f(6) - f(4) = -2$. Because we saw that $f(4) = 9$, we have $f(6) = -2 + 9 = 7$. Thus the three function values determined are $f(0) = 1$, $f(4) = 9$, and $f(6) = 7$. Any graph exhibiting these three values is correct. Here is one possible graph:

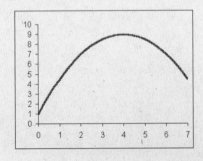

E-7. **The effect of adding a linear function**: The average rate of change for g is

$$\frac{g(b) - g(a)}{b - a} = \frac{(f(b) + 3b + 5) - (f(a) + 3a + 5)}{b - a}$$

$$= \frac{(f(b) - f(a)) + 3(b - a)}{b - a} = \frac{f(b) - f(a)}{b - a} + 3.$$

The first term on the far right is the average rate of change for f, so its value is 7; thus the average rate of change for g is $7 + 3 = 10$.

E-9. **Central difference quotients**: The average rate of change from $x = 2$ to $x = 3$ is

$$\frac{f(3) - f(2)}{3 - 2} = \frac{12 - 9}{1} = 3.$$

The average rate of change from $x = 3$ to $x = 5$ is

$$\frac{f(5) - f(3)}{5 - 3} = \frac{16 - 12}{2} = 2.$$

The average of these two rates is $\frac{3 + 2}{2} = 2.5$. Thus the central difference quotient at $x = 3$ is 2.5.

S-1. **A tabulated function**: Here $N(20)$ is 23.8 since that is the corresponding value in the table.

S-3. **Average rate of change**: The average rate of change in N from $t = 20$ to $t = 30$ is given by the change in N divided by the change in t:

$$\frac{N(30) - N(20)}{30 - 20} = \frac{44.6 - 23.8}{10} = 2.08.$$

Thus the average rate of change in N is 2.08.

S-5. **Averaging**: We can estimate the value of $N(35)$ by finding the average of $N(30)$ and $N(40)$ since 35 is the average of 30 and 40. The average is $\frac{N(30) + N(40)}{2} = \frac{44.6 + 51.3}{2}$, which equals 47.95, or about 48.0.

S-7. **Using average rates of change**: To estimate the value of $N(37)$, we calculate $N(37)$ as $N(30)$ plus 7 years of change at the average rate, that is, $N(37)$ is estimated to be $N(30) + 7 \times 0.67 = 44.6 + 7 \times 0.67$, which equals 49.29, or about 49.3.

S-9. **When limiting values occur**: We expect c to have a limiting value of 0. This is because the average speed c gets closer and closer to 0 as the time t required to travel 100 miles increases.

S-11. **Average rate of change**: The average rate of change in f from $x = 15$ to $x = 20$ is given by the change in f divided by the change in x:

$$\frac{f(20) - f(15)}{20 - 15} = \frac{-7.9 - (-3.6)}{5} = -0.86.$$

Thus the average rate of change in f is -0.86.

S-13. **A table**: Here $N(2000)$ is 427.0 because that is the corresponding value in the table.

S-15. **Using average rate of change**: To estimate the value of $N(2003)$, we calculate $N(2000)$ plus 3 years of change at the average rate. That is, $N(2003)$ is estimated to be

$$N(2000) + 3 \times -1.67 = 427.0 - 5.01 = 421.99,$$

or about 422.

1. **The American food dollar:**

   (a) Here $P(1980) = 27\%$. This means that in 1980 Americans spent 27% of their food dollars eating out.

   (b) The expression $P(1990)$ is the percent of the American food dollar spent eating away from home in 1990. Since 1990 falls halfway between 1980 and 2000, our best guess at $P(1990)$ is the average of $P(1980)$ and $P(2000)$, or

   $$\frac{P(1980) + P(2000)}{2} = \frac{27 + 37}{2} = 32.$$

   Approximately 32% of the American food dollar in 1990 was spent eating out.

   (c) The average rate of change per year in percentage of the food dollar spent away from home from 1980 to 2000 is

   $$\frac{P(2000) - P(1980)}{2000 - 1980} = \frac{37 - 27}{20} = 0.5,$$

   or 0.5 percentage point per year.

   (d) The expression $P(1997)$ is the percent of the American food dollar spent eating away from home in 1997. We estimate it as

   $$P(1997) = P(1980) + 17 \times \text{ yearly change } = 27 + 17 \times 0.5 = 35.5,$$

   or about 36%.

   (e) Assuming the increase in P continues at the same rate of about 0.5 percentage point per year as we calculated in Part (c), then

   $$P(2003) = P(2000) + 3 \times \text{ yearly change } = 37 + 3 \times 0.5 = 38.5,$$

   or about 39%.

3. **Cable TV:**

   (a) Here $C(1995) = 60$ million households. This means that 60 million American households had cable TV in 1995.

(b) The average rate of change per year from 1990 to 1995 is

$$\frac{\text{Change in } C}{\text{Time elapsed}} = \frac{C(1995) - C(1990)}{5} = \frac{60 - 52}{5} = 1.6 \text{ million households per year.}$$

(c) Since the average rate of change per year from 1990 to 1995 is 1.6 million households per year, we estimate

$$C(1992) = C(1990) + 2 \times \text{ yearly change } = 52 + 2 \times 1.6 = 55.2,$$

or about 55 million households.

5. **A troublesome snowball**: Here $W(t)$ is the volume of dirty water soaked into the carpet, so its limiting value is the total volume of water frozen in the snowball. The limiting value is reached when the snowball has completely melted.

7. **Carbon 14**:

(a) The average yearly rate of change for the first 5000 years is

$$\frac{\text{Amount of change}}{\text{Years elapsed}} = \frac{C(5) - C(0)}{5000} = \frac{2.73 - 5}{5000} = -4.54 \times 10^{-4} \text{ gram per year.}$$

That is -0.000454 gram per year.

(b) We use the average yearly rate of change from Part (a):

$$C(1.236) = C(0) + 1236 \times \text{ yearly rate of change } = 5 + 1236 \times -0.000454 = 4.44 \text{ grams.}$$

(c) The limiting value is zero since all of the carbon 14 will eventually decay.

9. **Effective percentage rate for various compounding periods**:

(a) We have that $n = 1$ represents compounding yearly, $n = 2$ represents compounding semiannually, $n = 12$ represents compounding monthly, $n = 365$ represents compounding daily, $n = 8760$ represents compounding hourly, and $n = 525,600$ represents compounding every minute.

(b) We have that $E(12)$ is the EAR when compounding monthly, and $E(12) = 12.683\%$.

(c) If interest is compounded daily then the EAR is $E(365)$. So the interest accrued in one year is

$$8000 \times E(365) = 8000 \times 0.12747 = \$1019.76.$$

(d) If interest were compounded continuously then the EAR would probably be about 12.750%. As the length of the compounding period decreases, the EAR given in the table appears to stabilize at this value.

11. **Growth in height:**

   (a) In functional notation, the height of the man at age 13 is given by $H(13)$.

   From ages 10 to 15, the average yearly growth rate in height is

   $$\frac{\text{Inches increased}}{\text{Years elapsed}} = \frac{67.0 - 55.0}{5} = 2.4 \text{ inches per year.}$$

   Since age 13 is 3 years after age 10, we can estimate $H(13)$ as

   $$H(10) + 3 \times \text{ yearly growth } = 55.0 + 3 \times 2.4 = 62.2 \text{ inches.}$$

   (b) i. We calculate the average yearly growth rate for each 5-year period just as we calculated 2.4 inches per year as the average yearly growth rate from ages 10 to 15 in Part (a). The average yearly growth rate is measured in inches per year.

Age change	0 to 5	5 to 10	10 to 15	15 to 20	20 to 25
Average yearly growth rate	4.2	2.5	2.4	1.3	0.1

   ii. The man grew the most from age 0 to age 5.

   iii. The trend is that as the man gets older, he grows more slowly.

   (c) It is reasonable to guess that 74 or 75 inches is the limiting value for the height of this man. He grew only 0.5 inches from ages 20 to 25, so it is reasonable to expect little or no further growth from age 25 on.

13. **Tax owed:**

   (a) The average rate of change over the first interval is

   $$\frac{T(16,200) - T(16,000)}{16,200 - 16,000} = \frac{888 - 870}{200} = 0.09 \text{ dollar per dollar.}$$

   Continuing in this way, we get the following table, where the rate of change is measured in dollars per dollar.

Interval	16,000 to 16,200	16,200 to 16,400	16,400 to 16,600
Rate of change	0.09	0.09	0.09

   (b) The average rate of change has a constant value of 0.09 dollar per dollar. This suggests that, at every income level in the table, for every increase of $1 in taxable income the tax owed increases by $0.09, or 9 cents.

   (c) Because the average rate of change is a nonzero constant and thus does not tend to 0, we would expect T not to have a limiting value but rather to increase at a constant rate as I increases.

15. **Yellowfin tuna**:

(a) The average rate of change in weight is

$$\frac{W(110) - W(100)}{110 - 100} = \frac{56.8 - 42.5}{10} = 1.43 \text{ pounds per centimeter.}$$

(b) The average rate of change in weight is

$$\frac{W(180) - W(160)}{180 - 160} = \frac{256 - 179}{20} = 3.85 \text{ pounds per centimeter.}$$

(c) Examining the table shows that the rate of change in weight is smaller for small tuna than it is for large tuna. Hence an extra centimeter of length makes more difference in weight for a large tuna.

(d) To estimate the weight of a yellowfin tuna that is 167 centimeters long we use the average rate of change we found in Part (b):

$$W(160) + 7 \times 3.85 = 179 + 7 \times 3.85 = 205.95 \text{ pounds.}$$

Hence the weight of a yellowfin tuna that is 167 centimeters long is 205.95, or about 206.0, pounds.

(e) Here we are thinking of the weight as the variable and the length as a function of the weight. The average rate of change in length is

$$\frac{\text{Length at 256 pounds} - \text{Length at 179 pounds}}{256 - 179} = \frac{180 - 160}{256 - 179} = 0.26,$$

so the average rate of change is 0.26 centimeter per pound. Note that this number is the reciprocal of the answer from Part (b).

(f) To estimate the length of a yellowfin tuna that weighs 225 pounds we use the average rate of change we found in Part (e):

Length at 179 pounds $+ (225 - 179) \times 0.26 = 160 + 46 \times 0.26 = 171.96$ centimeters.

Hence the length of a yellowfin tuna that weighs 225 pounds is 171.96, or about 172, centimeters.

17. **Widget production**:

   (a) The average rate of change over the first interval is

   $$\frac{W(20) - W(10)}{10} = \frac{37.5 - 25.0}{10} = 1.25 \text{ thousand widgets per worker.}$$

   Continuing in this way, we get the following table, where the rate of change is measured in thousands of widgets per worker.

Interval	10 to 20	20 to 30	30 to 40	40 to 50
Rate of change	1.25	0.63	0.31	0.15

   (b) The average rate of change decreases and approaches 0 as we go across the table. This means that the increase in production gained from adding another worker gets smaller and smaller as the level of workers employed moves higher and higher. Eventually there is very little benefit in employing an extra worker.

   (c) To estimate how many widgets will be produced if there are 55 full-time workers we use the entry from the table for the average rate of change over the last interval:

   $$W(50) + 5 \times 0.15 = 48.4 + 5 \times 0.15 = 49.15 \text{ thousand widgets.}$$

   Hence the number of widgets produced by 55 full-time workers is about 49.2 thousand, or 49,200.

   (d) Because the average rate of change is decreasing, the actual increase in production in going from 50 to 55 workers is likely to be less than what the average rate of change from 40 to 50 suggests. Thus our estimate is likely to be too high.

19. **The Margaria-Kalamen test**:

   (a) The average rate of change per year in excellence level from 25 years to 35 years old is

   $$\frac{168 - 210}{35 - 25} = -4.2 \text{ points per year.}$$

   (b) We estimate the power score needed for a 27-year-old man using the score for a 25-year-old man and the average rate of change from Part (a):

   $$210 + 2 \times -4.2 = 201.6 \text{ points.}$$

   Hence the power score that would merit an excellent rating for a 27-year-old man is 201.6, or about 202, points.

   (c) The decrease in power score for excellent rating over these three periods is the greatest in the second period (35 years to 45 years), so we would expect to see the greatest decrease in leg power from 35 to 45 years old.

21. **A home experiment**: Answers will vary greatly. In general, there will be initially a small percentage of bread surface covered with mold. That percentage will quickly rise as the mold covers much of the bread surface. There are usually a few small patches which the mold covers more slowly. Ultimately all the bread surface is covered with mold. Here is a typical data table:

Time	8 am	4 pm	12 am	8 am	4 pm	12 am
Mold	10%	25%	60%	98%	100%	100%

1.3 FUNCTIONS GIVEN BY GRAPHS

E-1. **Secant lines for graphs which are concave down**: One possible graph is the following.

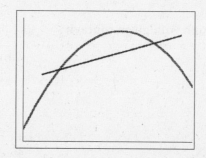

(a) The secant line is below the graph between the two points and above the graph outside those points.

(b) Recall that using average rates of change to approximate function values is the same as approximating the graph by its secant line. Using Part (a), we see that the average rate of change will give an estimate that is too small between the two points used and too large outside them.

S-1. **A function given by a graph**: The value of $f(1.8)$ is obtained from the graph by locating 1.8 on the horizontal axis, then moving vertically up to the point of the graph over 1.8, and then moving horizontally to locate the corresponding value on the vertical axis, Thus $f(1.8)$ is a little below 3.5, so about 3.3.

S-3. **The maximum**: The graph reaches a maximum at its highest point, which is when $x =$ 2.4 and f is about 4.0.

S-5. **Decreasing functions**: The graph is decreasing when it is falling, so that is when x is between 2.4 and 4.8.

S-7. **Concavity again**: If a graph is increasing, but at a decreasing rate, then it will look like the portion of the graph in Exercise S-1 for x between 1.8 and 2.4; in particular the graph is concave down.

S-9. **Maximum and zero**: Answers will vary. One solution is

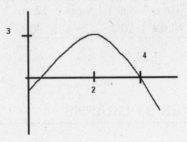

S-11. **Inflection points**: The points of inflection occur at about $x = 1.2$ and $x = 3.6$.

1. **Sketching a graph with given concavity**:

   (a)   (b)

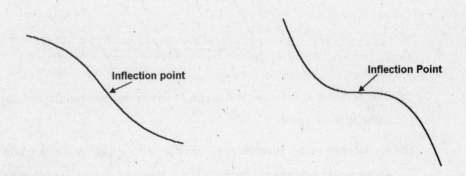

3. **A stock market investment**

   (a) According to the exercise our original investment made in 1970 was $10,000, so $v(1970) = \$10,000$. Our investment lost half its value in the 70's, so $v(1980) = \$5000$. The investment was worth $35,000 in 1990, so $v(1990) = \$35,000$. Since the stock remained stable after 1990, the value of the investment stays at the same level of $35,000. So $v(2000) = \$35,000$.

(b)

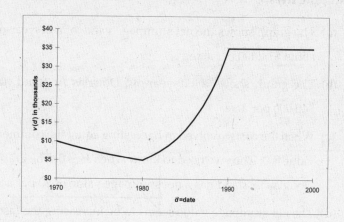

The graph should reflect the function values from Part (a), and it should increase from 1980 to 1990 and remain flat after that. There remains room for interpretation, and many graphs similar to the one above are acceptable.

(c) The stock was most rapidly increasing in the 1980's. For the graph above, 1985 is a good guess. (The answer will depend on what the graph in Part (b) looks like.)

5. **River flow**:

(a) Because the end of July is 7 months since the start of the year, the flow at that time is $F(7)$ in functional notation. According to the graph, the value is about 1500 cubic feet per second.

(b) The flow is at its greatest at $t = 6$, which corresponds to the end of June.

(c) The flow is increasing the fastest at $t = 5$, corresponding to an inflection point on the graph. The time is the end of May.

(d) The graph is practically level from $t = .0$ to $t = 2$, so the function is nearly constant there, and the average rate of change over this interval is about 0 cubic feet per second.

(e) Because the flow is measured near the river's headwaters in the Rocky Mountains, we expect any change in flow to come from melting snow primarily. This is consistent with almost no change in flow during the first two months of the year (as seen in Part (d)), a maximum increase in flow at the end of May (as seen in Part (c)), and a peak flow one month later (as seen in Part (b)).

7. **Cutting trees**:

   (a) The graph shows the net stumpage value of a 60-year-old Douglas fir stand to be about $14,000 per acre.

   (b) The graph shows a 110-year-old Douglas fir stand has net stumpage value of $40,000 per acre.

   (c) When the costs involved in harvesting equal the commercial value, the net stumpage value is 0. Thus we need to know when V is 0. The graph shows that a 30-year-old Douglas fir stand has a net stumpage value of $0 per acre.

   (d) The net stumpage value seems to be increasing the fastest in trees about 60 years old.

   (e) We would expect the trees to reach an age where they don't grow as much. When this happens the net stumpage value should level out. Here is one possible extended graph for the stumpage value of the Douglas fir.

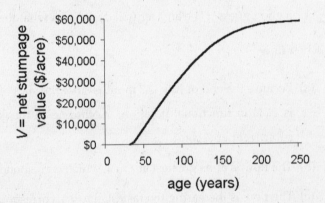

9. **Tornadoes in Oklahoma**:

   (a) The most tornadoes were reported in 1999. There were about 145 tornadoes reported that year.

   (b) The fewest tornadoes were reported in 2002. There were about 18 tornadoes reported then.

   (c) In 1995 there were about 79 tornadoes reported, and in 1996 there were about 46 tornadoes reported. Hence the average rate of decrease was

   $$\frac{\text{Decrease in } T}{\text{Years elapsed}} = \frac{79 - 46}{1} = 33 \text{ tornadoes per year.}$$

   (d) In 1997 the number of tornadoes reported was about 55, and in 1999 the number of tornadoes reported was about 145. Hence the average rate of increase was

   $$\frac{\text{Increase in } T}{\text{Years elapsed}} = \frac{145 - 55}{2} = 45 \text{ tornadoes per year.}$$

(e) The number of tornadoes reported in 1996 was about 46, and the number of tornadoes reported in 2000 was about 44. Hence the average rate of change was

$$\frac{\text{Change in } T}{\text{Years elapsed}} = \frac{44 - 46}{4} = -0.5 \text{ tornado per year.}$$

This is close to 0, and the answer can vary depending on how the graph is interpreted.

11. **Driving a car**: There are many possible stories. Common elements include that you start 6 miles from home and drive towards home, arriving there in about 12 minutes, staying at home for about 4 minutes, then driving about $2\frac{1}{2}$ miles away from home and staying there.

Additional features might be the velocity (rate of change) towards home as about 30 mph and the velocity away from home as about 15 mph.

13. **Photosynthesis**: In Parts (a) and (b) answers may vary somewhat.

   (a) We want to find where the graph corresponding to 80 degrees crosses the horizontal axis. The crossing point at about 0.7 thousand foot-candles, or about 700 foot-candles.

   (b) We want to find where the graph corresponding to 80 degrees crosses the graph corresponding to 40 degrees. The crossing point is at about 0.8 thousand foot-candles, or about 800 foot-candles.

   (c) Among the three graphs, the graph corresponding to 80 degrees meets the vertical axis at the lowest point, so a temperature of 80 degrees will result in the largest emission of carbon dioxide in the dark.

   (d) The graph corresponding to 40 degrees is the most level, so that temperature gives the net exchange that is least sensitive to light.

15. **Carbon dioxide concentrations**:

   (a) The minimum of the graph occurs at about 2 p.m.

   (b) The maximum concentration is attained over the entire interval from about 6 a.m. to about 9 a.m.

   (c) Net absorption is indicated by the interval where the graph is decreasing. This is the period from about 9 a.m. to about 2 p.m.

   (d) From 6 a.m. to 9 a.m. the graph is level, so the net carbon dioxide exchange is zero during that period.

17. **Profit from fertilizer:**

   (a) The maximum value of yield occurs at about 100 on the horizontal axis, so about 100 pounds of nitrogen per acre should be applied to produce maximum crop yield.

   (b) The profit can be read by measuring the difference between the yield and cost graphs.

   (c) The maximum difference between the graphs occurs at about 70 on the horizontal axis, so about 70 pounds of nitrogen per acre should be applied to produce maximum profit. (In this part answers may vary somewhat.)

19. **Laboratory experiment:** Answers will vary. Please see the website for further information.

1.4 FUNCTIONS GIVEN BY WORDS

E-1. **A fence next to a river:** Assume that the width w is measured in yards.

   (a) Let h be the height (in yards) of the rectangle. Three sides of the rectangle require fencing, with one having length w and two having length h. Because we are to use 200 yards of fence, we know that $w + 2h = 200$. We solve this equation for h:

$$
\begin{aligned}
w + 2h &= 200 \\
2h &= 200 - w \\
h &= \frac{200 - w}{2}.
\end{aligned}
$$

   Thus the height in yards is $h = \dfrac{200 - w}{2}$. (This can also be written as $h = 100 - \dfrac{w}{2}$.)

   (b) We find the area A of the rectangle (in square yards) in terms of w by using the formula for h from Part (a):

$$
A = wh = w\frac{200 - w}{2}.
$$

   Thus the area in square yards is $A = w\dfrac{200 - w}{2}$.

E-3. **Circumference in terms of area**: From the given formula we know that $A = \dfrac{C^2}{4\pi}$. We solve this equation for C:

$$
\begin{aligned}
A &= \frac{C^2}{4\pi} \\
4\pi A &= C^2 \\
\sqrt{4\pi A} &= C \\
2\sqrt{\pi A} &= C.
\end{aligned}
$$

Thus the circumference can be written as $C = 2\sqrt{\pi A}$.

E-5. **Making a box**:

(a) Two segments of size x are removed from the width of the original rectangle to obtain the width of the box. Thus the width of the box is $5 - 2x$.

(b) Two segments of size x are removed from the length of the original rectangle to obtain the length of the box. Thus the length of the box is $10 - 2x$.

(c) Because the tabs are folded up to form the box, the height of the box is x.

(d) The volume of the box is

$$
\text{width} \times \text{length} \times \text{height} = (5 - 2x)(10 - 2x)x.
$$

Thus the volume is $x(5 - 2x)(10 - 2x)$.

E-7. **A soda can with given volume**: Assume that the radius x and the height h are both measured in inches.

(a) From the statement of Part (c) of Exercise E-6 we know that the volume of the can is the area of the top times the height. Now the area of the top is πx^2 and the height is h, so the formula for the volume is $\pi x^2 h$. Since the volume is 25, we have the equation $\pi x^2 h = 25$. We solve this equation for h:

$$
\begin{aligned}
\pi x^2 h &= 25 \\
h &= \frac{25}{\pi x^2}.
\end{aligned}
$$

Thus the desired expression is $h = \dfrac{25}{\pi x^2}$.

(b) From the statement of Part (a) of Exercise E-6 we know that the total surface area is the area of the top plus the area of the bottom plus the area of the cylindrical part. The top and bottom are formed by circles of radius x, so each of these has area πx^2. We also know from the statement of Part (a) of Exercise E-6 that the area of the cylindrical part of the can is the circumference of the can times the height. Because the radius of the can is x, its circumference is $2\pi x$, and thus the area of the cylindrical part is $2\pi xh$. Hence the formula for the total surface area of the can is $2\pi x^2 + 2\pi xh$. Using the expression for h in terms of x from Part (a) gives the formula $2\pi x^2 + 2\pi x\dfrac{25}{\pi x^2}$. Simplifying gives the expression $2\pi x^2 + \dfrac{50}{x}$ for the surface area of the can in terms of x.

S-1. **A description**: If you have $5000 and spend half the balance each month, then the new balance will be

New balance after 1 month $= 5000 - \frac{1}{2} \times 5000 = 2500$

New balance after 2 months $= 2500 - \frac{1}{2} \times 2500 = 1250$

New balance after 3 months $= 1250 - \frac{1}{2} \times 1250 = 625$

New balance after 4 months $= 625 - \frac{1}{2} \times 625 = 312.5$

and so the balance left after 4 months is $312.50.

S-3. **A description**: We know that $f(0) = 5$ and that each time x increases by 1, the value of f triples, that is, it is three times its previous value. Therefore

$$f(1) = 3 \times f(0) = 3 \times 5 = 15$$
$$f(2) = 3 \times f(1) = 3 \times 15 = 45$$
$$f(3) = 3 \times f(2) = 3 \times 45 = 135$$
$$f(4) = 3 \times f(3) = 3 \times 135 = 405$$

so $f(4) = 405$. On the other hand, 5×3^4 is also 405.

S-5. **Getting a formula**: If the man loses 67 strands of hair each time he showers, then if he showers s times, he will lose $67 \times s$ strands of hair. Thus $N = 67s$.

S-7. **Getting a formula**: If you start with $500 and you add to that $37 each month, then the balance will be $500 plus $37 times the number of months. Thus the balance B is given by $B = 500 + 37t$.

S-9. **Proportionality**: If f is proportional to x and the constant of proportionality is 8, then $f = 8x$.

S-11. **Getting a formula for discounted items**: The cost per item decreases by $2 for each extra item rented. Because it costs $20 per item for 1 item, it costs $20 - 2 \times 4 = 12$ dollars per item if 5 items are rented. The total cost for 5 items is $5 \times 12 = 60$ dollars.

1. **United States population growth:**

   (a) Since t is the number of years since 1960, the year 1963 is represented by $t = 3$. In functional notation, the population of the U.S. in 1963 is given by $N(3)$. To calculate its value, we use the fact that the population increases by 1.2% per year. Since $N(0) = 180$, in millions,

   $$
   \begin{aligned}
   N(1) &= \text{Population in 1960} + 1.2\% \text{ growth} \\
        &= 180 + 0.012 \times 180 = 182.16 \text{ million} \\
   N(2) &= \text{Population in 1961} + 1.2\% \text{ increase} \\
        &= 182.16 + 0.012 \times 182.16 = 184.35 \text{ million} \\
   N(3) &= \text{Population in 1962} + 1.2\% \text{ growth} \\
        &= 184.35 + 0.012 \times 184.35 = 186.56 \text{ million.}
   \end{aligned}
   $$

   (b) The table is given below, calculating $N(4)$ and $N(5)$ as in Part (a) above.

Year	t	$N(t)$
1960	0	180.00
1961	1	182.16
1962	2	184.35
1963	3	186.56
1964	4	188.80
1965	5	191.06

   (c) A graph is shown below.

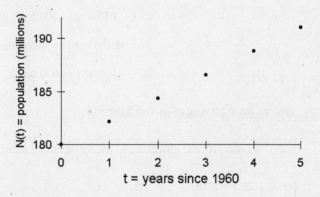

(d) The formula 180×1.012^t can be shown to give the same values as those found in Part (b) by substituting $t = 0$, $t = 1, \ldots$ into the formula:

t	180×1.012^t
0	$180 \times 1.012^0 = 180.00$
1	$180 \times 1.012^1 = 182.16$
2	$180 \times 1.012^2 = 184.35$
3	$180 \times 1.012^3 = 186.56$
4	$180 \times 1.012^4 = 188.80$
5	$180 \times 1.012^5 = 191.06$

(e) Using the formula and the fact that 2000 corresponds to $t = 40$, we find that the population in 2000 from this prediction is

$$N(40) = 180 \times 1.012^{40} = 290.06 \text{ million people.}$$

This estimate is about 9 million higher than the actual population.

3. **Altitude:**

(a) Let t be the time in minutes since takeoff and A the altitude in feet. Then

$$A(0) = \text{ Initial altitude } = 200 \text{ feet,}$$

and

$$A(1) = \text{ Initial altitude } + \text{ Increase over 1 minute}$$
$$= 200 + 150 \times 1 = 350 \text{ feet,}$$
$$A(2) = \text{ Initial altitude } + \text{ Increase over 2 minutes}$$
$$= 200 + 150 \times 2 = 500 \text{ feet,}$$
$$A(3) = \text{ Initial altitude } + \text{ Increase over 3 minutes}$$
$$= 200 + 150 \times 3 = 650 \text{ feet,}$$

and so on. This suggests the formula

$$A = \text{ Initial altitude } + \text{ Increase over } t \text{ minutes } = 200 + 150 \times t,$$

or $A = 200 + 150t$.

(b) In functional notation the altitude 90 seconds after takeoff is $A(1.5)$ because 90 seconds corresponds to 1.5 minutes. The value is $200 + 150 \times 1.5 = 425$ feet.

(c) Here is the graph.

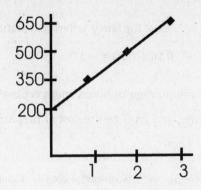

The altitude changes at a constant rate, and this is reflected in the fact that the graph is a straight line.

5. **A rental:**

   (a) If we rent the car for 2 days and drive 100 miles, it will cost us

   $$2 \text{ days rental } + \text{ charge for 100 miles } = 29 \times 2 + 0.06 \times 100 = \$64.$$

   (b) Let d be the number of days we rent the car, m the number of miles we drive the car, and C the cost in dollars of renting the car. Then the formula that gives us the cost of renting a car is

   $$C(d, m) = 29 \times \text{ days } + 0.06 \times \text{ miles } = 29d + 0.06m.$$

   (c) Since we drove from Dallas to Austin and back then we traveled a total of 500 miles. We kept the car for one week, or 7 days. Hence in functional notation the cost of the rental car is $C(7,500)$. This is calculated as

   $$C(7,500) = 29 \times 7 + 0.06 \times 500 = \$233.$$

7. **Preparing a letter, continued:**

   (a) There are 2 pages of regular stationery and 1 of fancy letterhead stationery, so the total cost is $2 \times 0.03 + 1 \times 0.16 = 0.22$ dollar, or 22 cents.

   (b) We know from Part (a) that the cost for stationery is 0.22 dollar. The secretarial cost is 2×6.25 dollars, and the cost for the envelope is 0.38 dollar. In total, the letter costs $0.22 + 2 \times 6.25 + 0.38 = 13.10$ dollars.

(c) Let c be the cost in dollars of the stationery and p the number of pages. There are $p - 1$ pages of regular stationery and 1 of fancy letterhead stationery. Then

$$
\begin{aligned}
c(p) &= \text{Cost for fancy letterhead stationery} + \text{Cost for regular stationery} \\
&= 0.16 + 0.03(p - 1).
\end{aligned}
$$

(d) Let h be the number of hours spent typing the letter, let p be the number of pages of the letter, and let C be the cost of preparing and mailing the letter (measured in dollars). Then

$$
\begin{aligned}
C(h, p) &= \text{Secretarial cost} + \text{Cost for paper} + \text{Cost for envelope} \\
&= 6.25h + 0.16 + 0.03(p - 1) + 0.38.
\end{aligned}
$$

(This can also be written as $C(h, p) = 6.25h + 0.03(p - 1) + 0.54$ or as $C(h, p) = 6.25h + 0.03p + 0.51$.)

(e) Here $h = \dfrac{25}{60}$ and $p = 2$, so the cost is

$$
C\left(\frac{25}{60}, 2\right) = 6.25 \times \frac{25}{60} + 0.16 + 0.03(2 - 1) + 0.38 = 3.17 \text{ dollars.}
$$

9. **Stock turnover rate:**

(a) If 350 shirts are sold annually then the number of orders of 50 shirts is $\dfrac{350}{50} = 7$.

(b) The annual stock turnover rate is 7, the number computed in Part (a), because this is the number of times that the average inventory of 50 shirts needs to be replaced if 350 shirts are sold in a year.

(c) If 500 shirts were sold in a year then the annual stock turnover rate would be $\dfrac{500}{50} = 10$.

(d) Let T be the annual stock turnover rate and S the number of shirts sold in a year. Then

$$
T = \frac{\text{Number of shirts sold}}{\text{Average inventory}} = \frac{S}{50}.
$$

Thus the formula is $T = \dfrac{S}{50}$.

11. **Total cost:**

(a) Let N be the number of widgets produced in a month and C the total cost in dollars. Because

$$
\text{Total cost} = \text{Variable cost} \times \text{Number of items} + \text{Fixed costs,}
$$

we have $C = 15N + 9000$.

(b) In functional notation the total cost is $C(250)$, and the value is $15 \times 250 + 9000 = 12{,}750$ dollars.

13. **More on revenue:**

   (a) We compute

$$p(100) = 50 - 0.01 \times 100 = 49$$

$$p(200) = 50 - 0.01 \times 200 = 48$$

$$p(300) = 50 - 0.01 \times 300 = 47$$

$$p(400) = 50 - 0.01 \times 100 = 46$$

$$p(500) = 50 - 0.01 \times 500 = 45.$$

   These values agree with those in the table.

   (b) We have $R = pN$, so $R = (50 - 0.01N)N$ dollars.

   (c) In functional notation the total revenue is $R(450)$, and the value is

$$(50 - 0.01 \times 450) \times 450 = 20{,}475 \text{ dollars.}$$

15. **Renting motel rooms:**

   (a) Since the group rents 2 extra rooms, we take off $4 from the base price of each room, which means that we charge $81 for each room. Since we rented 3 rooms altogether, we take in a total of $3 \times 81 = \$243$.

   (b) The formula that tells us how much to charge for each room is

$$\text{Rate} = \text{Base price} - \$2 \text{ per extra room dollars.}$$

   Now if n rooms are rented, then the number of extra rooms is $n - 1$. Since the base price is $85, the formula can be written as

$$\text{Rate} = 85 - 2 \times \text{ number of extra rooms} = 85 - 2(n - 1) \text{ dollars.}$$

   This can also be written as Rate $= 87 - 2n$ dollars.

   (c) The total revenue is the number of rooms times the rental per room:

$$R(n) = \text{Rooms rented} \times \text{Price per room} = n \times (85 - 2(n - 1)) \text{ dollars.}$$

   This can also be written as $R(n) = n(87 - 2n)$ dollars.

(d) The total cost of renting 9 rooms is expressed in functional notation as $R(9)$. This is calculated as $R(9) = 9(85 - 2 \times (9 - 1)) = \$621$.

17. **Catering a dinner:**

(a) i. If 50 people attend, then your cost is the rental fee plus the caterer's fee for each of the 50 people, or a total of $150 + 50 \times 10 = 650$ dollars. If 50 people attend, then to break even, each ticket should cost $\dfrac{650}{50} = 13$ dollars.

ii. Let n be the number of people attending and C the amount in dollars you should charge per person. Your total cost is the rental fee plus the caterer's fee for each of the n people, or a total of $150 + 10n$ dollars. Since n people attend, then to break even, each ticket should cost

$$C = \frac{150 + 10n}{n} \text{ dollars.}$$

This can also be written as $C = \dfrac{150}{n} + 10$, which can be thought of as each person's share of the \$150 rental fee plus the \$10 caterer's fee.

iii. If 65 people attend, then the amount to charge each is expressed in functional notation as $C(65)$. This is calculated as

$$C(65) = \frac{150}{65} + 10 = \$12.31 \text{ per ticket.}$$

(b) To make a profit of \$100, you should think of your cost as being \$100 more in determining your ticket price, so your new price would be

$$P = \frac{100 + 150 + 10n}{n} = \frac{250 + 10n}{n} \text{ dollars.}$$

This can also be written as $P = \dfrac{250}{n} + 10$ or as $P = \dfrac{100}{n} + \dfrac{150}{n} + 10$, which can be thought of as each person's share of the \$100 profit, plus each person's share of the \$150 rental fee, plus the \$10 caterer's fee.

19. **Production rate:**

(a) Let k be the constant of proportionality. Then $t = kn$.

(b) Since t is total number of items produced and n is the number of employees, it follows that k is the number of items produced per employee.

21. **Head and pressure:**

(a) Because p is proportional to h with constant of proportionality 0.434, the equation is $p = 0.434h$.

(b) The head of water at the mouth of the nozzle is $8 \times 12 = 96$ feet. The back pressure is the value of p given by the equation in Part (a) with the head of $h = 96$ feet: $p = 0.434 \times 96 = 41.66$. Thus the back pressure is 41.66 pounds per square inch.

(c) The head is the height of the nozzle above the pumper, and in this case that value is $185 - 40 = 145$ feet. The back pressure is $p = 0.434 \times 145 = 62.93$ pounds per square inch.

23. **Darcy's law:**

(a) Because V is proportional to S with constant of proportionality K, the equation is $V = KS$.

(b) The constant K equals the permeability of sandstone, 0.041 meter per day. The slope S is given as 0.03 meter per meter. We compute the velocity of the water flow using the equation in Part (a): $V = 0.041 \times 0.03 = 0.00123$. The units are found by multiplying the units for K with those for S, and we have that the velocity is 0.00123 meter per day.

(c) Now we take the constant K to be 41 meters per day. The slope S is still 0.03 meter per meter. The velocity of the water flow is $V = 41 \times 0.03 = 1.23$ meters per day.

25. **The $3x + 1$ problem:**

(a) We have

$$
\begin{aligned}
f(1) &= 3(1) + 1 = 4 \\
f(4) &= \frac{4}{2} = 2 \\
f(2) &= \frac{2}{2} = 1.
\end{aligned}
$$

The procedure repeats this cycle over and over.

(b) We have

$$
\begin{aligned}
f(5) &= 3(5) + 1 = 16 \\
f(16) &= \frac{16}{2} = 8 \\
f(8) &= \frac{8}{2} = 4 \\
f(4) &= 2 \\
f(2) &= 1.
\end{aligned}
$$

It took 5 steps to get to 1.

(c) We have

$$
\begin{aligned}
f(7) &= 3(7)+1 = 22 \\
f(22) &= 11 \\
f(11) &= 34 \\
f(34) &= 17 \\
f(17) &= 52 \\
f(52) &= 26 \\
f(26) &= 13 \\
f(13) &= 40 \\
f(40) &= 20 \\
f(20) &= 10 \\
f(10) &= 5.
\end{aligned}
$$

We followed the rest of this trail in Part (b). So it takes 16 steps to get to 1 starting with 7.

(d) These answers will vary. If a number is found that does not lead back to 1, the computation should be checked very carefully.

Chapter 1 Review Exercises

1. **Evaluating formulas**: To get the function value $M(9500, 0.01, 24)$, substitute $P = 9500$, $r = 0.01$, and $t = 24$ in the formula

$$
M(P, r, t) = \frac{Pr(1+r)^t}{(1+r)^t - 1}.
$$

The result is

$$
\frac{9500 \times 0.01 \times (1 + 0.01)^{24}}{(1 + 0.01)^{24} - 1},
$$

which equals 447.20.

2. **U.S. population**:

(a) Because 1790 corresponds to $t = 0$, the population in 1790 was $3.93 \times 1.03^0 = 3.93$ million.

(b) Because 1810 is 20 years after 1790, we take $t = 20$ and get that $N(20)$ is functional notation for the population in 1790.

(c) To find the population in 1810 we put $t = 20$ in the formula. The result is $3.93 \times 1.03^{20} = 7.10$, so the population in 1810 was 7.10 million according to the formula.

3. **Averages and average rate of change**:

(a) Because 5 is halfway between 4 and 6, we estimate $f(5)$ by

$$\frac{f(4) + f(6)}{2} = \frac{40.1 + 43.7}{2} = 41.9.$$

(b) The average rate of change is the change in f divided by the change in x, and that is

$$\frac{f(6) - f(4)}{2} = \frac{43.7 - 40.1}{2} = 1.8.$$

4. **High school graduates**:

(a) Here $N(1989)$ represents the number, in millions, graduating from high school in 1989. According to the table, its value is 2.47 million.

(b) In functional notation the number of graduates in 1988 is $N(1988)$. We estimate its value by averaging:

$$\frac{N(1987) + N(1989)}{2} = \frac{2.65 + 2.47}{2} = 2.56.$$

Thus there were about 2.56 million graduating in 1988.

(c) The average rate of change is the change in N divided by the change in t, and that is

$$\frac{N(1991) - N(1989)}{2} = \frac{2.29 - 2.47}{2} = -0.09 \text{ million per year.}$$

(d) To estimate the value of $N(1994)$, we calculate $N(1991)$ plus 3 years of change at the average rate found in the previous part. So, $N(1994)$ is estimated to be

$$N(1991) + 3 \times -0.09 = 2.29 - 0.27 = 2.02.$$

Our estimate for $N(1994)$ is 2.02 million.

5. **Increasing, decreasing, and concavity**:

(a) The function is increasing from 1990 to 2000.

(b) The function is concave down from 1996 to 2000 and concave up from 1990 to 1996.

(c) There is an inflection point at $d = 1996$.

6. **Logistic population growth**:

   (a) The population grows rapidly at first and then the growth slows. Eventually it levels off.

   (b) The population reaches 300 in mid-1997.

   (c) The population is increasing most rapidly in 1996.

   (d) The point of most rapid population increase is an inflection point.

7. **Getting a formula**: The balance (in dollars) is the initial balance of $780 minus $39 times the number of withdrawals. Thus $B = 780 - 39t$.

8. **Cell phone charges**:

   (a) Let t denote the number of text messages and C the charge, in dollars.

   (b) The charge (in dollars) is the flat monthly rate of $39.95 plus $0.10 times the number of messages in excess of 100. That excess is $t - 100$, so the formula is $C = 39.95 + 0.1(t - 100)$.

   (c) In functional notation the cost if you have 450 messages is $C(450)$. The value is $39.95 + 0.1(450 - 100) = 74.95$ dollars.

   (d) If the number of text messages is less than 100 then the only charge is the flat monthly rate of $39.95. So the formula is $C = 39.95$.

9. **Cell phone charges again**:

   (a) Let t denote the number of text messages, m the number of minutes, and C the charge, in dollars.

   (b) The charge (in dollars) is the flat monthly rate of $34.95, plus $0.35 times the number of minutes in excess of 4000, plus $0.10 times the number of messages in excess of 100. Thus the formula is $C = 34.95 + 0.35(m - 4000) + 0.1(t - 100)$.

   (c) The charges are $34.95 + 0.35(6000 - 4000) + 0.1(450 - 100) = 769.95$ dollars.

   (d) If the number of text messages is less than 100 then the only charges are the flat monthly rate of $34.95 plus $0.35 times the number of minutes in excess of 4000. So the formula is $C = 34.95 + 0.35(m - 4000)$.

   (e) We use the formula from Part (d). The charges are $34.95 + 0.35(4200 - 4000) = 104.95$ dollars.

10. **Practicing calculations:**

    (a) We calculate that $C(0) = 0.2 + 2.77e^{-0.37 \times 0} = 2.97$.

    (b) We calculate that $C(0) = \dfrac{12.36}{0.03 + 0.55^0} = 12$.

    (c) We calculate that $C(0) = \dfrac{0-1}{\sqrt{0+1}} = -1$.

    (d) We calculate that $C(0) = 5 \times 0.5^{0/5730} = 5$.

11. **Amortization:**

    (a) We calculate that

    $$M(5500, 0.01, 24) = \frac{5500 \times 0.01 \times (1+0.01)^{24}}{(1+0.01)^{24} - 1} = 258.90 \text{ dollars.}$$

    Your monthly payment if you borrow \$5500 at a monthly rate of 1% for 24 months is \$258.90.

    (b) In functional notation the payment is $M(8000, 0.006, 36)$. The value is

    $$\frac{8000 \times 0.006 \times (1+0.006)^{36}}{(1+0.006)^{36} - 1} = 247.75 \text{ dollars.}$$

12. **Using average rate of change:**

    (a) The average rate of change is the change in f divided by the change in x, and that is

    $$\frac{f(3) - f(0)}{3} = \frac{55 - 50}{3} = 1.67.$$

    (b) Again, the average rate of change is the change in f divided by the change in x, and in this case that is
    $$\frac{f(6) - f(3)}{3} = \frac{61 - 55}{3} = 2.$$

    (c) Because 4 is 1 unit more than 3 and the average rate of change is 2, we estimate $f(4)$ by

    $$f(3) + 1 \times 2 = 55 + 2 = 57.$$

13. **Timber values under Scribner scale:**

    (a) The average rate of change from \$20 to \$24 is $\dfrac{81.60 - 68.00}{24 - 20} = 3.4$. The units here are dollar value per MBF Scribner divided by dollar value per cord. Continuing in this way, we get the following table, where the rate of change has the units just given. Note that the change in the variable for the last interval is 8, not 4.

Interval	20 to 24	24 to 28	28 to 36
Rate of change	3.4	3.4	3.4

(b) No, the value per MBF Scribner should not have a limiting value: Its rate of change is a nonzero constant, so we expect it to increase at a constant rate.

(c) We use the average rate of change from $24 to $28 to estimate the value per MBF Scbribner when the value per cord is $25. That estimate is $81.60 + 1 \times 3.4 = 85$ dollars per MBF Scbribner. Because $85 is greater than $71, if you are selling then $25 per cord is a better value, but if you are buying then $71 per MBF Scribner is a better value. Another way to do this is to note by inspecting the table that the value of $25 per cord is greater than the value of $71 per MBF Scribner: The value of $25 per cord is greater than the value of $24 per cord listed in the table, so it is higher than the equivalent value of $81.60 per MBF Scribner listed in the table.

14. **Concavity**: If a graph is decreasing at an increasing rate then it is concave down. If it is decreasing at a decreasing rate then it is concave up.

15. **Longleaf pines**:

(a) The height of the tree increases quickly at first, but the growth rate decreases as the tree ages. It makes sense for a young tree to grow more quickly than an older tree.

(b) According to the graph the tree height for a 60-year-old tree is about 132 feet.

(c) Yes, there is a limiting value, since the graph eventually levels off.

(d) The graph is concave down. This means that the height is increasing at a decreasing rate, so each year the amount of growth decreases.

16. **Getting a formula**:

(a) If we rent 3 rooms we get a discount of $2 \times 2 = 4$ dollars, so each room will cost $56 - 4 = 52$ dollars.

(b) Because each room will cost $52 dollars, we will pay $3 \times 52 = 156$ dollars altogether.

(c) If we rent n rooms we get a discount of $2(n-1) = 2n-2$ dollars. If we let R denote the rental cost in dollars per room then $R = 56 - 2(n-1)$ or $R = 58 - 2n$ dollars.

(d) Let C denote the total cost in dollars. Then $C = n \times (56 - 2(n-1))$ or $C = n \times (58 - 2n)$.

17. **A wedding reception**:

(a) If you invite 100 guests then the cost is $3200 for the venue plus $31 times 50 (because 100 guests makes an excess of 50 over the number included). Thus the cost is $3200 + 31 \times 50 = 4750$ dollars.

(b) If you invite n guests then the cost is \$3200 for the venue plus \$31 times the number in excess of 50. That excess is $n - 100$, so if we let C denote the cost in dollars then $C = 3200 + 31(n - 50)$ or $C = 1650 + 31n$.

(c) We want to find n so that $C = 5500$. By the formula from Part (b) this says that $1650 + 31n = 5500$. By trial and error (or by solving for n using algebra) we find that we can invite 124 guests.

18. **Limiting values**:

   (a) No, not all tables show limiting values.

   (b) We can identify a limiting value from a table by checking whether the last few entries in the table show little change. If so, the limiting value is approximated by the trend established by the last few entries.

   (c) No, not all graphs show limiting values.

   (d) We can identify a limiting value from a graph by checking whether the last portion of the graph levels off. If so, the limiting value is approximated by that value where the graph is level.

Solution Guide for Chapter 2: Graphical and Tabular Analysis

2.1 TABLES AND TRENDS

E-1. **Verifying the basic exponential limit**: We show tables for two values of a. The table below on the left corresponds to $a = 0.5$, so the function is 0.5^t. The table on the right corresponds to $a = 0.3$, so the function is 0.3^t. In both cases the function values are approaching 0, in support of the statement of the basic exponential limit.

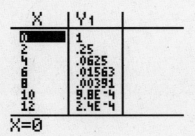 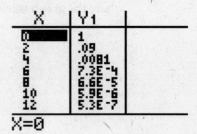

E-3. **Calculating with the basic exponential limit:**

(a) Because $\lim_{t \to \infty} 0.5^t = 0$ by the basic exponential limit, we have

$$\lim_{t \to \infty} 15(1 - 3 \times 0.5^t) = 15(1 - 3 \times 0) = 15.$$

(b) Because $\lim_{t \to \infty} 0.4^t = 0$ and $\lim_{t \to \infty} 0.7^t = 0$, we have

$$\lim_{t \to \infty} \frac{6 + 0.4^t}{3 - 0.7^t} = \frac{6 + 0}{3 - 0} = 2.$$

(c) Because $\lim_{t \to \infty} 0.5^t = 0$, we have

$$\lim_{t \to \infty} \sqrt{7 + 0.5^t} = \sqrt{7 + 0} = \sqrt{7}.$$

(d) Recall from the basic properties of exponents that $\dfrac{2^t}{3^t} = \left(\dfrac{2}{3}\right)^t$. Thus

$$\lim_{t \to \infty} \frac{2^t}{3^t} = \lim_{t \to \infty} \left(\frac{2}{3}\right)^t = 0;$$

here we have applied the basic exponential limit with $a = \dfrac{2}{3}$.

E-5. **Calculating using the division trick:**

(a) The highest power of t that appears is $t^1 = t$, so we divide top and bottom by t. We have

$$\lim_{t \to \infty} \frac{4t + 5}{t + 1} = \lim_{t \to \infty} \frac{4t + 5}{t + 1} \times \frac{\left(\frac{1}{t}\right)}{\left(\frac{1}{t}\right)} = \lim_{t \to \infty} \frac{4 + \frac{5}{t}}{1 + \frac{1}{t}} = \frac{4 + 5 \times 0}{1 + 0} = 4.$$

Here we have applied the basic power limit to $\dfrac{1}{t}$.

(b) The highest power of t that appears is t^3, so we divide top and bottom by t^3. We have

$$\lim_{t \to \infty} \frac{6t^3 + 5}{2t^3 + 3t + 1} = \lim_{t \to \infty} \frac{6t^3 + 5}{2t^3 + 3t + 1} \times \frac{\left(\frac{1}{t^3}\right)}{\left(\frac{1}{t^3}\right)} = \lim_{t \to \infty} \frac{6 + \frac{5}{t^3}}{2 + \frac{3}{t^2} + \frac{1}{t^3}} = \frac{6 + 5 \times 0}{2 + 3 \times 0 + 0} = 3.$$

Here we have applied the basic power limit to $\dfrac{1}{t^2}$ and $\dfrac{1}{t^3}$.

(c) The highest power of t that appears is t^3, so we divide top and bottom by t^3. We have

$$\lim_{t \to \infty} \frac{t^2 + 1}{t^3 + 1} = \lim_{t \to \infty} \frac{t^2 + 1}{t^3 + 1} \times \frac{\left(\frac{1}{t^3}\right)}{\left(\frac{1}{t^3}\right)} = \lim_{t \to \infty} \frac{\frac{1}{t} + \frac{1}{t^3}}{1 + \frac{1}{t^3}} = \frac{0 + 0}{1 + 0} = 0.$$

Here we have applied the basic power limit to $\dfrac{1}{t}$ and $\dfrac{1}{t^3}$.

E-7. **Calculating using both exponential and power limits:**

(a) We have

$$\lim_{t \to \infty} \frac{a + \frac{1}{t}}{b + 0.2^t} = \frac{a + 0}{b + 0} = \frac{a}{b}.$$

Here we have used the basic power limit for $\dfrac{1}{t}$ and the basic exponential limit for 0.2^t.

(b) We have

$$\lim_{t \to \infty} \left(b \times 0.8^t - \frac{a}{t^2} \right) = b \times 0 - a \times 0 = 0.$$

Here we have used the basic exponential limit for 0.8^t and the basic power limit for $\dfrac{1}{t^2}$.

(c) The highest power of t that appears is t^2, so we divide top and bottom by t^2. We have

$$\lim_{t \to \infty} \frac{at^2 + 0.3^t}{bt^2 + 1} = \lim_{t \to \infty} \frac{at^2 + 0.3^t \left(\frac{1}{t^2}\right)}{bt^2 + 1 \left(\frac{1}{t^2}\right)} = \lim_{t \to \infty} \frac{a + 0.3^t \times \frac{1}{t^2}}{b + \frac{1}{t^2}} = \frac{a + 0 \times 0}{b + 0} = \frac{a}{b}.$$

Here we have used the basic exponential limit for 0.3^t and the basic power limit for $\dfrac{1}{t^2}$.

E-9. Concentration of salt: The eventual concentration of salt is given by the following limit:

$$\lim_{t \to \infty} \left(a + \frac{b}{t} \right) = a + b \times 0 = a.$$

(Here we have used the basic power limit for $\frac{1}{t}$.) Thus the eventual concentration of salt in the solution is a pounds per gallon.

S-1. Making a table: To make a table for $f(x) = x^2 - 1$ showing function values for $x = 4, 6, 8, \ldots$, we enter the function as $Y1 = X^2 - 1$ and use a table starting value of 4 and a table increment value of 2, resulting in the following table.

X	Y1
4	15
6	35
8	63
10	99
12	143
14	195
16	255

X=4

S-3. Making a table: To make a table for $f(x) = 16 - x^3$ showing function values for $x = 3, 7, 11, \ldots$, we enter the function as $Y1 = 16 - X^3$ and use a table starting value of 3 and a table increment value of 4, resulting in the following table.

X	Y1
3	-11
7	-327
11	-1315
15	-3359
19	-6843
23	-12151
27	-19667

X=3

S-5. Finding a limiting value: We enter the function as $Y1 = (4X^2 - 1)/(7X^2 + 1)$ and use a table starting value of 0 and a table increment value of 20, resulting in the following table. The table suggests a limiting value of about 0.5714.

X	Y1
0	-1
20	.57087
40	.57129
60	.57137
80	.57139
100	.57141
120	.57141

X=0

S-7. **Finding a minimum**: To find the minimum value of f, we make a table. We enter the function as $Y1 = X^2 - 8X + 21$ and use a table starting value of 0 and a table increment value of 1, resulting in the following table.

X	Y1
0	21
1	14
2	9
3	6
4	5
5	6
6	9

X=0

The table shows that f has a minimum value of 5 at $x = 4$, at least for values of x up to 6. Scrolling down the table through $x = 20$, we see that the function increases, so indeed the minimum value of f is 5 at $x = 4$.

S-9. **Finding a maximum**: To find the maximum value of f, we make a table. We enter the function as $Y1 = 9X^2 - 2^X + 1$ and use a table starting value of 0 and a table increment value of 1. Scrolling farther down the table yields the following table.

X	Y1
4	129
5	194
6	261
7	314
8	321
9	218
10	-123

X=8

The table shows that f reaches a maximum of 321 at $x = 8$.

S-11. **Making a table**: To make tables for $f(x) = 13/(0.93^x + 0.05)$, we enter the function as $Y1 = 13/(0.93^X + 0.05)$. First we use a table starting value of 1 and a table increment value of 1, resulting in the table on the left. Then we use a table starting value of 10 and a table increment value of 10, resulting in the table on the right.

X	Y₁
1	13.265
2	14.209
3	15.216
4	16.29
5	17.434
6	18.652
7	19.948

X=1

X	Y₁
10	24.345
20	45.736
30	79.575
40	123.97
50	169.81
60	206.83
70	231.23

X=10

1. **Harvard Step Test**: Here is a table of values for E as a function of P:

X	Y₁
150	100
160	93.75
170	88.235
180	83.333
190	78.947
200	75
210	71.429

X=150

(a) From the table above, we see that the index E decreases with increasing values of the variable P. This should hold for all values of P because increasing the denominator in the fraction defining E makes the fraction smaller. This means that a person having a larger total pulse count after the exercise than another person has a lower physical efficiency index and hence is not as physically fit.

(b) The index of someone with a total pulse count of 200 is expressed in functional notation by $E(200)$. From the table we generated we find the value to be 75.

(c) We saw in Part (b) that the index of someone with a total pulse count of 200 is 75. This is in the interval from 65 to 79 of the table in the text, and that table indicates that the physical condition of such a person is high average.

(d) According to the table in the text, for an excellent rating the index must be 90 or above. From the table we generated we see that this transition occurs between $P = 160$ and $P = 170$. Here is a table examining the function values more closely:

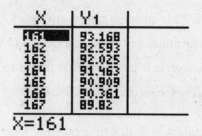

$$X=161$$

We see that the index is 90 or above when the total pulse count is 166 or lower.

3. **Earlier public high school enrollment**: Here are two tables of values for N as a function of t:

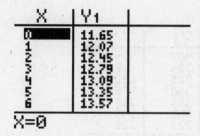

$$X=0 \qquad\qquad X=7$$

(a) From the second table we generated we find the value of $N(7)$ to be 13.75 million students. This is the number (in millions) of students enrolled in U.S. public high schools in the year 1972 (because that is 7 years after 1965).

(b) Examining the tables shown above suggests that the maximum value of the function occurs when $t = 11$. Extending the table to the end of the period (when $t = 20$) shows that the function values continue to decrease after $t = 11$. Thus the enrollment was the largest in 1976 (11 years after 1965). From the second table we see that the largest enrollment was 14.07 million students.

(c) Here 1965 corresponds to $t = 0$, and 1985 corresponds to $t = 20$. From the first table above we find that $N(0) = 11.65$, and examining further the table on our calculator yields the value $N(20) = 12.45$. The average yearly rate of change is then

$$\frac{N(20) - N(0)}{20} = \frac{12.45 - 11.65}{20} = 0.04 \text{ million students per year.}$$

Thus the average yearly rate of change from 1965 to 1985 is 0.04 million students per year, or 40,000 students per year. This relatively small result is misleading because, as we saw in Part (b), the enrollment actually increased significantly over this period (reaching 14.07 million students in 1972) before it decreased to a level near the original value.

5. **Competition**: This question can be answered by trial and error, but we will solve it by finding two formulas and making a table. Let n be the number of races run, $F = F(n)$ the time in seconds it takes the first friend to run a mile, and $S = S(n)$ the time in seconds it takes the second friend. We have

Time taken = Initial time taken − Decrease in time × Number of races.

Now for the first friend the initial time taken is 7 minutes, or 420 seconds, and the decrease in time is 13 seconds. This gives the formula $F = 420 - 13n$. Similar reasoning gives the formula $S = 440 - 16n$ for the second friend. Here is a partial table of values of these two functions:

X	Y1	Y2
1	407	424
2	394	408
3	381	392
4	368	376
5	355	360
6	342	344
7	329	328

X=1

From the table we see that the first race in which the second friend beats the first is the seventh race.

7. **Counting when order matters**: Here is a table of values for the factorial function:

X	Y1	
1	1	
2	2	
3	6	
4	24	
5	120	
6	720	
7	5040	

X=7

(a) The number of ways you can arrange 5 people in a line is 5!. From the table we see that the value is 120. (This can also be computed directly: $5! = 5 \times 4 \times 3 \times 2 \times 1 = 120$.) Thus there are 120 ways to arrange 5 people in a line.

(b) From the table we see that the factorial function first exceeds 1000 when the variable is 7, so 7 (or more) people will result in more than 1000 possible arrangements for a line.

(c) The number of guesses is the same as the number of ways to arrange 4 people in a line because there are 4 different digits given. Thus the number is 4!, which is 24. Hence we would need at most 24 guesses.

(d) The number of shufflings is the same as the number of ways to arrange 52 people in a line. Thus the number is 52!. The calculator gives that $52! = 8.07 \times 10^{67}$. Thus there are 8.07×10^{67} possible shufflings of a deck of cards.

9. **APR and EAR:**

(a) We would expect the EAR to be larger if interest is compounded more often because once interest is compounded it accrues additional interest.

(b) We proceed by making a table of values for

$$\text{EAR} = \left(1 + \frac{\text{APR}}{n}\right)^n - 1 = \left(1 + \frac{0.1}{n}\right)^n - 1$$

The table below on the left shows that when $n = 1$ (yearly compounding), the EAR is exactly 10%. The table on the right has been extended to include $n = 12$, monthly compounding, and it shows the EAR to be 10.471%.

X	Y1
1	.1
2	.1025
3	.10337
4	.10381
5	.10408
6	.10426
7	.10439

X=1

X	Y1
6	.10426
7	.10439
8	.10449
9	.10456
10	.10462
11	.10467
12	.10471

X=12

The table below includes $n = 365$ (daily compounding), and it shows an EAR of 10.516%.

X	Y1
360	.10516
361	.10516
362	.10516
363	.10516
364	.10516
365	.10516
366	.10516

X=365

(c) On a loan of \$5000 compounded monthly, after one year the interest owed would be

$$5000 \times \text{ monthly EAR} = 5000 \times 0.10471 = 523.55 \text{ dollars.}$$

The total amount owed is the loan amount plus the interest, which is $5000 + 523.55 = 5523.55$ dollars.

If we compound daily, after one year the interest owed would be

$$5000 \times \text{ daily EAR} = 5000 \times 0.10516 = 525.80 \text{ dollars.}$$

That gives a total amount owed of \$5525.80.

(d) The EAR when we compound continuously is

$$r = e^{\text{APR}} - 1 = e^{0.10} - 1 = 0.10517.$$

This is 10.517%, which is only 0.0046 percentage point (less than 0.05 percentage point) higher than monthly compounding.

11. **An amortization table for continuous compounding:**

(a) The monthly interest rate for this exercise is $r = \dfrac{\text{APR}}{12} = \dfrac{0.09}{12} = 0.0075$, we borrowed $P = \$3500$, and $t = 24$ months is the life of the loan. Our monthly payment for compounding continuously would be

$$M = \frac{3500(e^{0.0075} - 1)}{1 - e^{-0.0075 \times 24}} = \$159.95.$$

This is only 5 cents more than the payment for compounding monthly.

(b) The amortization table is a table of values for

$$B = \frac{P\left(e^{rt} - e^{rk}\right)}{e^{rt} - 1} = \frac{3500 \times \left(e^{0.0075 \times 24} - e^{0.0075 \times k}\right)}{e^{0.0075 \times 24} - 1}.$$

The amortization table is below. Comparing the entries here with those in Part (c) of Exercise 10 above, we see that the current entries are larger, but the difference is less than a dollar.

Number of payments made	Amount still owed	Number of payments made	Amount still owed
0	3500.00	13	1682.51
1	3366.40	14	1535.23
2	3231.79	15	1386.83
3	3096.17	16	1237.32
4	2959.53	17	1086.69
5	2821.85	18	934.92
6	2683.15	19	782.01
7	2543.40	20	627.94
8	2402.59	21	472.72
9	2260.73	22	316.33
10	2117.80	23	158.76
11	1973.79	24	0
12	1828.70		

13. **Inventory**:

   (a) The exercise tells us that $N = 36$, $c = 850$, and $f = 230$. So inventory expense is given by

$$E(Q) = \left(\frac{Q}{2}\right) \times 850 + \left(\frac{36}{Q}\right) \times 230.$$

   (b) For 3 cars, the yearly inventory expense is

$$E(3) = \left(\frac{3}{2}\right) \times 850 + \left(\frac{36}{3}\right) \times 230 = \$4035.$$

   (c) We can find this by generating a table of values for the inventory cost function given in Part (a). As we scroll down the table we see that the cost decreases until we get to \$3770, which corresponds to ordering 4 cars at a time. The table is shown below for $Q = 1$ to $Q = 7$ cars per order.

X	Y1
1	8705
2	4990
3	4035
4	3770
5	3781
6	3930
7	4157.9

X=4

   (d) Since we expect to sell 36 cars this year and we order 4 cars at a time to minimize our inventory cost, then we will place $\frac{36}{4} = 9$ orders to Detroit this year.

   (e) The average rate of increase in yearly inventory cost from ordering four cars to ordering six cars is

$$\frac{\text{Change in cost}}{\text{Change in cars}} = \frac{E(6) - E(4)}{6 - 4} = \frac{3930 - 3770}{2} = \$80 \text{ per year per additional car.}$$

15. **Falling with a parachute:**

(a) The velocity 2 seconds into the fall is expressed in functional notation as $v(2)$. Its value is
$$v(2) = 20(1 - 0.2^2) = 19.2 \text{ feet per second.}$$

(b) The average change in velocity during the first second of the fall is
$$\frac{\text{Change in velocity}}{\text{Seconds}} = \frac{v(1) - v(0)}{1} = \frac{16 - 0}{1} = 16 \text{ feet per second per second.}$$

The average velocity from the fifth to sixth second is
$$\frac{\text{Change in velocity}}{\text{Seconds}} = \frac{v(6) - v(5)}{1} = \frac{19.999 - 19.994}{1} = 0.005 \text{ foot per second per second.}$$

The velocity is increasing as the seconds go by (as can be seen from a table of values), but the average rate of increase is decreasing.

(c) We make a table of values for velocity. The table below, which shows velocity from the $t = 2$ through $t = 8$, indicates that velocity levels off at about 20 feet per second.

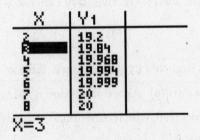

(d) Now 99% of terminal velocity is 99% of 20 feet per second, which is $0.99 \times 20 = 19.8$ feet per second. The table above shows that this occurs approximately 3 seconds into the fall.

In Example 2.1 it took 25 seconds to reach 99% of terminal velocity, while in this exercise it took only 3 seconds. This indicates that objects (such as parachutes) for which air resistance is large will reach terminal velocity faster. Thus, we would expect the feather to reach terminal velocity more quickly than a cannonball.

17. **Profit:**

(a) Assume that the total cost C is measured in dollars. Because
$$\text{Total cost} = \text{Variable cost} \times \text{Number of items} + \text{Fixed costs,}$$
we have $C = 50N + 150$.

(b) Assume that the total revenue R is measured in dollars. Because

$$\text{Total revenue} = \text{Selling price} \times \text{Number of items,}$$

we have $R = 65N$.

(c) Assume that the profit P is measured in dollars. Because

$$\text{Profit} = \text{Total revenue} - \text{Total cost,}$$

we have $P = R - C$. Using the formulas from Parts (a) and (b), we obtain

$$P = 65N - (50N + 150).$$

(This can also be written as $P = 15N - 150$.)

(d) We want to find for what value of N we have $P(N) = 0$. Examining a table of values for P shows that this occurs when $N = 10$. Thus a break-even point occurs at a production level of 10 widgets per month.

19. **A precocious child and her blocks:**

(a) If we think of going around the rectangle and adding up the sides as we encounter them, we see that the perimeter is $h + w + h + w$. That is two h's and two w's, so $P = 2h + 2w$ inches.

(b) We substitute $w = \dfrac{64}{h}$ in the above formula and get

$$P = 2h + 2 \times \frac{64}{h} \text{ inches.}$$

(c) To solve this exercise we first generate a table for $P(h)$. This is shown below for heights of 5 through 11 blocks. The smallest value is 32, and this happens when the height is 8. If the height is 8 then the width is $w = \dfrac{64}{8} = 8$ inches. So the blocks should be arranged in an 8 by 8 square to get a minimum perimeter.

X	Y1
5	35.6
6	33.333
7	32.286
8	32
9	32.222
10	32.8
11	33.636

X=8

(d) In this case the perimeter would still be $P = 2h + 2w$, and then $P = 2h + 2 \times \dfrac{60}{h}$ (measuring P in inches). We make a table of values as before. The table below shows a minimum perimeter of 31 inches, and this occurs when the height is 8 inches. However, a height of 8 inches gives a width of $w = \dfrac{60}{8} = 7.5$ inches, which cannot be accomplished without cutting the blocks into pieces.

X	Y1
5	34
6	32
7	31.143
8	31
9	31.333
10	32
11	32.909

X=8

We know that $hw = 60$ and so h must divide evenly into 60. This means that h can only be one of 1, 2, 3, 4, 5, 6, 10, 12, 15, 20, 30, or 60. If we look at these values in the table we see that a height of 6 or 10 leads us to a perimeter of 32, which is the minimum perimeter among these values. If the height is 6, then the width must be 10, and so the child should lay the blocks in a 6 by 10 pattern if she wants to get a minimum perimeter.

21. **Growth in length of haddock:**

(a) The value of $L(4)$ is

$$L(4) = 53 - 42.82 \times 0.82^4 = 33.64 \text{ centimeters.}$$

(This can also be found from a table of values for L.) This means that a haddock that is 4 years old is approximately 33.64 centimeters long.

(b) Using a table for the function L gives the values $L(5) = 37.12$, $L(10) = 47.11$, $L(15) = 50.82$, and $L(20) = 52.19$. Now the average yearly rate of growth in length from age 5 years to age 10 years is

$$\frac{L(10) - L(5)}{10 - 5} = \frac{47.11 - 37.12}{5} = 2.00 \text{ centimeters per year.}$$

The average yearly rate of growth in length from age 15 years to age 20 years is

$$\frac{L(20) - L(15)}{20 - 15} = \frac{52.19 - 50.82}{5} = 0.27 \text{ centimeter per year.}$$

The growth rate is lower over the later period, and this suggests that haddock grow more rapidly when they are young than when they are older.

(c) Scanning down a table on the calculator indicates that the function L increases to a limiting value of 53 as t gets larger and larger. Thus the longest haddock is 53 centimeters long.

23. **California earthquakes**: Here is a table of values for the function p:

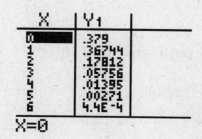

(a) We are given that $n = 3$, and from the table we see that the function value is $p(3) = 0.058$. Thus the probability is 0.058 or 5.8%.

(b) Scanning down the table suggests that the function values approach 0 as the variable n increases. Thus the limiting value is 0. This means that it is very unlikely that a California home will be affected by a large number of earthquakes over a 10-year period.

(c) We are given that $n = 0$, and from the table we see that the function value is $p(0) = 0.379$. Thus the probability of a California home being affected by no major earthquakes over a 10-year period is 0.379 or 37.9%.

(d) According to the hint, we have

> Probability of no major earthquake occurring
> + Probability of at least one major earthquake occurring $= 1$,

which can be rearranged to give

> Probability of at least one major earthquake occurring
> $= 1 -$ Probability of no major earthquake occurring.

We saw in Part (c) that the probability on the right has the value 0.379. Thus the probability of at least one major earthquake occurring is $1 - 0.379 = 0.621$ or 62.1%.

25. **Research project**: Answers will vary.

2.2 GRAPHS

E-1. **Shifting:**

(a) We shift the graph 2 units to the left:

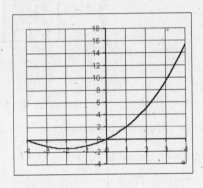

(b) We shift the original graph up 2 units:

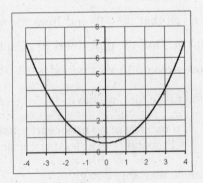

(c) We shift the original graph 2 units to the right:

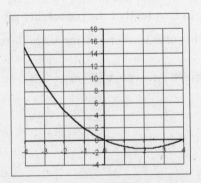

(d) We shift the original graph down 2 units:

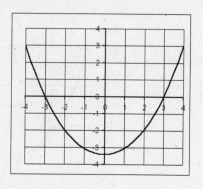

E-3. **Combinations of shifting and stretching**:

(a) We stretch the graph horizontally by a factor of 2:

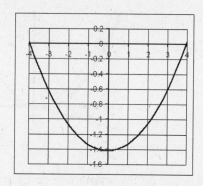

(b) We stretch the graph from Part (a) vertically by a factor of 2:

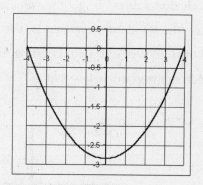

(c) We shift the graph from Part (b) up 1 unit:

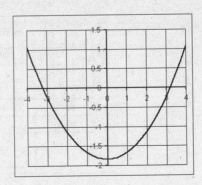

E-5. **Adding functions**: We show the graph with a horizontal span of -5 to 5 and a vertical span of -5 to 20. The darker graph is the sum.

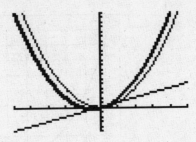

The heights of points on the graphs of f and g are added to get the heights of points on the graph of $f + g$.

S-1. **Graphs and function values**: To get the value of $f(3)$, we enter the function as $\text{Y1} = 2 - X^2$ and use the standard view, that is, the window with a horizontal span from -10 to 10 and the same vertical span. Locating the point on the graph corresponding to $X = 3$, we see from the graph below that $f(3) = -7$.

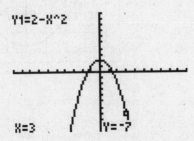

S-3. **Graphs and function values**: To get the value of $f(3)$, we enter the function as Y1 = $(X^2 + 2^X)/(X + 10)$ and use the standard view, that is, the window with a horizontal span from -10 to 10 and the same vertical span. Locating the point on the graph corresponding to $X = 3$, we see from the graph below that $f(3) = 1.31$.

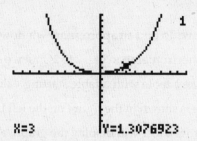

S-5. **Finding a window**: To find an appropriate window setup which will show a good graph of $\dfrac{x^3}{500}$ with a horizontal span of -3 to 3, first we enter the function as Y1 = $(X^3)/500$, and then we make a table with a table starting value of -3 and a table increment value of 1. Such a table is shown in the figure on the left below. The table of values shows that a vertical span from -0.06 to 0.06 will display the graph, as shown on the right below.

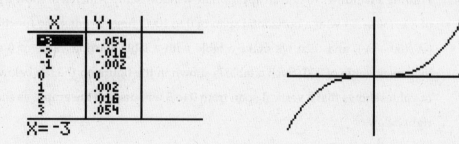

S-7. **Finding a window**: To find an appropriate window setup which will show a good graph of $\dfrac{x^4 + 1}{x^2 + 1}$ with a horizontal span of 0 to 300, first we enter the function as Y1 = $(X^4 + 1)/(X^2 + 1)$, and then we make a table with a table starting value of 0 and a table increment value of 50. Such a table is shown in the figure on the left below. The table of values shows that a vertical span from 0 to $90,000$ will display the graph, as shown on the right below.

S-9. **Finding a window**: To find an appropriate window setup which will show a good graph of $\dfrac{1}{x^2+1}$ with a horizontal span of -2 to 2, first we enter the function as $Y1 = 1/(X^2+1)$, and then we make a table with a table starting value of -2 and a table increment value of 1. Such a table is shown in the figure on the left below. The table of values shows that a vertical span from 0 to 1 will display the graph, as shown on the right below.

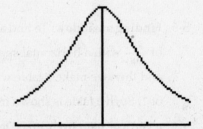

S-11. **Finding a window**: To find an appropriate window setup which will show a good graph of $3X/(50+X)$ with a horizontal span of 0 to 1000, first we enter the function as $Y1 = 3X/(50+X)$, and then we make a table with a table starting value of 0 and a table increment value of 200. Such a table is shown in the figure on the left below. The table of values shows that a vertical span from 0 to 3 will display the graph, as shown on the right below.

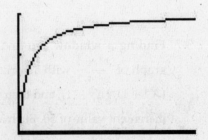

1. **Weekly cost:**

   (a) The weekly cost if there are 3 employees is $2500 + 350 \times 3 = 3550$ dollars.

   (b) Let n be the number of employees and C the weekly cost in dollars. Because

   $$\text{Weekly cost} = \text{Fixed cost} + \text{Cost per employee} \times \text{Number of employees},$$

   we have $C = 2500 + 350n$.

   (c) We show the graph with a horizontal span of 0 to 10 and a vertical span of 2500 to 7000. In choosing this vertical span we were guided by the table below.

   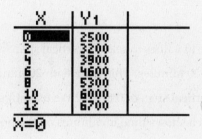 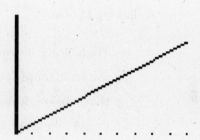

   (d) We want to find what value of the variable n gives the function value $C = 4250$. Looking at a table of values or the graph shows that this occurs when $n = 5$. Thus if there are 5 employees the weekly cost will be \$4250.

3. **Resale value:**

   (a) The resale value in the year 2001 is

   $$\begin{aligned} \text{Resale value} \quad &= \quad \text{Value in the year 2000} - \text{Decrease over 1 year} \\ &= \quad 18{,}000 - 1700 \\ &= \quad 16{,}300 \text{ dollars.} \end{aligned}$$

   Thus the resale value in the year 2001 is \$16,300.

   (b) Because

   $$\text{Resale value} = \text{Value in the year 2000} - \text{Decrease each year} \times \text{Number of years since 2000},$$

   we have $V = 18{,}000 - 1700t$.

   (c) We show the graph with a horizontal span of 0 to 4 and a vertical span of 10,000 to 20,000. In choosing this vertical span we were guided by the table below.

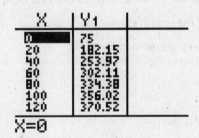

(d) Since 2003 is 3 years since 2000, the resale value in that year is $V(3)$ in functional notation. The value is $18,000 - 1700 \times 3 = 12,900$ dollars. (This can also be found from a table of values or the graph.)

5. **Baking a potato**:

(a) The first step is to make a table of values to select a vertical span. As suggested, we use a horizontal span of 0 to 120 minutes. This table with 20 minute increments is shown below. It suggests a vertical span of 0 to 420. We used this window setting to make the graph below, which shows time in minutes on the horizontal axis and temperature on the vertical axis.

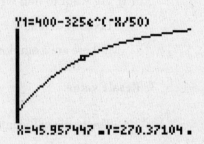

(b) The initial temperature of the potato is its temperature when placed in the oven. That is the value of P when t is zero:

$$P(0) = 400 - 325e^{-0/50} = 75 \text{ degrees.}$$

(c) The change in the potato's temperature in the first 30 minutes is

$$
\begin{aligned}
\text{Temperature at 30 minutes } - \text{ Initial temperature} \quad &= \quad P(30) - P(0) \\
&= \quad 221.636 - 75 \\
&= \quad 146.636 \text{ degrees.}
\end{aligned}
$$

The change in the potato's temperature in the second 30 minutes is

Temperature at 60 minutes $-$ Temperature at 30 minutes $= P(60) - P(30)$

$$= 302.111 - 221.636$$

$$= 80.475 \text{ degrees.}$$

The potato's temperature rose the most in the first 30 minutes.

The average rate of change during the first 30 minutes of baking is

$$\frac{\text{Change in temp}}{\text{Minutes elapsed}} = \frac{P(30) - P(0)}{30} = \frac{146.636}{30} = 4.89 \text{ degrees per minute.}$$

The average rate of change during the second 30 minutes of baking is

$$\frac{\text{Change in temp}}{\text{Minutes elapsed}} = \frac{P(60) - P(30)}{30} = \frac{80.475}{30} = 2.68 \text{ degrees per minute.}$$

We used three decimal accuracy for the various values of P to ensure that our final answer would still be accurate to two decimal places after division.

(d) The graph is concave down, and this tells us that, although the temperature is rising, the rate of increase in temperature is decreasing. In other words, the temperature is not rising as fast at later times as it was at first. Note that in Part (c) the average increase from 0 to 30 minutes was 4.89 degrees per minute, whereas the average increase from 30 to 60 minutes was only 2.68 degrees per minute.

(e) The potato will reach a temperature of 270 degrees after about 46 minutes. This can be seen by tracing the graph, as we have done in the graph above.

(f) If the potato is left in the oven a long time, its temperature will match that of the oven. To see what happens to the potato after a long time, we made a new table with a table starting value of 0 and a table increment value of 120. It appears that the limiting value is 400. We see that the temperature of the oven is about 400 degrees.

X	Y1
0	75
120	370.52
240	397.33
360	399.76
480	399.98
600	400
720	400

X=0

7. **Population growth:**

   (a) We show the graph with a horizontal span of 0 to 25 and a vertical span of −35 to 35. In choosing this vertical span we were guided by the table below.

   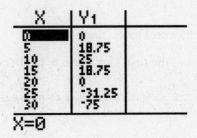 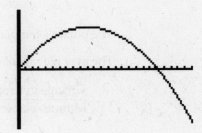

   (b) The growth over a week if the population at the beginning is 4 thousand animals is expressed as $G(4)$ in functional notation. From a table of values or the graph we find that the value is 16 thousand animals.

   (c) From a table of values or the graph we find that the value of $G(22)$ is −11 thousand animals. This means that if the population at the beginning of a week is 22 thousand animals then the population will decrease by 11 thousand over that week.

   (d) Tracing the graph shows that the function increases from $n = 0$ to about $n = 10$. Over this interval the graph is concave down. This means that, if the population at the beginning of the week is at most 10 thousand, for larger initial populations the growth over a week is larger, but the rate of increase in this growth actually decreases.

9. **The economic order quantity model:**

   (a)   i. Since the demand is $N = 400$ units per year and the carrying cost is $h = 24$ dollars per year, the function we want is

   $$Q(c) = \sqrt{\frac{800c}{24}} \text{ items per order.}$$

   Since we do not expect the fixed ordering cost to exceed $25, we use 0 to 25 as the horizontal span. The table below suggests a vertical span of 0 to 30. The graph is shown below. The horizontal axis corresponds to fixed order cost, and the vertical span is number of items per order.

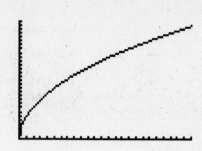

ii. If we evaluate the function above at 6 (using a table of values or the graph), we get that the number of items we need to order at a time is 14.14, or around 14.

iii. From the graph we can see that as the fixed order cost increases so does the number of items we need to order at a time.

(b) i. The exercise tells us that $N = 400$ and $c = 14$. So the function we need is

$$Q = \sqrt{\frac{2 \times 400 \times 14}{h}} = \sqrt{\frac{11200}{h}} \text{ items per order.}$$

Since h ranges from 0 to 25 dollars, we use that for the horizontal span. We used the table of values below to get the vertical span. We used 0 to 150 to make the graph below. The horizontal axis corresponds to carrying cost, and the vertical axis shows number of items per order.

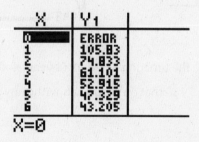 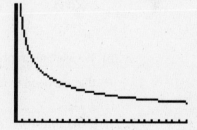

ii. If we evaluate the above function at 15 we get 27.33. So the optimum order size for a carrying cost of $15 is about 27 items.

iii. From the graph we can see that an increase in the carrying cost decreases the number of items ordered.

iv. Thinking of Q as a function of h alone and evaluating the function at 15 and 18, we see that $Q(15) = 27.325$ and $Q(18) = 24.944$. So the average rate of change from $15 to $18 is

$$\frac{\text{Change in number}}{\text{Change in carrying cost}} = \frac{Q(18) - Q(15)}{3} = -0.79 \text{ item per dollar.}$$

Here again we used three decimal places to ensure that our answer is still accurate to two decimal places after division.

v. The graph is concave up, and this tells us that the rate of decrease in the number of items ordered is decreasing as the carrying cost increases.

11. **An annuity**:

(a) The monthly interest rate is $r = 0.01$ and we want a monthly withdrawal of $M = \$200$, so the function is

$$P(t) = 200 \times \frac{1}{0.01} \times \left(1 - \frac{1}{(1 + 0.01)^t}\right) \text{ dollars.}$$

Note that t is measured in months.

We use a horizontal span of 0 to 500 months and we chose a vertical span of 0 to 25,000 dollars from the table below. The horizontal axis is number of months of withdrawal, and the vertical axis is dollars invested.

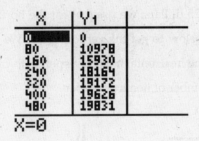
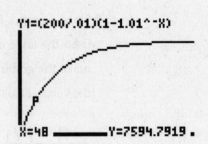

(b) We use the graph to evaluate the function at $t = 48$ (4 years is 48 months) and find that we need to invest \$7594.79 so that our child can withdraw \$200 per month for 4 years.

(c) We evaluate the function at $t = 120$ (10 years is 120 months) and find that we need to invest \$13,940.10 so that our child can withdraw \$200 per month for 10 years.

(d) The graph levels off and appears to have a limiting value. If we trace out toward the tail end of the graph, we see that an investment of \$20,000 will be sufficient to establish a \$200 per month perpetuity.

13. **Artificial gravity**:

(a) i. We want an acceleration of $a = 9.8$ meters per second per second, so we use the function

$$N = \frac{30}{\pi} \times \sqrt{\frac{9.8}{r}} \text{ rotations per minute.}$$

ii. We use a horizontal span of 10 to 200 meters, and from the table below choose a vertical span of 0 to 10 revolutions per minute. The horizontal axis corresponds to the radius of the station, and the vertical axis to revolutions per minute.

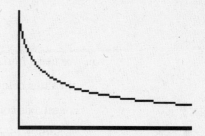

iii. As the distance r increases, the number of rotations decreases, although more and more slowly. In practical terms, the larger the space station, the fewer rotations per minute are needed to produce artificial gravity.

iv. If we evaluate the function using the graph, or a table of values, we see that in order to produce Earth gravity for a space station of radius $r = 150$ meters, we need $N = 2.44$ rotations per minute.

(b) i. Now we are looking at a space station of radius $r = 150$ meters. Then gravity as a function of revolutions per minute is given by

$$N = \frac{30}{\pi} \times \sqrt{\frac{a}{150}} \text{ revolutions per minute.}$$

ii. We want the horizontal span to be from 2.45 meters per second per second to 9.8 meters per second per second. The table of values below suggests a vertical span of 1 to 3 revolutions per minute. The horizontal axis shows acceleration due to gravity, and the vertical axis shows revolutions per second.

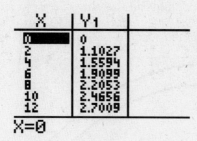 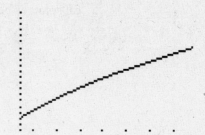

iii. As the desired acceleration increases, the number of rotations needed also increases, although more and more slowly.

15. **More on plant growth**:

(a)   i. We are assuming a rainfall of $R = 100$ millimeters, so our function, measured in kilograms per hectare, is

$$Y = -55.12 - 0.01535N - 0.00056N^2 + 3.946 \times 100$$
$$= 339.48 - 0.01535N - 0.00056N^2.$$

ii. We use a horizontal span of 0 to 800 kilograms per hectare and from the table of values below, a vertical span of -50 to 400 kilograms per hectare. The horizontal axis corresponds to initial biomass and the vertical axis corresponds to growth in biomass.

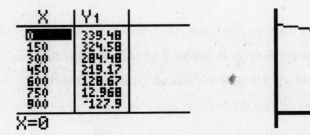

iii. As the amount of initial plant biomass, N, increases, the amount of growth, Y, decreases. Practically, there is a limit to how much biomass the land can support, and each kilogram present leaves less room for more plants to grow.

(b)   i. We are now looking at a rainfall of $R = 80$ millimeters, so our function now, measured in kilograms per hectare, is

$$Y = -55.12 - 0.01535N - 0.00056N^2 + 3.946 \times 80 = 260.56 - 0.01535N - 0.00056N^2.$$

ii. We add this new graph to the one above using the same horizontal and vertical spans as before.

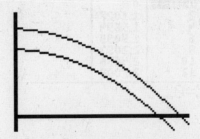

iii. We see that lowered rainfall decreases biomass growth. Yes, this is quite consistent with the prediction made in the previous exercise.

17. **Magazine circulation**:

(a) We use a horizontal span from 0 to 6 years, and we used the table of values shown below to choose a vertical span from 0 to 55 thousand circulation. The horizontal axis is years since 1992 and the vertical axis is circulation in thousands.

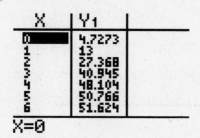 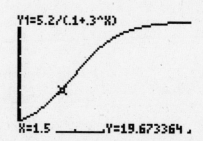

(b) Since 18 months is 1.5 years and t is measured in years, $C(1.5)$ expresses in functional notation the circulation of the magazine 18 months after it was started. The value of $C(1.5)$ can be found using the graph, as shown in the figure above on the right. We see that 18 months after it started the magazine has a circulation of 19.67 thousand, or 19,670 magazines.

(c) The graph is concave up for about the first 2 years. (Answers will vary here.) In practical terms, the circulation was increasing more and more quickly during the first two years.

(d) The circulation was increasing the fastest where the graph is steepest. The graph appears to be steepest near $t = 2$, or after 2 years. (Answers will vary here but should be consistent with the answer to Part (c).)

(e) To see the limiting value for C, which is in practical terms the level of market saturation, we increase the horizontal span to 20 years and trace the tail end of the graph. We see from the graph below that the limiting value is about 52 thousand magazines. (This can also be seen by scanning down a table.)

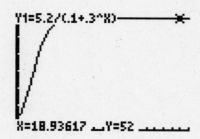

19. **Buffalo**:

(a) We show the graph with a horizontal span of 0 to 30 and a vertical span of 0 to 330.

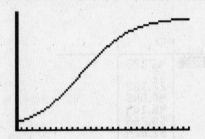

(b) The year 2002 corresponds to $t = 0$. From a table of values or the graph we find that the value of the function at $t = 0$ is 21. Thus there were 21 buffalo in the herd in 2002.

(c) We want to find what value of the variable t gives the function value $N = 300$. This can be done by tracing the graph, and the value is between $t = 24$ and $t = 25$. Since t is the number of years since 2002, the number of buffalo will first exceed 300 in the year 2026.

(d) Scrolling down a table on the calculator (or tracing the graph), we see that the limiting value of the function is 315. Thus the population will eventually increase to 315 buffalo.

(e) Tracing the graph shows that it is concave up from about $t = 0$ to about $t = 11$. It is concave down afterwards. This means that the herd is growing at an increasing rate from 2002 to 2013 and that the rate of growth begins to decrease after that time.

21. **Research project**: Answers will vary.

2.3 SOLVING LINEAR EQUATIONS

E-1. **Solving equations linear in one variable**: First we gather the terms involving x to the left side of the equation and those not involving x to the right side, then we factor out x, and then we divide by the coefficient of x.

(a) We have

$$
\begin{aligned}
xy^3 &= xy + y \\
xy^3 - xy &= y \\
(y^3 - y)x &= y \\
x &= \frac{y}{y^3 - y}.
\end{aligned}
$$

Thus the solution is $x = \dfrac{y}{y^3 - y}$. This can be simplified to $x = \dfrac{1}{y^2 - 1}$.

(b) We have

$$
\begin{aligned}
3x\sqrt{y} &= 2x + \sqrt{y} \\
3x\sqrt{y} - 2x &= \sqrt{y} \\
(3\sqrt{y} - 2)x &= \sqrt{y} \\
x &= \frac{\sqrt{y}}{3\sqrt{y} - 2}.
\end{aligned}
$$

Thus the solution is $x = \dfrac{\sqrt{y}}{3\sqrt{y} - 2}$.

(c) We have

$$
\begin{aligned}
yx + z^2 x &= z - y^2 x \\
yx + z^2 x + y^2 x &= z \\
(y + z^2 + y^2)x &= z \\
x &= \frac{z}{y + z^2 + y^2}.
\end{aligned}
$$

Thus the solution is $x = \dfrac{z}{y + z^2 + y^2}$.

(d) We have

$$
\begin{aligned}
\sqrt{9 + y} + x\sqrt{7 + y} &= yz + xz \\
x\sqrt{7 + y} - xz &= yz - \sqrt{9 + y} \\
(\sqrt{7 + y} - z)x &= yz - \sqrt{9 + y} \\
x &= \frac{yz - \sqrt{9 + y}}{\sqrt{7 + y} - z}.
\end{aligned}
$$

Thus the solution is $x = \dfrac{yz - \sqrt{9 + y}}{\sqrt{7 + y} - z}$.

E-3. **Finding inverse functions**: In each case we solve the equation $f(y) = x$ to find an expression y giving the inverse.

(a) We have

$$7y + 5 = x$$
$$7y = x - 5$$
$$y = \frac{x - 5}{7}.$$

Thus the inverse function of f is $y = \frac{x - 5}{7}$.

(b) We have

$$\frac{3y + 1}{2y - 5} = x$$
$$3y + 1 = x(2y - 5)$$
$$3y + 1 = 2xy - 5x$$
$$3y - 2xy = -5x - 1$$
$$(3 - 2x)y = -5x - 1$$
$$y = \frac{-5x - 1}{3 - 2x}.$$

Thus the inverse function of f is $y = \frac{-5x - 1}{3 - 2x}$. This can also be written as $y = \frac{5x + 1}{2x - 3}$.

(c) We have

$$\frac{2y - 4}{y} = x$$
$$2y - 4 = xy$$
$$2y - xy = 4$$
$$(2 - x)y = 4$$
$$y = \frac{4}{2 - x}.$$

Thus the inverse function of f is $y = \frac{4}{2 - x}$. This can also be written as $y = \frac{-4}{x - 2}$.

(d) We have

$$\frac{y}{4y - 3} = x$$
$$y = x(4y - 3)$$
$$y = 4xy - 3x$$
$$y - 4xy = -3x$$
$$(1 - 4x)y = -3x$$
$$y = \frac{-3x}{1 - 4x}.$$

Thus the inverse function of f is $y = \dfrac{-3x}{1-4x}$. This can also be written as $y = \dfrac{3x}{4x-1}$.

E-5. Making equations linear:

(a) Multiplying both sides of the equation $\dfrac{7}{x+1} = 3$ by $x+1$ gives

$$\dfrac{7}{x+1} \times (x+1) = 3(x+1)$$
$$7 = 3(x+1)$$
$$7 = 3x + 3.$$

Now we solve for x:

$$7 = 3x + 3$$
$$7 - 3 = 3x$$
$$4 = 3x$$
$$\dfrac{4}{3} = x.$$

Thus the solution is $x = \dfrac{4}{3}$.

(b) Multiplying both sides of the equation $\dfrac{9}{2x} = \dfrac{7}{3}$ by $6x$ gives

$$\dfrac{9}{2x} \times 6x = \dfrac{7}{3} \times 6x$$
$$\dfrac{9 \times 6}{2} = \dfrac{7 \times 6}{3} \times x$$
$$27 = 14x.$$

Now we solve for x:

$$27 = 14x$$
$$\dfrac{27}{14} = x.$$

Thus the solution is $x = \dfrac{27}{14}$.

(c) Squaring both sides of the equation

$$3\sqrt{x} = a$$

gives

$$(3\sqrt{x})^2 = a^2$$
$$9x = a^2.$$

Now we solve for x:

$$x = \dfrac{a^2}{9}.$$

Thus the solution is $x = \dfrac{a^2}{9}$.

(d) Multiplying both sides of the equation $3\sqrt{x} = 2\sqrt{x} + \dfrac{4}{\sqrt{x}}$ by $\sqrt{x}$ gives

$$
\begin{aligned}
3\sqrt{x} \times \sqrt{x} &= \left(2\sqrt{x} + \dfrac{4}{\sqrt{x}}\right) \times \sqrt{x} \\
3x &= 2\sqrt{x} \times \sqrt{x} + \dfrac{4}{\sqrt{x}} \times \sqrt{x} \\
3x &= 2x + 4.
\end{aligned}
$$

Now we solve for x:

$$
\begin{aligned}
3x &= 2x + 4 \\
x &= 4.
\end{aligned}
$$

Thus the solution is $x = 4$.

S-1. Linear equations: To solve $3x + 7 = x + 21$ for x, we follow the procedure described in the solution of Exercise E-1 above: First we gather the terms involving the variable to the left side of the equation and those not involving the variable to the right side, then we factor out the variable, and then finally we divide by the coefficient of the variable.

$$
\begin{aligned}
3x + 7 &= x + 21 \\
3x - x &= 21 - 7 \\
2x &= 14 \\
x &= 7.
\end{aligned}
$$

S-3. Linear equations: To solve $3 - 5x = 23 + 4x$ for x, we follow the procedure described above, just as for Exercise S-1:

$$
\begin{aligned}
3 - 5x &= 23 + 4x \\
-5x - 4x &= 23 - 3 \\
-9x &= 20 \\
x &= \dfrac{20}{-9} = -\dfrac{20}{9}.
\end{aligned}
$$

S-5. Linear equations: To solve $12x + 4 = 55x + 42$ for x, we follow the procedure described above, just as for Exercise S-1:

$$
\begin{aligned}
12x + 4 &= 55x + 42 \\
12x - 55x &= 42 - 4 \\
-43x &= 38 \\
x &= \dfrac{38}{-43} = -\dfrac{38}{43}.
\end{aligned}
$$

S-7. **Linear equations**: To solve for $cx + d = 12$ for x, we follow the procedure described above, just as for Exercise S-1:

$$
\begin{aligned}
cx + d &= 12 \\
cx &= 12 - d \\
x &= \frac{12 - d}{c}.
\end{aligned}
$$

S-9. **Linear equations**: To solve $2k + m = 5k + n$ for k, we follow the procedure described above, just as for Exercise S-1, only in this case the variable is named k instead of x:

$$
\begin{aligned}
2k + m &= 5k + n \\
2k - 5k &= n - m \\
-3k &= n - m \\
k &= \frac{n - m}{-3} = -\frac{n - m}{3}.
\end{aligned}
$$

This expression can also be written as $k = \dfrac{m - n}{3}$.

S-11. **Linear equations**: To solve $2x - 3 = 4 - 2x$ for x we follow the procedure described above, just as for Exercise S-1:

$$
\begin{aligned}
2x - 3 &= 4 - 2x \\
2x + 2x &= 4 + 3 \\
4x &= 7 \\
x &= \frac{7}{4}.
\end{aligned}
$$

1. **Gross domestic product**:

   (a) Because P is calculated as the sum of C, I, G, and E, we have $P = C + I + G + E$.

   (b) We are given that $C = 7303.7$, $I = 1593.2$, $G = 1972.9$, and $E = -423.6$, all in billions of dollars.

       i. Recall that E is exports minus imports. Because E is negative, imports were larger.

       ii. From the given values we have

$$
P = 7303.7 + 1593.2 + 1972.9 - 423.6 = 10{,}446.2,
$$

   so the gross domestic product was 10,446.2 billion dollars.

   (c) We solve the equation $P = C + I + G + E$ for E by subtracting $C + I + G$ from both sides. The result is $E = P - C - I - G$.

(d) We are given the same values for C, I, and G as in Part (b), and we know that $P = 10{,}886$ (in billions of dollars). The equation from Part (c) gives

$$E = 10{,}886 - 7303.7 - 1593.2 - 1972.9 = 16.2,$$

so the net sales to foreigners is 16.2 billion dollars.

3. **Resale value:**

(a) The formula gives $V(3) = 12.5 - 1.1 \times 3 = 9.2$ thousand dollars. This means that the resale value of the boat will be 9.2 thousand dollars, or \$9200, at the start of the year 2004 (because that is 3 years after the start of 2001).

(b) We want to find what value of the variable t gives the value 7 for the function V. This can be done either by examining a table of values (or a graph) for the function or by solving the linear equation $12.5 - 1.1t = 7$. We take the second approach:

$$
\begin{aligned}
12.5 - 1.1t &= 7 \\
-1.1t &= 7 - 12.5 && \textbf{Subtract } 12.5 \textbf{ from both sides.} \\
t &= \frac{7 - 12.5}{-1.1} && \textbf{Divide both sides by } -1.1. \\
t &= 5.
\end{aligned}
$$

Thus the resale value will be 7 thousand dollars at the start of the year 2006 (because that is 5 years after the start of 2001).

(c) We solve the equation $V = 12.5 - 1.1t$ for t as follows:

$$
\begin{aligned}
V &= 12.5 - 1.1t \\
V - 12.5 &= -1.1t && \textbf{Subtract } 12.5 \textbf{ from both sides.} \\
\frac{V - 12.5}{-1.1} &= t && \textbf{Divide both sides by } -1.1.
\end{aligned}
$$

Thus the formula is $t = \dfrac{V - 12.5}{-1.1}$. This can also be written as $t = \dfrac{12.5 - V}{1.1}$ or as $t = 11.36 - 0.91V$.

(d) When we use the value $V = 4.8$ in the answer to Part (c), we obtain $t = \dfrac{4.8 - 12.5}{-1.1} = 7$. Thus the resale value will be 4.8 thousand dollars at the start of the year 2008 (because that is 7 years after the start of 2001).

5. **Gas mileage:**

(a) The exercise tells us that we have $g = 12$ gallons of gas in the tank and that our car gets $m = 24$ miles per gallon. So we can travel

$$d = gm = 12 \times 24 = 288 \text{ miles.}$$

(b) To solve for m, we divide both sides by g:

$$d = gm$$

$$\frac{d}{g} = m \text{ miles per gallon.}$$

    i. This equation tells us that the mileage our car gets is the distance traveled divided by the amount of gas used.

    ii. If we drive $d = 335$ miles using $g = 13$ gallons of gas, then our car got

$$m = \frac{d}{g} = \frac{335}{13} = 25.77 \text{ miles per gallon.}$$

(c) i. Since $d = 425$ miles, we have $425 = gm$, so $g = \dfrac{425}{m}$.

    ii. To choose a window size we need first to pick a reasonable range for gas mileage. We choose 0 to 30 for the horizontal span. The table below suggests a vertical span from 0 to 75. The horizontal axis corresponds to gas mileage, and the vertical axis corresponds to tank capacity.

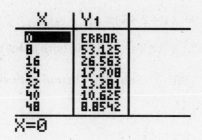
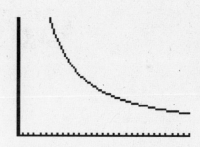

The graph is not a straight line.

7. **Supply and demand**: At the equilibrium price we have $S = D$, so the expression $1.9S - 0.7$ has the same value as the expression $2.8 - 0.6S$; the equilibrium price is then the common value of these expressions. Thus we want to solve the equation $1.9S - 0.7 = 2.8 - 0.6S$ for S. Here are the steps:

$$1.9S - 0.7 = 2.8 - 0.6S$$

$$1.9S + 0.6S = 2.8 + 0.7 \qquad \textbf{Add } 0.7 \textbf{ and } 0.6S \textbf{ to both sides.}$$

$$2.5S = 3.5 \qquad \textbf{Combine terms.}$$

$$S = \frac{3.5}{2.5} \qquad \textbf{Divide both sides by } 2.5.$$

$$S = 1.4.$$

Thus the equilibrium price occurs at the quantity 1.4 billion bushels; using the formula $P = 1.9S - 0.7$ shows that that price is $1.9 \times 1.4 - 0.7 = 1.96$ dollars per bushel.

9. **Fire engine pressure**:

(a) We are given that $NP = 80$ and $K = 0.51$. Because the length of the hose is 80 feet, we have $L = \dfrac{80}{50} = 1.6$. Thus $EP = 80(1.1 + 0.51 \times 1.6) = 153.28$. The engine pressure is 153.28 psi.

(b) We are given that $K = 0.73$ and $EP = 150$. Because the length of the hose is 45 feet, we have $L = \dfrac{45}{50} = 0.9$. Thus $150 = NP(1.1 + 0.73 \times 0.9)$. Dividing both sides by $1.1 + 0.73 \times 0.9$ gives $NP = \dfrac{150}{1.1 + 0.73 \times 0.9} = 85.37$. The nozzle pressure is 85.37 psi.

(c) To solve the equation $EP = NP(1.1 + K \times L)$ for NP, we divide both sides by $1.1 + K \times L$. The result is $NP = \dfrac{EP}{1.1 + K \times L}$.

(d) We are given that $EP = 160$ and $NP = 90$. Because the length of the hose is 190 feet, we have $L = \dfrac{190}{50} = 3.8$. Thus $160 = 90(1.1 + K \times 3.8)$. We solve for K as follows:

$$
\begin{aligned}
160 &= 90(1.1 + 3.8K) \\
\frac{160}{90} &= 1.1 + 3.8K \qquad &&\textbf{Divide both sides by } 90. \\
\frac{160}{90} - 1.1 &= 3.8K \qquad &&\textbf{Subtract } 1.1 \textbf{ from both sides.} \\
\frac{\frac{160}{90} - 1.1}{3.8} &= K \qquad &&\textbf{Divide both sides by } 3.8. \\
0.18 &= K.
\end{aligned}
$$

Thus the "K" factor is 0.18.

(e) We solve for K as follows:

$$
\begin{aligned}
EP &= NP(1.1 + K \times L) \\
\frac{EP}{NP} &= 1.1 + K \times L \qquad &&\textbf{Divide both sides by } NP. \\
\frac{EP}{NP} - 1.1 &= K \times L \qquad &&\textbf{Subtract } 1.1 \textbf{ from both sides.} \\
\frac{\frac{EP}{NP} - 1.1}{L} &= K \qquad &&\textbf{Divide both sides by } L.
\end{aligned}
$$

Thus $K = \dfrac{\frac{EP}{NP} - 1.1}{L}$.

11. **Net profit:**

(a) Our gross profit is the number of dolls sold times the selling price. Then we have to subtract the cost of making the dolls, which is the number sold times the cost per doll. We also have to subtract the cost of rent. We find that

$$\text{Net profit} = \text{Sales price} \times \#\,\text{sold} - \text{Cost per doll} \times \#\,\text{sold} - \text{Rent}$$
$$p = dn - cn - R.$$

This can also be written as $P = (d - c)n - R$.

(b) The exercise tells us that the rent is $R = \$1280$, it costs $c = \$2$ each to make the dolls, the selling price is $d = \$6.85$, and we sell $n = 826$ dolls. So our net profit would be

$$p = dn - cn - R = 6.85 \times 826 - 2 \times 826 - 1280 = \$2726.10.$$

(c) We have

$$p = dn - cn - R$$
$$p + cn = dn - R \qquad \textbf{Add } cn \textbf{ to both sides.}$$
$$p + cn + R = dn \qquad \textbf{Add } R \textbf{ to both sides.}$$

$$\frac{p + cn + R}{n} = d \qquad \textbf{Divide both sides by } n.$$

Thus the formula is $d = \dfrac{p + cn + R}{n}$. This can also be written as $d = \dfrac{p + R}{n} + c$.

(d) The required net profit is $p = \$4000$, the rent is $R = \$1200$, the cost per doll is $c = \$2$, and we expect to sell $n = 700$ dolls. So under these conditions the price we need to get for each doll is

$$d = \frac{p + cn + R}{n} = \frac{4000 + 700 \times 2 + 1200}{700} = \$9.43.$$

13. **Temperature conversion:**

(a) Here $K(30)$ is the temperature in kelvins when the temperature on the Celsius scale is 30 degrees. The value is $30 + 273.15 = 303.15$ kelvins.

(b) To express the Celsius temperature C in terms of kelvins K we solve the given equation relating the two:

$$K = C + 273.15$$
$$K - 273.15 = C \qquad \textbf{Subtract 273.15 from each side.}$$

Thus we have

$$C = K - 273.15.$$

(c) We use the expression from Part (b) to replace C in $F = 1.8C + 32$. The result is

$$F = 1.8(K - 273.15) + 32.$$

This can also be written as $F = 1.8K - 459.67$.

(d) If the temperature is $K = 310$ kelvins, then by Part (c) the Fahrenheit temperature is

$$F = 1.8(310 - 273.15) + 32 = 98.33 \text{ degrees.}$$

15. **Running ants**:

(a) The ambient temperature is $T = 30$ degrees Celsius, so in functional notation $S(30)$ is the speed of the ants. This is calculated as

$$S(30) = 0.2 \times 30 - 2.7 = 3.3 \text{ centimeters per second.}$$

(b) We have

$$
\begin{aligned}
S &= 0.2T - 2.7 \\
S + 2.7 &= 0.2T \quad \textbf{Add 2.7 to each side.}
\end{aligned}
$$

$$\frac{S + 2.7}{0.2} = T \quad \textbf{Divide each side by 0.2.}$$

Thus the formula is $T = \dfrac{S + 2.7}{0.2}$.

(c) The speed is $S = 3$ centimeters per second, so the ambient temperature is

$$T = \frac{S + 2.7}{0.2} = \frac{3 + 2.7}{0.2} = 28.5 \text{ degrees Celsius.}$$

17. **Profit**:

(a) Assume that the total cost C is in dollars. Because

$$\text{Total cost} = \text{Variable cost} \times \text{Number of items} + \text{Fixed costs,}$$

we have $C = 55N + 200$.

(b) Assume that the total revenue R is in dollars. Because

$$\text{Total revenue} = \text{Selling price} \times \text{Number of items,}$$

we have $R = 58N$.

(c) Assume that the profit P is in dollars. Because

$$\text{Profit} = \text{Total revenue} - \text{Total cost,}$$

we have $P = R - C$. Using the formulas from Parts (a) and (b), we obtain

$$P = 58N - (55N + 200).$$

Because

$$58N - (55N + 200) = 58N - 55N - 200 = 3N - 200,$$

this can also be written as $P = 3N - 200$. The latter is the formula we use in Part (d).

(d) We want to find what value of the variable N gives the function value $P = 0$. Thus we need to solve the equation $3N - 200 = 0$. We solve it as follows:

$$
\begin{aligned}
3N - 200 &= 0 \\
3N &= 200 \qquad \textbf{Add 200 to both sides.} \\
N &= \frac{200}{3} \qquad \textbf{Divide both sides by 3.} \\
N &= 66.67.
\end{aligned}
$$

Thus the break-even point occurs at a production level of about 66.67 thousand widgets per month. This is $66,670$ widgets per month.

19. **Plant growth:**

(a) We have

$$Y = -55.12 - 0.01535N - 0.00056N^2 + 3.946R$$

$$Y + 55.12 = -0.01535N - 0.00056N^2 + 3.946R \qquad \textbf{Add 55.12 to each side.}$$

$$Y + 55.12 + 0.01535N = -0.00056N^2 + 3.946R \qquad \textbf{Add } 0.01535N \textbf{ to each side.}$$

$$Y + 55.12 + 0.01535N + 0.00056N^2 = 3.946R \qquad \textbf{Add } 0.00056N^2 \textbf{ to each side.}$$

$$\frac{Y + 55.12 + 0.01535N + 0.00056N^2}{3.946} = R \qquad \textbf{Divide both sides by 3.946.}$$

Thus the formula is

$$R = \frac{Y + 55.12 + 0.01535N + 0.00056N^2}{3.946}.$$

(b) If there is no plant growth, then $Y = 0$, and so

$$R = \frac{55.12 + 0.01535N + 0.00056N^2}{3.946} \text{ millimeters.}$$

(c) We use a horizontal span of 0 to 800 kilograms per hectare. Using the table below, we choose a vertical span of 0 to 120 millimeters. The horizontal axis corresponds to initial plant biomass, and the vertical axis corresponds to rainfall.

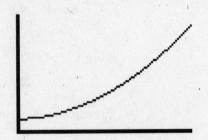

(d) As N increases, R also increases. In practical terms, the greater the initial plant biomass, the more rainfall it needs just to maintain its size.

(e) If the initial plant biomass is $N = 400$ kilograms per hectare, then we can evaluate the function using the graph to get $R = 38.23$ millimeters.

(f) If the initial plant biomass N is 500 kilograms per hectare, then, by using the graph, we find that the corresponding zero isocline rainfall R needed to sustain that biomass is $R(500) = 51.39$ millimeters of rain. Unfortunately, there are only 40 millimeters of rain, so the plants will die back.

21. **Competition between populations**:

(a) We need to solve the equation $5(1 - m - n) = 0$ for n. We solve it as follows:

$$
\begin{aligned}
5(1 - m - n) &= 0 \\
1 - m - n &= 0 \quad \textbf{Divide both sides by } 5. \\
1 - m &= n \quad \textbf{Add } n \textbf{ to both sides.}
\end{aligned}
$$

Thus $n = 1 - m$ describes this isocline.

(b) We need to solve the equation $6(1 - 0.7m - 1.2n) = 0$ for n. We solve it as follows:

$$
\begin{aligned}
6(1 - 0.7m - 1.2n) &= 0 \\
1 - 0.7m - 1.2n &= 0 \quad \textbf{Divide both sides by } 6. \\
1 - 0.7m &= 1.2n \quad \textbf{Add } 1.2n \textbf{ to both sides.} \\
\frac{1 - 0.7m}{1.2} &= n \quad \textbf{Divide both sides by } 1.2.
\end{aligned}
$$

Thus $n = \dfrac{1 - 0.7m}{1.2}$ describes this isocline.

(c) By Parts (a) and (b), the per capita growth rates will both be zero if $n = 1 - m$ and $n = \dfrac{1 - 0.7m}{1.2}$. This gives the equation $1 - m = \dfrac{1 - 0.7m}{1.2}$. We solve this equation as follows:

$$
\begin{aligned}
1 - m &= \frac{1 - 0.7m}{1.2} \\
1.2(1 - m) &= 1 - 0.7m \qquad \textbf{Multiply both sides by } 1.2. \\
1.2 - 1.2m &= 1 - 0.7m \qquad \textbf{Expand.} \\
1.2 - 1 &= -0.7m + 1.2m \qquad \textbf{Add } 1.2m \textbf{ and subtract } 1 \textbf{ on both sides.} \\
0.2 &= 0.5m \qquad \textbf{Combine terms.} \\
\frac{0.2}{0.5} &= m \qquad \textbf{Divide both sides by } 0.5. \\
0.4 &= m.
\end{aligned}
$$

Thus $m = 0.4$, and from the equation in Part (a) we find that $n = 1 - m = 1 - 0.4 = 0.6$. Hence the equilibrium point occurs at $m = 0.4, n = 0.6$ thousand animals.

2.4 SOLVING NONLINEAR EQUATIONS

E-1. **Factoring quadratics**:

(a) We know from the signs that we are looking for factors of the form $(x + a)(x - b)$ with $0 < a < b$. We want $ab = 9$ and $b - a = 8$. Thus we should choose $a = 1$ and $b = 9$. Hence $x^2 - 8x - 9 = (x + 1)(x - 9)$. Of course this can also be written as $x^2 - 8x - 9 = (x - 9)(x + 1)$.

(b) We know from the signs that we are looking for factors of the form $(x + a)(x + b)$ with a, b positive. We want $ab = 9$ and $a + b = 6$. Thus we should choose $a = 3$ and $b = 3$. Hence $x^2 + 6x + 9 = (x + 3)(x + 3)$, or $x^2 + 6x + 9 = (x + 3)^2$. *Note*: Here we could also use the perfect-square formula.

(c) We use the following basic formula from the text:

$$
x^2 + (a + b)x + ab = (x + a)(x + b).
$$

Since there is no x term in $x^2 - 16$, we have $a + b = 0$; further, from the constant term we have $ab = -16$. We choose $a = -4$ and $b = 4$. (Of course, the choice $a = 4$ and $b = -4$ is also valid.) Hence $x^2 - 16 = (x - 4)(x + 4)$.

(d) We know from the signs that we are looking for factors of the form $(x - a)(x - b)$ with a, b positive. We want $ab = 35$ and $a + b = 12$. We choose $a = 7$ and $b = 5$. (Of course, the choice $a = 5$ and $b = 7$ is also valid.) Hence $x^2 - 12x + 35 = (x - 7)(x - 5)$.

E-3. **Solving higher-order equations**:

(a) We first bring everything to the left-hand side of the equation: $x^3 - 3x^2 - 4x = 0$. Next we note that each term has a common factor of x, and we factor this out: $x(x^2 - 3x - 4) = 0$. The quadratic term factors as $(x + a)(x - b)$ with $0 < a < b$. We want $ab = 4$ and $b - a = 3$. Thus we should choose $a = 1$ and $b = 4$. Hence $x^2 - 3x - 4 = (x + 1)(x - 4)$, and the original equation is equivalent to $x(x + 1)(x - 4) = 0$. Thus we get three solutions: $x = 0$, $x = -1$, and $x = 4$.

(b) The left-hand side is already factored, and at least one factor must be zero. Thus either $2x - 3 = 0$ or $3x + 4 = 0$ or $2x - 5 = 0$. Solving each of these linear equations gives the three solutions $x = \dfrac{3}{2}$, $x = -\dfrac{4}{3}$, and $x = \dfrac{5}{2}$.

(c) The first step is to factor the quadratic term $x^2 + 2x - 15$. This factors as $(x + a)(x - b)$ with $0 < b < a$. We want $ab = 15$ and $a - b = 2$. Thus we should choose $a = 5$ and $b = 3$. Hence $x^2 + 2x - 15 = (x + 5)(x - 3)$. Thus the original equation can be written as $(x - 4)(x + 5)(x - 3) = 0$. Thus we get three solutions: $x = 4$, $x = -5$, and $x = 3$.

(d) First we factor each quadratic term. On page 148 of the text, the factorization $x^2 - 5x + 6 = (x - 3)(x - 2)$ is found. The second quadratic, namely $x^2 + 7x + 12$, factors as $(x + a)(x + b)$ with a, b positive. We want $ab = 12$ and $a + b = 7$. We choose $a = 4$ and $b = 3$. Hence $x^2 + 7x + 12 = (x + 4)(x + 3)$. Thus the original equation can be written as $(x - 3)(x - 2)(x + 4)(x + 3) = 0$. Thus we get four solutions: $x = 3$, $x = 2$, $x = -4$, and $x = -3$.

E-5. **Rational solutions**:

(a) Here the constant term a_0 and the leading term $a_n = a_4$ are both 3. The only divisors of 3 are ± 1 and ± 3. Hence the only possible rational solutions are $\pm\dfrac{1}{3}$, ± 1, and ± 3. By plugging these into the equation, we find that $\dfrac{1}{3}$ and 3 are solutions and that the others are not. Thus the only rational solutions are $x = \dfrac{1}{3}$ and $x = 3$.

(b) Here the constant term a_0 is -1 and the leading term $a_n = a_3$ is 2. The only divisors of -1 are ± 1, and the only divisors of 2 are ± 1 and ± 2. Hence the only possible rational solutions are $\pm\dfrac{1}{2}$ and ± 1. By plugging these into the equation, we find that $\dfrac{1}{2}$ is a solution and that the others are not. Thus the only rational solution is $x = \dfrac{1}{2}$.

(c) Here the constant term a_0 is 1 and the leading term $a_n = a_4$ is 2. The only divisors of 1 are ± 1, and the only divisors of 2 are ± 1 and ± 2. Hence the only possible rational solutions are $\pm\dfrac{1}{2}$ and ± 1. By plugging these into the equation, we find that none of these is a solution. Thus there are no rational solutions of the equation. (In fact, examining the equation shows that there are no real solutions: $2x^4 + 3x^2 + 1$ is at least 1 if x is real because x^4 and x^2 are nonnegative in that case.)

S-1. **The crossing-graphs method**: To solve $\dfrac{20}{1 + 2^x} = x$ using the crossing-graphs method, we enter the $\dfrac{20}{1 + 2^X}$ as Y1 and X as Y2. Using the standard viewing window, we obtain the graph below, which shows that the solution is $x = 2.69$.

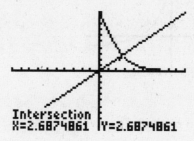

S-3. **The crossing-graphs method**: To solve $3^x + x = 2^x + 1$ using the crossing-graphs method, we enter the $3^X + X$ as Y1 and $2^X + 1$ as Y2. Using a horizontal span of -2 to 2 and a vertical span of 0 to 5, we obtain the graph below, which shows that the solution is $x = 0.59$.

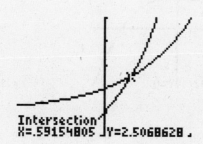

S-5. **The crossing-graphs method**: To solve $\dfrac{5}{x^2 + x + 1} = 1$ using the crossing-graphs method, we enter the $5/(X^2 + X + 1)$ as Y1 and 1 as Y2. Using a horizontal span of -2 to 2 and a vertical span of 0 to 7, we obtain the graphs below, which show that the solutions are $x = -2.56$ and $x = 1.56$.

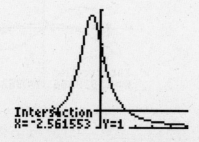
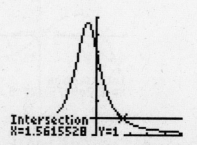

S-7. **The single-graph method**: To solve $\dfrac{5}{x^2 + x + 1} - 1 = 0$ using the single-graph method, we enter the function as Y1. Using a horizontal span of -4 to 4 and a vertical span from -3 to 6, we obtain the graphs below, which show that the solutions are $x = -2.56$ and $x = 1.56$.

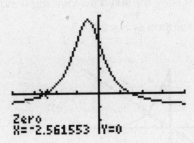
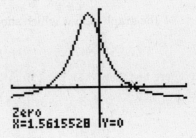

S-9. **The single-graph method**: To solve $\dfrac{-x^4}{x^2 + 1} + 1 = 0$ using the single-graph method, we enter the function as Y1. Using a horizontal span of -2 to 2 and a vertical span from -2 to 2, we obtain the graphs below, which show that the solutions are $x = -1.27$ and $x = 1.27$.

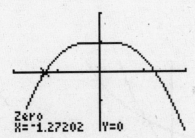
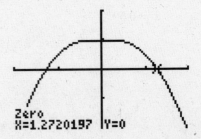

S-11. **The crossing-graphs method**: To solve $x^3 - 5x = 1 - x^2$ using the crossing-graphs method, we enter the $X^3 - 5X$ as Y1 and $1 - X^2$ as Y2. Using a horizontal span of -3 to 3 and a vertical span of -12 to 12, we obtain the graphs below, which show that the solutions are $x = -2.71$, $x = -0.19$, and $x = 1.90$.

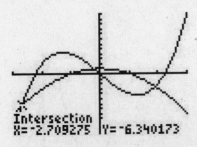
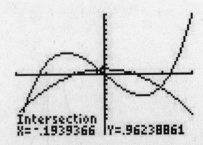

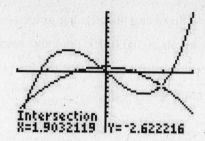

Intersection
X=1.9032119   Y=-2.622216

PLEASE NOTE: The exercises in this section involve solving nonlinear equations. There are many ways to do this, so it is important that students, graders, and instructors be aware of the various ways to solve equations. For example, some people prefer to use tables and get better and better approximations by refining the table; some will graph and zoom in repeatedly; other will use two graphs and find their intersection; others may graph and trace; and still others will rearrange the expression and find the appropriate root of an equation.

1. **A population of foxes**:

    (a) The number of foxes introduced is the value of N when t is zero:

    $$N(0) = \frac{37.5}{0.25 + 0.76^0} = 30 \text{ foxes.}$$

    (b) The exercise gave no information about the horizontal span. We have chosen to look at the fox population over a 25-year period. The following table of values led us to choose a vertical span of 0 to 160 foxes. The horizontal axis is time in years, and the vertical axis is number of foxes.

X	Y1
0	30
5	74.471
10	119.32
15	140.82
20	147.56
25	149.37
30	149.84

    X=0

    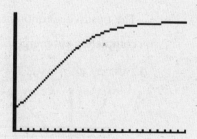

    The population of foxes grows rapidly at first, and then the growth slows down until there is almost no growth at all.

    (c) The fox population reaches 100 individuals when $N(t) = 100$. Thus we need to solve the equation

    $$\frac{37.5}{0.25 + 0.76^t} = 100$$

for t. We do this using the crossing graphs method. In the figure below we have added the graph of 100 (thick line) and calculated the crossing point. We read from the bottom of the screen that the fox population will reach 100 individuals after 7.58 years.

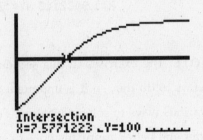

Intersection
X=7.5771223 Y=100

3. **Revisiting the floating ball**: We need to make a graph with a horizontal span of -2 to 7. We used the table of values on the left below to choose a vertical span of -3300 to 2100. The graph is on the right below.

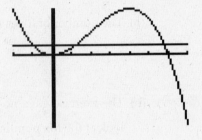

In the figures below we have found the other two solutions, $d = -0.98$ foot and $d = 5.80$ feet. The negative solution would have the ball floating above the surface of the water. The positive solution is larger than the diameter of the ball, and so the ball would be completely submerged.

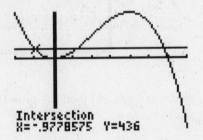

Intersection
X=-.9778575 Y=436

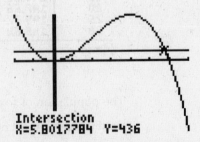

Intersection
X=5.8017784 Y=436

5. **Falling with a parachute**: The man has fallen 140 feet when

$$12.5(0.2^t - 1) + 20t = 140.$$

Thus we need to solve this equation. We use the crossing graphs method. Even with a parachute, it doesn't take long to fall 140 feet—surely less than 20 seconds. Thus we take a horizontal span of 0 to 20. The distance he falls increases from 0 up to 140, and so we use a slightly larger vertical span of 0 to 160. Below we have graphed $12.5(0.2^t - 1) + 20t$, and we have added the graph of 140 (thick line). We get the crossing point at $t = 7.62$. Thus the man falls 140 feet in 7.62 seconds.

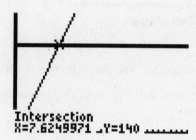

7. **Reaction rates**:

   (a) The exercise suggests a horizontal span of 0 to 100. The table of values on the left below led us to choose a vertical span of -5 to 25. The graph is on the right below.

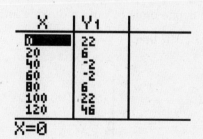 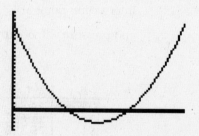

   (b) The reaction rate when the concentration is 15 moles per cubic meter is represented by $R(15)$ in functional notation. Using a table of values or the graph gives that the value is 9.25 moles per cubic meter per second.

(c) We want to find what two values of the variable x give the value 0 for the function R. That is, we want to find the two solutions of the equation $0.01x^2 - x + 22 = 0$. We do this using the single-graph method. As the graphs below indicate, the solutions are $x = 32.68$ and $x = 67.32$. Thus the reaction is in equilibrium at the concentrations of 32.68 moles per cubic meter and 67.32 moles per cubic meter.

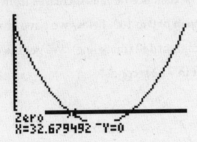
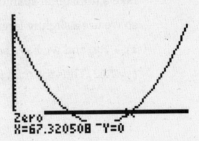

9. **Van der Waals equation**: The pressure is $p = 100$ atm, and the temperature is $T = 500$ kelvins. Putting these values into the equation, we have

$$100 = \frac{0.082 \times 500}{V - 0.043} - \frac{3.592}{V^2}.$$

We want to know the value of V. That is, we need to solve the equation above.

For the graph of $\frac{0.082 \times 500}{V - 0.043} - \frac{3.592}{V^2}$ the exercise suggests we use a horizontal span from 0.1 to 1 liter. The table of values below leads us to choose a vertical span of 0 to 400 atm. We have added the graph of 100 (thick line) to the picture and calculated the intersection point at $V = 0.37$. Thus one mole of carbon dioxide at this temperature and pressure will occupy a volume of 0.37 liter.

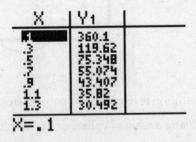
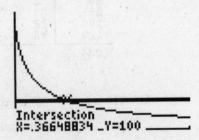

11. **Radiocarbon dating**: Since the half-life of C_{14} is $H = 5.77$ thousand years, the formula that gives the fraction of the original amount present is

$$\frac{A}{A_0} = 0.5^{t/5.77}.$$

(Here t is measured in thousands of years.) We want to know then this value is $\frac{1}{3}$. That is, we want to solve the equation

$$\frac{1}{3} = 0.5^{t/5.77}.$$

We use the crossing graphs method to do that. Since half of the C_{14} will be gone in 5.77 thousand years, and half of that will be gone in another 5.77 thousand years, we know that there will be only a quarter of the original amount of carbon 14 left after $2 \times 5.77 = 11.54$ thousand years. Thus the fraction reaches $\frac{1}{3}$ sometime before 12 thousand years. Hence to graph $\frac{A}{A_0}$ we use a horizontal span of 0 to 12. The fraction starts at 1 and decreases. Thus we use a vertical span of 0 to 1. In the picture below, we have added the graph of $\frac{1}{3}$ and calculated the intersection point at $t = 9.15$. Thus the tree died about 9.15 thousand years ago.

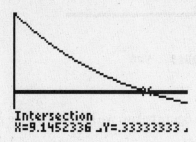

```
Intersection
X=9.1452336 _Y=.33333333 .
```

13. **Grazing kangaroos:**

    (a) i. The exercise suggests a horizontal span of 0 to 2000 pounds per acre. The table of values below for G led us to choose a vertical span of -1 to 3. The graph shows available vegetation biomass on the horizontal axis and daily intake on the vertical axis.

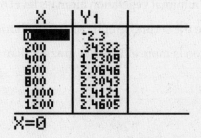
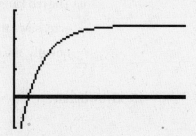

X	Y1
0	-2.3
200	.34322
400	1.5309
600	2.0646
800	2.3043
1000	2.4121
1200	2.4605

X=0

      ii. The graph is concave down. Since the graph is increasing, the kangaroo eats more when the vegetation biomass is greater. Small changes in V at the lower biomass levels cause the kangaroo to eat much more, while small changes in V at higher levels do not change the eating habits of the kangaroo very much.

iii. The minimal vegetation biomass is the amount that will cause the kangaroo to consume zero pounds. Thus we want to find the value of V where the graph crosses the horizontal axis. In the left-hand picture below, we have calculated where the graph crosses the horizontal axis as $V = 163.08$ pounds per acre.

iv. The satiation level is the limit of the amount the kangaroo will eat. That is where the curve levels off and becomes horizontal. Tracing toward the end of the graph, we see that G values approach 2.5. So the satiation level for the western grey kangaroo is a daily intake of about 2.5 pounds.

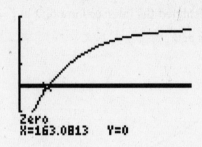

Zero
X=163.0813   Y=0

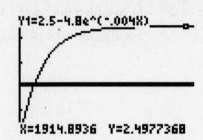

Y1=2.5-4.8e^(-.004X)

X=1914.8936   Y=2.4977368

(b)   i. In the figure below, we have added the graph of R.

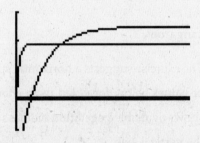

ii. The red kangaroo has a minimal vegetation biomass level of $V = 0$ pounds per acre, since that is where the second graph crosses the horizontal axis. Consequently, the red kangaroo is more efficient at grazing than the grey kangaroo.

15. **Growth rate:**

(a) As suggested in the exercise, we use a horizontal span of 0 to 1000 pounds per acre. The table below leads us to choose a vertical span of -1 to 1. The horizontal axis on the graph below corresponds to vegetation biomass and the vertical axis to the per capita growth rate.

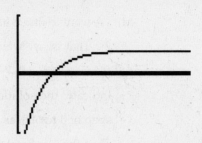

(b) The per capita growth rate is zero where the graph crosses the horizontal axis. This point is calculated in the graph below and shows that $r = 0$ when $V = 201.18$. The population size is stable when the vegetation level is 201.18 pounds per acre.

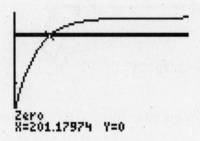

17. **Breaking even**:

(a) Assume that C is measured in dollars. Because

$$\text{Total cost} = \text{Variable cost} \times \text{Number} + \text{Fixed costs},$$

we have

$$C = 65N + 700.$$

(b) Assume that R is measured in dollars. Because

$$\text{Total revenue} = \text{Price} \times \text{Number},$$

we have $R = pN$, and thus

$$R = (75 - 0.02N)N.$$

(c) Assume that P is measured in dollars. Because

$$\text{Profit} = \text{Total revenue} - \text{Total cost},$$

we have $P = R - C$. Using Parts (a) and (b) gives

$$P = (75 - 0.02N)N - (65N + 700).$$

(This can also be written as $P = -0.02n^2 + 10N - 700$.)

(d) We want to find what two values of the variable N give the value 0 for the function P. That is, we want to find the two solutions of the equation $(75 - 0.02N)N - (65N + 700) = 0$. We do this using the single-graph method. As the graphs below indicate, the solutions are $N = 84.17$ and $N = 415.83$. (We used a horizontal span of 0 to 500, as suggested in the exercise, and a vertical span of -800 to 650.) Thus the two break-even points are 84.17 thousand widgets per month and 415.83 thousand widgets per month.

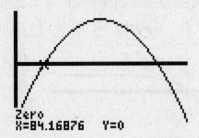

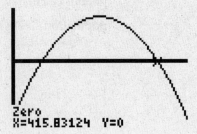

19. **Water flea**: We are given that $N_0 = 50$, so the relation is

$$e^{0.44t} = \frac{N}{50}\left(\frac{228 - 50}{228 - N}\right)^{4.46}.$$

We want to find what value of t corresponds to the value 125 for N. That is, we want to find the solution of the equation

$$e^{0.44t} = \frac{125}{50}\left(\frac{228 - 50}{228 - 125}\right)^{4.46}.$$

Note that the right-hand side of this equation is a constant. We solve this equation using the crossing-graphs method, putting the function $e^{0.44t}$ in **Y1** and the constant $\frac{125}{50}\left(\frac{228 - 50}{228 - 125}\right)^{4.46}$ in **Y2**. As the graph below indicates, the solution is $t = 7.63$. (After examining a table of values, we used a horizontal span of 0 to 8 and a vertical span of 0 to 40.) Thus it takes 7.63 days for the population to grow to 125.

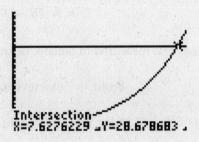

21. **Radius of a shock wave:**

   (a) We are given that $E = 10^{15}$, and we want to find what value of the variable t gives the value 4000 for the function R. That is, we want to find the solution of the equation $4.16(10^{15})^{0.2}t^{0.4} = 4000$. We do this using the crossing-graphs method. As the graph on the left below indicates, the solution is $t = 0.91$. (From the statement about the period over which the relation is valid, we expect the time to be around 1 second or less. After examining a table of values, we used a horizontal span of 0 to 1 and a vertical span of 0 to 4500.) Thus 0.91 second is required for the shock wave to reach a point 40 meters away.

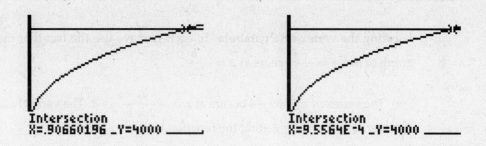

   (b) We are given that $E = 9 \times 10^{20}$, and we want to find what value of the variable t gives the value 4000 for the function R. That is, we want to find the solution of the equation $4.16(9 \times 10^{20})^{0.2}t^{0.4} = 4000$. We do this using the crossing-graphs method. As the graph on the right above indicates, the solution is $t = 9.6 \times 10^{-4} = 0.00096$. (Because the energy is much greater here than in Part (a), we expect the time required to be much smaller. After examining a table of values, we used a horizontal span of 0 to 0.001 and a vertical span of 0 to 4500.) Thus 0.00096 second is required for the shock wave to reach a point 40 meters away.

   (c) We are given that $t = 1.2$, and we want to find what value of the variable E gives the value 5000 for the function R. That is, we want to find the solution of the equation $4.16E^{0.2}1.2^{0.4} = 5000$. We do this using the crossing-graphs method. As the graph below indicates, the solution is $E = 1.74 \times 10^{15}$. (Because the time required in Part (a) is close to the time required here, we expect the energy to be on the order of 10^{15} here also. After examining a table of values, we used a horizontal span of 0 to 2×10^{15} and a vertical span of 0 to 5500.) Thus an energy of 1.74×10^{15} ergs was released by the explosion.

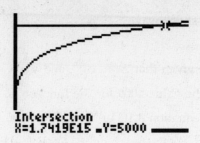

```
Intersection
X=1.7419E15 ▪Y=5000 ▬▬▬▬
```

23. **Research project**: Answers will vary.

2.5 OPTIMIZATION

E-1. **Locating the vertex of a parabola**: In each part we use the fact that the vertex of the graph of $ax^2 + bx + c$ occurs at $x = -\dfrac{b}{2a}$.

(a) The vertex of $x^2 + 6x - 4$ occurs at $x = -\dfrac{6}{2 \times 1} = -3$. The vertical coordinate of the vertex is found by getting the function value at $x = -3$:

$$\text{Vertical coordinate of vertex} = (-3)^2 + 6 \times (-3) - 4 = -13.$$

Thus the vertex of the parabola is $(-3, -13)$. Because the leading coefficient 1 is positive, this is a minimum.

(b) The vertex of $3x^2 - 30x + 1$ occurs at $x = -\dfrac{-30}{2 \times 3} = 5$. The vertical coordinate of the vertex is found by getting the function value at $x = 5$, which is $3 \times 5^2 - 30 \times 5 + 1 = -74$. Thus the vertex of the parabola is $(5, -74)$. Because the leading coefficient 3 is positive, this is a minimum.

(c) The vertex of $-2x^2 + 12x - 3$ occurs at $x = -\dfrac{12}{2 \times -2} = 3$. The vertical coordinate of the vertex is found by getting the function value at $x = 3$, which is $-2 \times 3^2 + 12 \times 3 - 3 = 15$. Thus the vertex of the parabola is $(3, 15)$. Because the leading coefficient -2 is negative, this is a maximum.

E-3. **Determining if the horizontal axis is crossed**:

(a) The vertex of $x^2 - 2x + 5$ occurs at $x = -\dfrac{-2}{2 \times 1} = 1$. The vertical coordinate of the vertex is found by getting the function value at $x = 1$, which is $1^2 - 2 \times 1 + 5 = 4$. Thus the vertex is above the horizontal axis. Because the leading coefficient 1 is positive, the vertex is a minimum. This means that the parabola opens upward from the vertex at $(1, 4)$, so it does not cross the horizontal axis. Another way to say this is that the vertical coordinate 4 of the vertex is the minimum value of the function, so the graph does not cross the horizontal axis.

(b) The vertex of $x^2 + 4x - 1$ occurs at $x = -\dfrac{4}{2 \times 1} = -2$. The vertical coordinate of the vertex is found by getting the function value at $x = -2$, which is $(-2)^2 + 4 \times (-2) - 1 = -5$. Thus the vertex is below the horizontal axis. Because the leading coefficient 1 is positive, the parabola opens upward from the vertex at $(-2, -5)$, so it crosses the horizontal axis twice.

E-5. **Distance from a line**: To find the x value giving the point on the line nearest the origin, we find where the minimum of D, the square of the distance function, occurs. To analyze D we expand the expression and collect terms:

$$D = x^2 + (x+1)^2 = x^2 + (x^2 + 2x + 1) = 2x^2 + 2x + 1.$$

Now $2x^2 + 2x + 1$ is a quadratic function in standard form, and the vertex occurs at $x = -\dfrac{2}{2 \times 2} = -\dfrac{1}{2}$. Because the leading coefficient 2 is positive, this is a minimum. Thus the desired x value is $x = -\dfrac{1}{2}$.

S-1. **Maximum**: To find the maximum value of $5x + 4 - x^2$ on the horizontal span of 0 to 5, we graph the function. A table of values leads us to choose a vertical span of 0 to 15. The graph, below, shows a maximum value of 10.25 at $x = 2.5$.

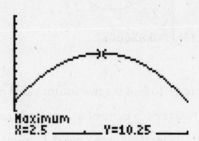

S-3. **Minimum**: To find the minimum value of $x + \dfrac{x+5}{x^2+1}$ on the horizontal span of 0 to 5, we graph the function. A table of values leads us to choose a vertical span of 0 to 7. The graph, below, shows a minimum value of 3.40 at $x = 1.92$.

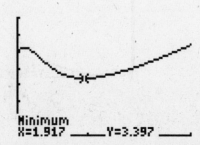

S-5. **Maximum**: To find the maximum value of $x^{1/x}$ on the horizontal span of 0 to 10, we graph the function. A table of values leads us to choose a vertical span of 0 to 3. The graph, below, shows a maximum value of 1.44 at $x = 2.72$.

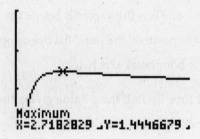

S-7. **Maxima and minima**: To find maxima and minima of $f = x^3 - 6x + 1$ with a horizontal span from -2 to 2 and a vertical span from -10 to 10, we graph the function. The graph on the left below shows the maximum is at $x = -1.41$, $y = 6.66$, while the graph on the right below shows the minimum is at $x = 1.41$, $y = -4.66$.

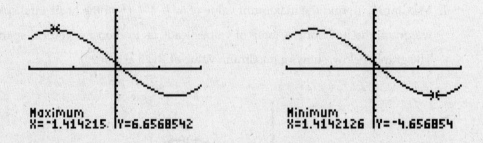

S-9. **Endpoint maximum**: To find the maximum value of $x^3 + x$ on the horizontal span of 0 to 5, we graph the function. A table of values leads us to choose a vertical span of 0 to 130. The graph, below, is increasing, so the maximum value of 130 occurs at the endpoint $x = 5$.

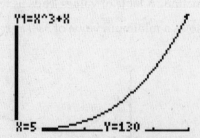

S-11. **Maxima and minima**: To find maxima and minima of $f = x(100 - 2x)$ with a horizontal span of 0 to 50, we graph the function. A table of values leads us to choose a vertical span of 0 to 1500. The graph on the left below shows that the maximum is at $x = 25$, $y = 1250$, while the other graphs show that the two minima are at $x = 0$, $y = 0$ and at $x = 50$, $y = 0$, both of which are endpoint minima.

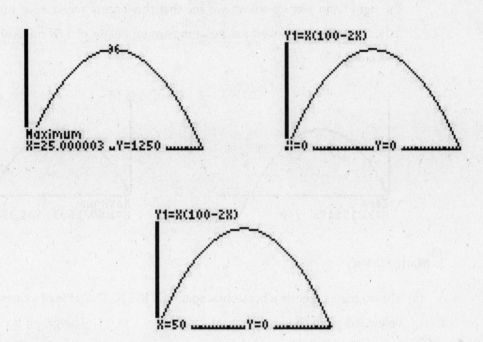

1. **The cannon at a different angle**:

   (a) We can look at the table of values below to get both horizontal and vertical spans. The cannonball follows the graph of y until it hits the ground, which corresponds to the horizontal axis. Thus we are only interested in the graph where it is positive. The table below shows that the cannonball strikes the ground just short of 4 miles, so we use 0 to 4 as a horizontal span. The table also shows that the cannonball doesn't get higher than about 1.5 miles. We use 0 to 2 miles for the vertical span.

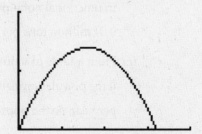

(b) The cannonball strikes the ground where the graph crosses the horizontal axis. In the left-hand figure below, we have expanded the vertical span a little so that the crossing point is clearly visible, and the calculator shows that this crossing point occurs at $x = 3.22$. Thus the cannonball travels 3.22 miles downrange.

(c) The cannonball reaches its maximum height where the graph reaches a peak. In the right-hand picture below, we see that this occurs when $x = 1.61$ and $y = 1.39$. Thus the cannonball reaches a maximum height of 1.39 miles at 1.61 miles downrange.

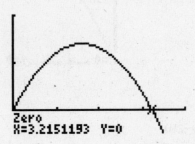
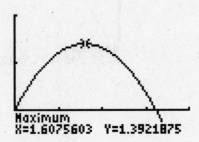

3. **Marine fishery:**

(a) The exercise suggests a horizontal span of 0 to 1.5. The table of values on the left below led us to choose a vertical span of -0.1 to 0.1. The graph is on the right below.

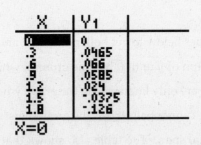
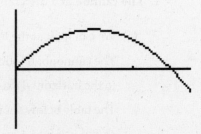

(b) The growth rate if the population size is 0.24 million tons is represented by $G(0.24)$ in functional notation. From a table of values or the graph we see that the value is 0.04 million tons per year.

(c) From a table of values or the graph we see that $G(1.42) = -0.02$. This means that if the population size is 1.42 million tons then the growth rate is -0.02 million tons per year, so the population is decreasing at a rate of 0.02 million tons per year.

(d) To find where the largest growth rate occurs, we locate the peak of the graph. In the figure below, we see that the peak occurs when $n = 0.67$ and $G = 0.07$. Thus the growth rate is the largest at a population size of 0.67 million tons.

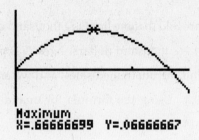

Maximum
X=.66666699 Y=.06666667

5. **Forming a pen**:

(a) We use two sides that are W feet long and one that is L feet long, so the total amount that we need is $2W + L$ feet of fence.

(b) The area of a rectangle is the width times the length, and because that area is to be 100 square feet we have $WL = 100$.

(c) To solve $WL = 100$ for L we divide both sides by W. The result is $L = \dfrac{100}{W}$.

(d) By Part (a) the total amount of fence needed is $2W + L$ feet, and by the equation in Part (c) this gives $F = 2W + \dfrac{100}{W}$.

(e) After examining a table of values, we used a horizontal span of 1 to 15 and a vertical span of 0 to 110. The graph is on the left below.

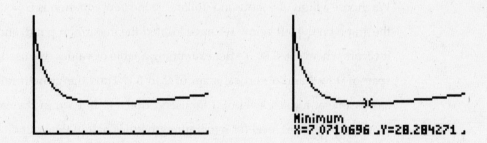

Minimum
X=7.0710696 Y=28.284271

(f) In the graph on the right above we have located the minimum point, and we see that it occurs where $W = 7.07$ feet. From the equation in Part (c) we find that the corresponding length is $L = \dfrac{100}{7.07} = 14.14$. Thus using 7.07 feet perpendicular to the building and 14.14 feet parallel requires a minimum amount of fence.

7. **Maximum sales growth**:

   (a) First we find a formula for unattained sales. Now

$$\text{Unattained sales} = \text{Limit} - \text{Sales level},$$

   and we are told that the limit is 4 thousand dollars. Thus the formula for unattained sales is $4 - s$ thousand dollars. Now G is proportional to the product of the sales level s and the unattained sales, and we are told that the constant of proportionality is 0.3. Using the formula for unattained sales, we see that the equation is $G = 0.3s(4 - s)$.

   (b) After examining a table of values, we used a horizontal span of 0 to 4 and a vertical span of 0 to 1.5. The graph is on the left below.

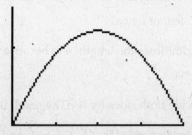 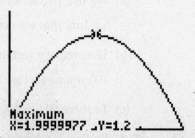

   (c) In the graph on the right above we have located the maximum point, and we see that it occurs where $s = 2.00$. Thus the growth rate is as large as possible at a sales level of 2 thousand dollars.

   (d) We choose a limit of 6 thousand dollars, so the new equation is $G = 0.3s(6 - s)$. In the graph on the left below we have located the maximum point, and we see that it occurs where $s = 3.00$. (After examining a table of values, we used a horizontal span of 0 to 6 and a vertical span of 0 to 3.) Thus the growth rate is as large as possible at a sales level of 3 thousand dollars. In each of the cases we have examined, the sales level for maximum growth is half of the limit on sales.

   If we change the constant of proportionality, say to 0.5, and keep a limit of 6 thousand dollars, then the formula for G changes to $G = 0.5s(6 - s)$. In the graph on the right below we have located the maximum point, and we see that it still occurs where $s = 3.00$. (After examining a table of values, we used a horizontal span of 0 to 6 and a vertical span of 0 to 5.) Thus the growth rate is again as large as possible at a sales level of 3 thousand dollars. This is still half of the limit on sales, which suggests that the relationship doesn't change if the proportionality constant is changed.

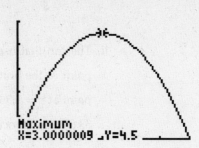

(e) Because the sales level for maximum growth is always half of the limit on sales, we can calculate this limit from the data by taking twice the sales level for maximum growth—that is, by doubling the sales level where the growth rate changes from increasing to decreasing.

9. **An aluminum can, continued**:

(a) If the radius is 1 inch then the height is

$$h(1) = \frac{15}{\pi} = 4.77 \text{ inches.}$$

The area is

$$A(1) = 2\pi + 30 = 36.28 \text{ square inches.}$$

(b) If the radius is 5 inches then the height is

$$h(5) = \frac{15}{\pi 5^2} = 0.19 \text{ inch.}$$

The area is

$$A(5) = 2\pi 5^2 + \frac{30}{5} = 163.08 \text{ square inches.}$$

(c) i. The aluminum will have a radius of only a few inches, and so we use a horizontal span of 0 to 4 inches. The table below suggests a vertical span of 0 to 50. In the graph below, the radius is on the horizontal axis, and the surface area is on the vertical axis. The graph shows the surface area needed to make a can holding 15 cubic inches as a function of the radius. From the graph it is evident that, as the radius increases, first the amount of aluminum needed decreases to a minimum, and then it increases.

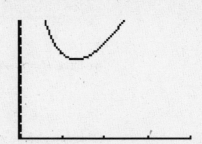

ii. The can that uses the minimum amount of aluminum is represented by the point at the bottom of the graph. In the figure below, we have located that point at $r = 1.34$ and $A = 33.67$. Thus we get a can of minimum surface area, 33.67 square inches, if we use a radius of 1.34 inches.

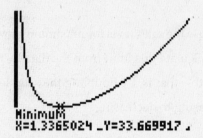

Minimum
X=1.3365024 _Y=33.669917 .

iii. We know that the radius should be about 1.34 to minimize the amount of aluminum used, so the height in that case would be

$$h(1.34) = \frac{15}{\pi 1.34^2} = 2.66 \text{ inches.}$$

11. **Profit**:

(a) Assume that C is measured in dollars. Because

$$\text{Total cost} = \text{Variable cost} \times \text{Number} + \text{Fixed costs,}$$

we have

$$C = 60N + 600.$$

(b) Assume that R is measured in dollars. Because

$$\text{Total revenue} = \text{Price} \times \text{Number,}$$

we have $R = pN$, and thus, using $p = 70 - 0.03N$, we find

$$R = (70 - 0.03N)N.$$

(c) Assume that P is measured in dollars. Because

$$\text{Profit} = \text{Total revenue} - \text{Total cost,}$$

we have $P = R - C$. Using Parts (a) and (b) gives

$$P = (70 - 0.03N)N - (60N + 600).$$

(This can also be written as $P = -0.03n^2 + 10N - 600$.)

(d) We graph the function P. In the graph below we used a horizontal span of 0 to 300, as suggested in the exercise, and a vertical span of -600 to 300. The graph below shows the maximum, and we see that it occurs at $N = 166.67$. Thus profit is maximized at a production level of 166.67 thousand widgets per month.

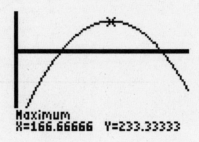

Maximum
X=166.66666 Y=233.33333

13. **Laying phone cable, continued:**

(a) Now

$$\text{Total cost} = \text{Cost for cable on shore} + \text{Cost for cable under water.}$$

Also,

$$\text{Cost for cable on shore} = 300 \times \text{Length of cable on shore}$$
$$= 300L,$$

and

$$\text{Cost for cable under water} = 500 \times \text{Length of cable under water}$$
$$= 500W.$$

Thus

$$C = 300L + 500W.$$

(b) We use the formula for W in terms of L from Part (a) of Exercise 12: Because $C = 300L + 500W$, that formula for W gives

$$C = 300L + 500\sqrt{1 + (5 - L)^2}.$$

(c) Again, we use the formula for W in terms of L from Part (a) of Exercise 12. Because $L = 1$, the amount of cable under water is

$$W = \sqrt{1 + (5 - 1)^2} = 4.12 \text{ miles.}$$

By Part (b) of this exercise, the total cost of the project is

$$C = 300 \times 1 + 500\sqrt{1 + (5 - 1)^2} = 2361.55 \text{ dollars.}$$

Here is the picture:

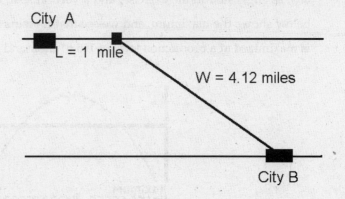

(d) Because $L = 3$, the amount of cable under water is

$$W = \sqrt{1 + (5 - 3)^2} = 2.24 \text{ miles.}$$

The total cost is

$$C = 300 \times 3 + 500\sqrt{1 + (5 - 3)^2} = 2018.03 \text{ dollars.}$$

Here is the picture:

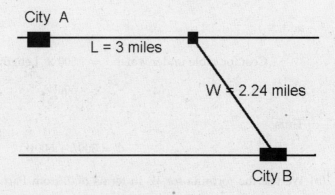

(e) Because City B is only 5 miles downriver from City A, the value of L will be no more than 5. Thus we use a horizontal span of 0 to 5. The table below led us to choose a vertical span from 1000 to 3000. The horizontal axis is miles of cable laid on shore, and the vertical axis is cost in dollars. The graph shows how much it costs to run cable from City A to City B as a function of miles of cable on shore. From the graph it is evident that, as the amount of cable on shore increases, first the cost decreases to a minimum, and then it increases.

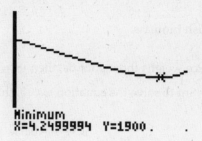

(f) In the picture below, we have located the minimum at $L = 4.25$ and $C = 1900$. Thus the minimum cost plan uses 4.25 miles on shore (and costs 1900).

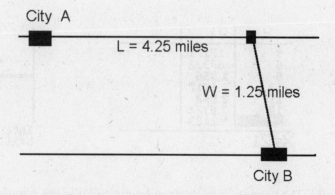

Minimum
X=4.2499994 Y=1900.

(g) Because $L = 4.25$ miles,

$$W = \sqrt{1 + (5 - 4.25)^2} = 1.25 \text{ miles.}$$

Here is the picture:

City A

L = 4.25 miles

W = 1.25 miles

City B

(h) The effect of the increase on the formulas from Parts (a) and (b) of this exercise is to replace 300 by 700. The new formula is

$$C = 700L + 500\sqrt{1 + (5 - L)^2}.$$

Because the land cost of $700 per mile is greater than the water cost and the water route is the shortest, the least cost project would lay all the cable under water. This can also be seen by making a graph (shown below) of the new cost function. (We used a vertical span of 1500 to 4500.) As expected (see Parts (c) and (d) of

Exercise 12), this function is always increasing, so the least cost is at the beginning, when $L = 0$.

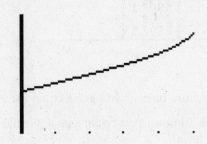

15. **Growth of fish biomass:**

(a) If a plaice weighs three pounds, then $w = 3$, and we want to know the age. That is, we want to solve the equation $w = 3$, that is

$$6.32(1 - 0.93e^{-0.095t})^3 = 3.$$

The table of values below for w suggests a horizontal span of 0 to 20 years and a vertical span of 0 to 5 pounds. In the right-hand picture, we have plotted the graph of w, along with that of the constant function 3, and calculated the intersection point. This shows that a 3-pound plaice is about 15.18 years old.

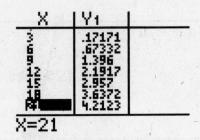
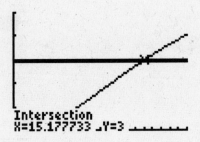

(b) The biomass B is obtained by multiplying the population size by the weight of a fish:

$$B = N \times w$$
$$B = 1000e^{-0.1t} \times 6.32(1 - 0.93e^{-0.095t})^3.$$

To make the graph, we use a horizontal span of 0 to 20 years as suggested. From the table below, we chose a vertical span of 0 to 700 pounds.

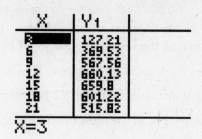

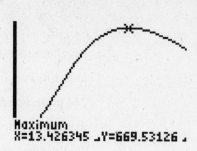

The horizontal axis is age, and the vertical axis is biomass.

(c) The maximum biomass occurs at the peak of the graph above, and we see from there that it occurs at $t = 13.43$ years.

(d) i. From Part (a), we know that a 3-pound plaice is 15.18 years old, so we can only harvest plaice 15.18 years old or older. In the graph below, we have placed the cursor at $t = 15.18$, and we are looking for the maximum considering the graph only from that point on. But biomass is decreasing beyond 15.18, so 15.18 years is the age giving the largest biomass.

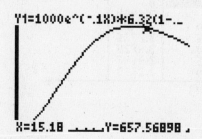

  ii. As in Part (a), we can find the age of a 2-pound plaice as the solution to $6.32(1 - 0.93e^{-0.095t})^3 = 2$. We have done this by the crossing graphs method in the left-hand picture below and find that a 2-pound plaice is 11.28 years old. We can harvest plaice 11.28 years old or older. We have marked the graph in the right-hand picture below at this point, and we see that the part of the graph from there on includes the maximum at $t = 13.43$. Thus, if we are able to catch 2-pound plaice, we harvest at the maximum biomass which occurs at 13.43 years.

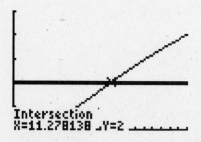

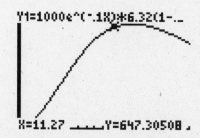

An alternative solution is to find the weight of a plaice at the age $t = 13.43$, found in Part (c), that gives the maximum biomass. This is $w(13.43)$, about 2.56 pounds. Since the weight increases with age, a 2-pound plaice is younger than 13.43 years, and so (as marked in the right-hand picture above) the relevant portion of the graph of biomass will include the maximum at $t = 13.43$ years.

17. **Rate of growth**:

(a) Since the maximum size of a North Sea cod is about 40 pounds, we use a horizontal span of 0 to 40 pounds. From the table of values below, we choose a vertical span of 0 to 5 pounds per year. The graph shows weight on the horizontal axis and rate of growth on the vertical axis.

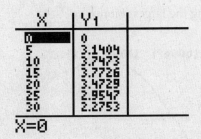 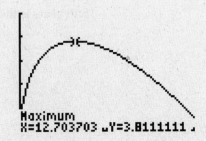

(b) The peak of the graph is at $w = 12.70$ pounds and $G = 3.81$ pounds per year. For this part of the exercise, we are interested in cod weighing 5 pounds or more. Thus we should be looking at the graph from $w = 5$ on. This includes the peak of the graph, and so the maximum growth rate is $G = 3.81$ pounds per year.

(c) For cod weighing at least 25 pounds, we should look at the graph of G against w from $w = 25$ on. This is beyond the peak of the graph, and from this point on the graph is decreasing. Thus the maximum value of G occurs when the weight is $w = 25$ pounds. That maximum is 2.95 pounds per year.

19. **Size of high schools:**

    (a) The study was for schools with a maximum enrollment of 2400 students, and so we use a horizontal span of 0 to 2400. The table below suggests a horizontal span of 0 to 750 dollars per pupil. The horizontal axis on the graph corresponds to enrollment, and the vertical axis corresponds to cost per pupil.

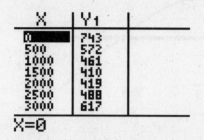
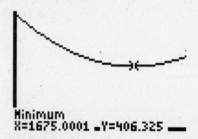

    (b) The enrollment giving the minimum per-pupil cost occurs at the bottom of the graph, and from the picture above, this is at $n = 1675$ students (with a cost of $C = \$406.33$ per pupil).

    (c) If a high school had an enrollment of 1200, then its cost would be $C(1200) = \$433.40$ per pupil. Increasing the school size to the optimum size of 1675 would lower the cost to $\$406.33$ per pupil, a savings of $\$27.07$ per pupil.

21. **Water flea:**

    (a) The exercise suggests a horizontal span of 0 to 350, and a table of values led us to choose a vertical span of -15 to 15. The graph is on the left below.

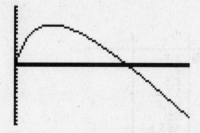
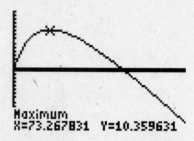

    (b) To find where the greatest rate of growth occurs, we locate the peak of the graph. In the figure on the right above, we see that the peak occurs when $N = 73.27$ and $G = 10.36$. Thus the greatest rate of growth occurs at a population level of 73.27.

    (c) It is evident from the graph or a table of values that $G = 0$ when $N = 0$. We find the other solution to the equation $G(N) = 0$ using the single-graph method. As the graph below indicates, that solution is $N = 228$. Thus the desired values are $N = 0$ and $N = 228$. (Another way to do this part is to notice that in the formula for G the numerator is factored, and this makes it easy to find the desired solutions.) At the two population levels of 0 and 228, the growth rate is 0, that is, the size of the population is not changing.

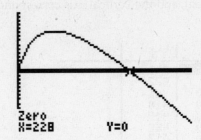

Zero
X=228   Y=0

(d) From the graph or a table of values we see that if the population size is 300 then the rate of population growth is -7.51 per day. Because the growth rate is negative, at this level the population is decreasing in size (at a rate of 7.51 per day).

23. **Research project**: Answers will vary.

Chapter 2 Review Exercises

1. **Finding a minimum**: To find the minimum value of f, we make a table. We enter the function as $Y1 = X^3 - 9X^2/2 + 6X + 1$ and use a table starting value of 1 and a table increment value of 1, resulting in the following table.

X	Y1
1	3.5
2	3
3	5.5
4	17
5	43.5
6	91
7	165.5

X=1

The table shows that f has a minimum value of 3 at $x = 2$.

2. **A population of foxes:**

   (a) We show the graph with a horizontal span of 0 to 20 and a vertical span of 0 to 150.

   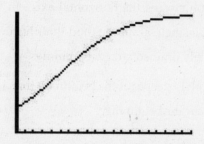

   (b) From a table of values or the graph we find that the value of the function at $t = 9$ is 112.08, which we round to 112. Thus there are 112 foxes 9 years after introduction.

   (c) Tracing the graph shows that it is concave up from $t = 0$ to about $t = 5$. It is concave down afterwards.

   (d) Scrolling down a table on the calculator (or tracing the graph), we see that the limiting value of the function is 150.

3. **Linear equations:** To solve $11 - x = 3 + x$ for x, we follow the usual procedure:

$$
\begin{aligned}
11 - x &= 3 + x \\
-x - x &= 3 - 11 \\
-2x &= -8 \\
x &= \frac{-8}{-2} = 4.
\end{aligned}
$$

4. **Water jug:**

   (a) On the left below we show the graph with a horizontal span of 0 to 7.5 and a vertical span of 0 to 16.

   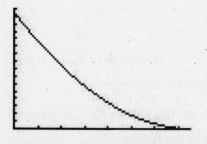 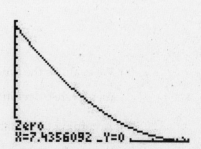

(b) From a table of values or the graph we find that $D(3)$, the value of the function at $t = 3$, is 5.72. Thus the depth of the water is 5.72 inches above the spigot after the spigot has been open 3 minutes.

(c) The time when the jug will be completely drained corresponds to the point when the graph crosses the horizontal axis. In the graph on the right above we find using the single-graph method that this occurs at $t = 7.44$. Thus the jug will be completely drained after 7.44 minutes.

(d) The graph is steeper at the beginning than it is toward the end, so the water drains faster near the beginning.

5. **Maxima and minima**: To find the maximum and minimum values of $x^2 + 100/x$ with a horizontal span of 1 to 5, we graph the function. A table of values leads us to choose a vertical span of 0 to 115. The graph on the left below shows that the maximum is at the left endpoint $x = 1$, where the value is $y = 101$. The graph on the right shows that the minimum is at $x = 3.68$, where the value is $y = 40.72$.

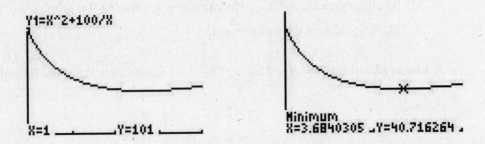

6. **George Deer Reserve population**: Below is a table of values for the function N.

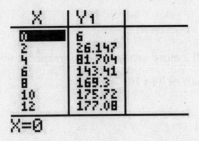

(a) We evaluate the function N at $t = 0$. According to the table that value is 6. Hence there were 6 deer introduced into the deer reserve.

(b) According to the table $N(4)$ is 81.70, or about 82. This means that there were 82 deer in the reserve 4 years after introduction.

(c) Scrolling down the table on the calculator, we see that the limiting value of the function is 177.43, or about 177, and this is the carrying capacity.

(d) The average rate of increase from $t = 0$ to $t = 2$ is

$$\frac{N(2) - N(0)}{2 - 0} = \frac{26.147 - 6}{2} = 10.07 \text{ deer per year.}$$

Continuing in this way, we get the following table, where the rate of change is measured in deer per year.

Interval	0 to 2	2 to 4	4 to 6	6 to 8
Rate of change	10.07	27.78	30.85	12.94

The population increases more and more rapidly, and then the rate of growth decreases.

7. **Making a graph**: To find an appropriate window that will show a good graph of the function, first we enter the function as Y1 = $315/(1 + 14e^{(-0.23X)})$, and then we make a table of values. The table shows that a vertical span from 0 to 330 will display the graph, and the graph is shown below.

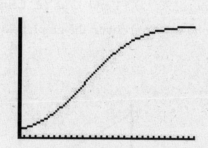

8. **Forming a pen**:

(a) We use two sides that are W feet long and two that are L feet long, so the total amount that we need is $F = 2W + 2L$ feet of fence.

(b) The area of a rectangle is the width times the length, and because that area is to be 144 square feet we have $144 = W \times L$.

(c) To solve $144 = W \times L$ for W we divide both sides by L. The result is $W = \dfrac{144}{L}$.

(d) By Part (a) the total amount of fence needed is $F = 2W + 2L$ feet, and by the equation in Part (c) this gives $F = 2 \times \dfrac{144}{L} + 2L$ or $F = \dfrac{288}{L} + 2L$.

(e) After examining a table of values, we used a horizontal span of 0 to 24 and a vertical span of 0 to 100. The graph is on the left below.

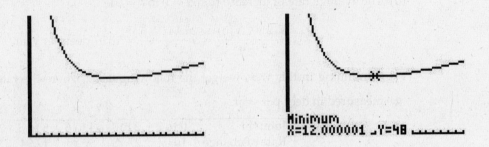

(f) In the graph on the right above we have located the minimum point, and we see that it occurs where $L = 12$ feet. From the equation in Part (c) we find that the corresponding width is $W = \dfrac{144}{12} = 12$ feet. Thus a square (12 by 12) pen requires a minimum amount of fence, which makes sense. The graph illustrates that, as the length increases, first the area decreases to a minimum, and then it increases.

9. **The crossing-graphs method**: To solve the equation using the crossing-graphs method, we enter $6 + 69 \times 0.96^X$ as Y1 and 32 as Y2. Using a horizontal span of 0 to 30 and a vertical span of 0 to 80, we obtain the graph below, which shows that the solution is $t = 23.91$.

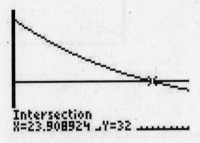

10. **Gliding pigeons**:

   (a) We show the graph on the left below with a horizontal span of 0 to 20 and a vertical span of 0 to 10.

   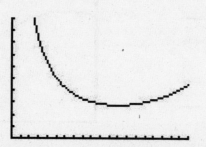
   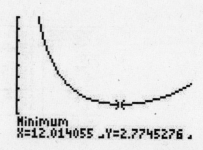

   (b) In the graph on the right above we have located the minimum point, and we see that it occurs where $u = 12.01$ meters per second.

   (c) The graph shows that if u is very small then s is very large. Hence if the airspeed is very slow the pigeon sinks quickly.

11. **Finding a maximum**: To locate the maximum of f, we make a table. We enter the function as $Y1 = 12X - X^2 - 25$ and use a table starting value of 3 and a table increment value of 1, resulting in the following table.

   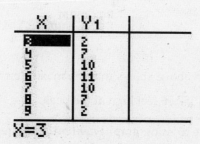

   The table shows that f has a maximum value of 11 at $x = 6$.

12. **Thrown ball:**

   (a) On the left below we show the table only from $t = 0$ to $t = 1.5$. Here we use a table starting value of 0 and a table increment value of 0.25.

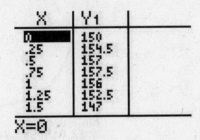
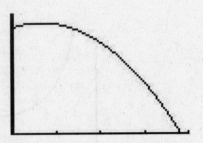

   (b) Using a horizontal span of 0 to 4 and a vertical span of 0 to 170, we obtain the graph on the right above.

   (c) In the graph below we have located the maximum point, and we see that it occurs at $t = 0.69$ second.

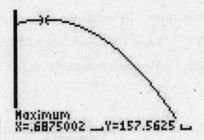

   (d) The graph above shows that the maximum value is 157.56 feet. This means that the ball is 157.56 feet high at the peak.

   (e) We use the crossing-graphs method. We enter Y2 = 155 and, to get a better picture of the intersection points, we change the vertical span to go from 150 to 160. The graphs below show that the two times are $t = 0.29$ second and $t = 1.09$ seconds.

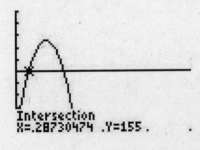
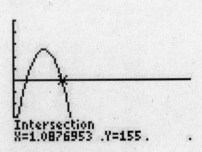

13. **Linear equations**: To solve $L = 98.42 + 1.08W - 4.14A$ for W, we follow the usual procedure:

$$
\begin{aligned}
L &= 98.42 + 1.08W - 4.14A \\
L - 98.42 + 4.14A &= 1.08W \\
W &= \frac{L - 98.42 + 4.14A}{1.08}.
\end{aligned}
$$

14. **Growth of North Sea sole**:

    (a) We show the graph on the left below with a horizontal span of 1 to 10 and a vertical span of 0 to 15.

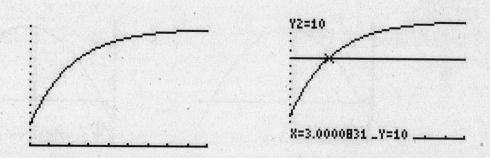

    (b) In functional notation the length of a 4-year-old sole is $L(4)$. From a table of values or the graph we find that the value is 11.77 inches.

    (c) To use the crossing-graphs method we enter Y2 = 10. The graph on the right above shows that the age is $t = 3$ years.

    (d) Scrolling down a table on the calculator (or tracing the graph), we see that the limiting value of the function is 14.8. This means that the limiting length for a North Sea sole is 14.8 inches.

15. **Minimum**: To find the minimum value of $x^2 + 20/(x + 1)$ with a horizontal span of 0 to 10, we graph the function. A table of values leads us to choose a vertical span of 0 to 105. The graph below shows that the minimum is at $x = 1.54$, where the value is $y = 10.25$.

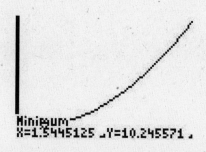

16. **Profit**:

(a) In functional notation the profit at a production level of 7 million items is $P(7)$.

(b) When we put $N = 0$ into the formula we find $P = -6.34$. This means that the loss at a production level of $N = 0$ million items is 6.34 million dollars.

(c) The break-even points occur where the graph crosses the horizontal axis. To make the graph we use a horizontal span of 0 to 10 and a vertical span of -10 to 20. We find by means of the single-graph method that the graph crosses the horizontal axis at $N = 0.68$ and at $N = 9.32$, as shown in the graphs below. Hence the two break-even points are at 0.68 million items and at 9.32 million items.

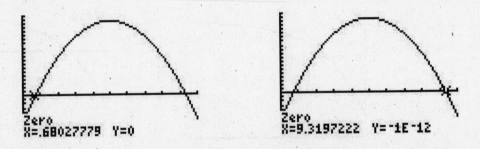

(d) In the graph below we have located the maximum point, and we see that it occurs where $N = 5$ and $P = 18.66$. Thus the production level that gives maximum profit is 5 million items, and the amount of the maximum profit is 18.66 million dollars.

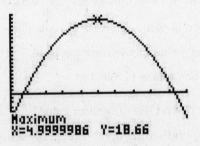

17. **Making a graph**: To find an appropriate window setup which will show a good graph of the function, first we enter the function as $Y1 = 0.5X/(6380 + X)$, and then we make a table of values. The table shows that a vertical span from 0 to 0.5 will display the graph, and the graph is shown below.

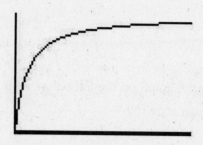

18. **Temperature conversions**:

(a) To solve these equations for C, we follow the usual procedure. For the first, we have

$$F = 1.8C + 32$$
$$F - 32 = 1.8C$$
$$C = \frac{F - 32}{1.8}.$$

For the second, we have

$$K = C + 273.15$$
$$C = K - 273.15.$$

(b) We put the formula for C in terms of K from Part (a) into the original formula for F in terms of C. The result is

$$F = 1.8(K - 273.15) + 32.$$

(c) We put the formula for C in terms of F from Part (a) into the original formula for K in terms of C. The result is

$$K = \frac{F - 32}{1.8} + 273.15.$$

This can also be done by solving the formula from Part (b) to get K in terms of F.

(d) We put $C = 0$ into the two original formulas. The results are $F = 32$ and $K = 273.15$. Thus 0 degrees Celsius corresponds to 32 degrees Fahrenheit and 273.15 kelvins.

(e) First we put $F = 72$ into the formula $C = \dfrac{F - 32}{1.8}$ from Part (a). The result is $C = 22.22$. Thus 72 degrees Fahrenheit corresponds to 22.22 degrees Celsius. Next we put this result, $C = 22.22$, into the original formula $K = C + 273.15$. The result is $K = 295.37$. (This can also be done directly by using the formula from Part (c).) Thus 72 degrees Fahrenheit corresponds to 295.37 kelvins.

19. **Lidocaine:**

(a) We show the graph on the left below with a horizontal span of 0 to 240 and a vertical span of 0 to 30.

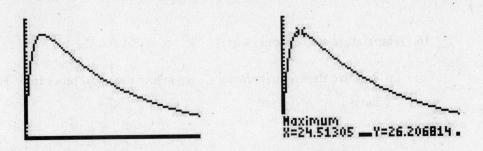

(b) In the graph on the right above we have located the maximum point, and we see that it occurs at $t = 24.51$. Thus the drug reaches its maximum level after 24.51 minutes. For future reference, note from this graph that the maximum level is 26.21 mg.

(c) We use the crossing-graphs method and enter Y2 = 7.5. The graphs below show that the two intersection points are at $t = 2.277$ minutes and at $t = 198.295$ minutes. The difference is 196.02 minutes, and that is how long the drug is at or above the level of 7.5 mg.

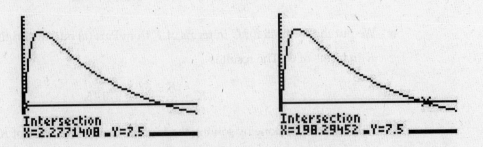

(d) As we noted in Part (b), the maximum level is 26.21 mg. Thus the amount never exceeds 30 mg, so this dose is not lethal for a person of typical size.

(e) We use the crossing-graphs method and enter Y2 = 15. The graph below shows that the first intersection point is at $t = 5.54$ minutes. Thus this dose is lethal for a small person, and it will be lethal after 5.54 minutes.

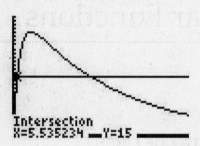

20. **The single-graph method**: To solve the equation using the single-graph method, we enter the function on the left as Y1. Using a horizontal span of -2 to 2 and a vertical span of -1 to 1, we obtain the graphs below, which show that the solutions are $x = -1.22$, $x = 0$, and $x = 1.22$.

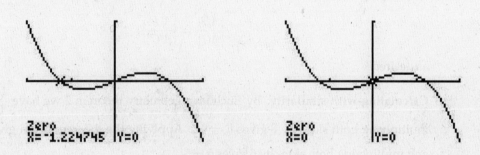

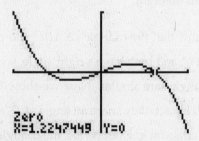

Solution Guide for Chapter 3: Straight Lines and Linear Functions

3.1 THE GEOMETRY OF LINES

E-1 **An alternative version of similarity**: We know from Euclidean geometry theorem 2 that if $\triangle ABC$ is similar to $\triangle A'B'C'$ then

$$\frac{|AB|}{|A'B'|} = \frac{|AC|}{|A'C'|}.$$

Multiplying both sides by $|A'B'|$ and dividing both sides by $|AC|$ gives

$$\frac{|AB|}{|AC|} = \frac{|A'B'|}{|A'C'|},$$

as desired.

E-3 **Calculating with similarity**: By Euclidean geometry theorem 2 we have $\frac{10}{5} = \frac{a}{6}$, and multiplying both sides by 6 gives $a = 12$. Applying the theorem again gives $\frac{5}{10} = \frac{b}{6}$, and multiplying both sides by 6 gives $b = 3$.

E-5 **The Pythagorean theorem**:

(a) First we show that the triangles $\triangle ABC$ and $\triangle DAC$ are similar. In fact, they share an angle at C, and both have a right angle, so by the two-angle criterion in Exercise E-2 the triangles are similar. Next we show that the triangles $\triangle ABC$ and $\triangle DBA$ are similar. In fact, they share an angle at B, and both have a right angle, so by the two-angle criterion in Exercise E-2 the triangles are similar.

(b) Because $\triangle ABC$ and $\triangle DAC$ are similar, by Euclidean geometry theorem 2 we have

$$\frac{|BC|}{|AC|} = \frac{|AC|}{|DC|},$$

which says that

$$\frac{c}{a} = \frac{a}{d}.$$

Multiplying both sides by ad gives $cd = a^2$, as desired.

(c) Because $\triangle ABC$ and $\triangle DBA$ are similar, by Euclidean geometry theorem 2 we have

$$\frac{|BC|}{|BA|} = \frac{|AB|}{|DB|},$$

which says that

$$\frac{c}{b} = \frac{b}{e}.$$

Multiplying both sides by be gives $ce = b^2$, as desired.

(d) By Parts (b) and (c) we have the equations $cd = a^2$ and $ce = b^2$. Adding these equations gives

$$a^2 + b^2 = cd + ce.$$

Now $d + e = c$ because the length of the line segment BC is the sum of the lengths of DC and BD. Thus

$$cd + ce = c(d + e) = c(c) = c^2.$$

Combining the last two displayed equations yields $a^2 + b^2 = c^2$, as desired.

E-7 **A right triangle**: First we find the slope of the line passing through $(1, 1)$ and $(3, 4)$. We have

$$\text{Slope} = \frac{\text{Vertical change}}{\text{Horizontal change}} = \frac{4 - 1}{3 - 1} = \frac{3}{2}.$$

Next we find the slope of the line passing through $(1, 1)$ and $(4, -1)$. We have

$$\text{Slope} = \frac{\text{Vertical change}}{\text{Horizontal change}} = \frac{-1 - 1}{4 - 1} = \frac{-2}{3}.$$

Because the slope $-\dfrac{2}{3}$ is the negative reciprocal of the slope $\dfrac{3}{2}$, the two lines are perpendicular. Hence the given points do form the vertices of a right triangle.

E-9 **A system of equations with no solution**: Solving the first equation for y gives $y = -\dfrac{2}{7}x + \dfrac{9}{7}$, so the slope of the first line is $-\dfrac{2}{7}$. Solving the second equation for y gives $y = -\dfrac{2}{7}x + \dfrac{5}{14}$, so the slope of the second line is $-\dfrac{2}{7}$ also. Because the slopes are the same, the lines are parallel. Evidently, though, the lines have different vertical intercepts, so they are not coincident (that is, not the same). Thus the graphs represented by the two linear equations do not cross. Hence the system has no solution.

S-1. **Slope from rise and run**: We have $\text{Slope} = \dfrac{\text{Rise}}{\text{Run}}$. In this case, the rise is 8 feet, the height of the wall, and the run is 2 feet, since that is the horizontal distance from the wall. Thus the slope is $\dfrac{8}{2} = 4$ feet per foot.

S-3. **Height from slope and horizontal distance**: We have

$$\text{Vertical change} = \text{Slope} \times \text{Horizontal change}.$$

In this case, the slope of the ladder is 2.5, while the horizontal distance is 3 feet, so the vertical height is $2.5 \times 3 = 7.5$ feet.

S-5. **Horizontal distance from height and slope**: We have

$$\text{Vertical change} = \text{Slope} \times \text{Horizontal change}.$$

In this case, the slope of the ladder is 1.75, while the vertical distance is 9 feet, so we have $9 = 1.75 \times$ Horizontal change. Thus the horizontal distance is $\dfrac{9}{1.75} = 5.14$ feet.

S-7. **Slope from two points**: We have Slope $= \dfrac{\text{Vertical change}}{\text{Horizontal change}}$. In this case, the vertical change is -2 feet (12 feet dropped to 10 feet) while the horizontal change is 3 feet (since west is the positive direction). Thus the slope is $\dfrac{-2}{3} = -\dfrac{2}{3}$ foot per foot.

S-9. **A circus tent**: We have Vertical change $=$ Slope $\times$ Horizontal change. In this case, the slope is -0.8, while the horizontal change is 7 feet, so the vertical change is $-0.8 \times 7 = -5.6$ feet. Thus the height of the tent if I walk 7 feet west is $22 - 5.6 = 16.4$ feet.

S-11. **Slope**: We have Slope $= \dfrac{\text{Rise}}{\text{Run}}$. In this case, the rise is 100 feet, the height of the building, and the run is 70 feet, since that is the horizontal distance to the building. Thus the slope is $\dfrac{100}{70} = 1.43$ feet per foot.

1. **A line with given intercepts**: The picture is drawn below. Since the line is falling from left to right, we expect the slope of the line to be negative. The slope is

$$m = \frac{\text{Vertical change}}{\text{Horizontal change}} = \frac{0-4}{3-0} = -\frac{4}{3}.$$

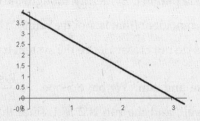

3. **Another line with given vertical intercept and slope**: Because the graph falls by 2 units for each unit of run, we need to move $\dfrac{8}{2} = 4$ units to the right of the vertical axis

to get to the horizontal intercept. Thus the horizontal intercept is 4. More formally, if we move from the location of the horizontal intercept to that of the vertical, we have

$$\text{Slope} = \frac{\text{Vertical change}}{\text{Horizontal change}} = \frac{\text{Vertical intercept} - 0}{0 - \text{Horizontal intercept}},$$

so

$$-2 = \frac{8}{-\text{Horizontal intercept}},$$

and solving gives the value of 4 for the horizontal intercept.

5. **Lines with the same slope**: The first line should have a vertical intercept at 3 and rise 2 units for each unit of run. The second should have a vertical intercept of 1 and rise 2 units for each unit of run. Our picture is shown below. The lines do not cross. Different lines with the same slope are parallel.

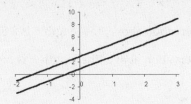

7. **A ramp to a building**:

(a) The slope of the graph is 0.4, so one foot of run results in 0.4 foot of rise. Thus, one foot from the base of the ramp, it is 0.4 foot high.

(b) To get to the steps we have moved horizontal 15 feet from the base of the ramp. Since the slope is 0.4, the ramp rises

$$\text{Rise} = \text{Run} \times \text{Slope} = 15 \times 0.4 = 6 \text{ feet.}$$

9. **A cathedral ceiling**:

(a) When we move 3 feet east from the west wall, the ceiling rises $10.5 - 8 = 2.5$. So the slope is

$$\text{Slope} = \frac{\text{Rise}}{\text{Run}} = \frac{2.5}{3} = 0.83 \text{ foot per foot.}$$

(b) If we move 17 feet to the right of the west wall then the vertical change is

$$\text{Rise} = \text{Slope} \times \text{Run} = 0.83 \times 17 = 14.11 \text{ feet.}$$

So the height of the ceiling at that point is $8 + 14.11 = 22.11$ feet.

(c) Since we start at 8 feet high on the west wall and we want to reach a 12-foot height, then the vertical change is 4 feet. Now

$$
\begin{aligned}
\text{Vertical change} &= \text{Horizontal change} \times \text{Slope} \\
4 &= \text{Horizontal change} \times 0.83 \\
\frac{4}{0.83} &= \text{Horizontal change} \qquad \textbf{Divide both sides by 0.83.} \\
4.82 &= \text{Horizontal change.}
\end{aligned}
$$

So if we place the light 4.82 feet from the west wall, then we will still be able to change the bulb.

11. **Cutting plywood siding**: The shape of the first piece shows that for a horizontal change (to the left) of 4 feet the vertical change along the roof line is $2.5 - 1 = 1.5$ feet. Now each piece has width 4 feet, so we observe that for each piece the longer side and the shorter side differ in length by 1.5 feet.

(a) By the observation above, the length h is $2.5 + 1.5 = 4$ feet.

(b) By the observation above, the length k is $h + 1.5 = 4 + 1.5 = 5.5$ feet, or 5 feet 6 inches.

Alternative approach: In both parts we focus on the part of the roof line sloping upward from right to left, and we take movement toward the center as going in the positive direction. Moving along this line from the outer wall to the far corner of the first piece of siding, we have a horizontal change of 4 feet and a vertical change of $2.5 - 1 = 1.5$ feet. Thus the slope of this line is $\frac{1.5}{4} = 0.375$ foot per foot. For Part(a): Moving along upper edge of the second piece of siding toward the peak, we have a horizontal change of 4 feet. Because the slope is 0.375, from

$$\text{Vertical change} = \text{Slope} \times \text{Horizontal change},$$

we find

$$\text{Vertical change} = 0.375 \times 4 = 1.5 \text{ feet.}$$

To find h we add this vertical change to the length 2.5 feet of the short side of the second piece. Thus the length h is $2.5 + 1.5 = 4$ feet.

For Part(b): Moving along upper edge of the third piece of siding toward the peak, we have a horizontal change of 4 feet. Because the slope is 0.375, from

$$\text{Vertical change} = \text{Slope} \times \text{Horizontal change},$$

we again find

$$\text{Vertical change} = 0.375 \times 4 = 1.5 \text{ feet.}$$

To find k we add this vertical change to the length 4 feet of the short side of the third piece. Thus the length h is $4 + 1.5 = 5.5$ feet.

13. **Looking over a wall**: Assuming that the man can just see the top of the building over the wall, we focus on the line of sight from the man to the top of the building. Taking south as the positive direction, we compute the slope of the line by considering the change from the top of the wall to the top of the building. We find that the slope of the line is

$$\text{Slope} = \frac{\text{Vertical change}}{\text{Horizontal change}} = \frac{50 - 35}{20} = \frac{15}{20} = 0.75 \text{ foot per foot.}$$

Now we find the horizontal distance from the wall to the man by considering the change along the line from the man to the top of the wall. We use

$$\text{Vertical change} = \text{Slope} \times \text{Horizontal change,}$$

which says

$$35 - 6 = 0.75 \times \text{Horizontal distance.}$$

Dividing both sides by 0.75 shows that the horizontal distance is $\dfrac{35 - 6}{0.75} = 38.67$ feet. Thus the man must be at least 38.67 feet north of the wall.

15. **A road up a mountain**:

(a) The vertical change is $4960 - 4130 = 830$ feet. The horizontal change is 3 miles. So the slope is

$$\frac{\text{Vertical change}}{\text{Horizontal change}} = \frac{830}{3} = 276.67 \text{ feet per mile.}$$

(b) The vertical change in moving 5 miles from the first sign is

$$\text{Rise} = \text{Run} \times \text{Slope} = 5 \times 276.67 = 1383.35 \text{ feet.}$$

Since we started at 4130 feet at the first sign, the elevation is $4130 + 1383.35 = 5513.35$ feet when we are five miles from the first sign.

(c) The vertical change from the first sign to the peak is $10,300 - 4130 = 6170$ feet.

$$
\begin{aligned}
\text{Vertical change} &= \text{Horizontal change} \times \text{Slope} \\
6170 &= \text{Horizontal change} \times 276.67 \\
\frac{6170}{276.67} &= \text{Horizontal change} \qquad \textbf{Divide both sides by 276.67.} \\
22.30 &= \text{Horizontal change.}
\end{aligned}
$$

Thus the peak is 22.30 horizontal miles away.

17. **Earth's umbra**: We take the positive horizontal axis to pass from the point where the umbra ends through the center of Earth. We find the slope of the upper line shown in the figure by considering the change from the point where the umbra ends to the surface of Earth. We find that the slope is

$$\text{Slope} = \frac{\text{Vertical change}}{\text{Horizontal change}} = \frac{\text{Radius of Earth}}{\text{Distance from Earth to end of umbra}} = \frac{3960}{860,000} \text{ mile per mile.}$$

Now we find the radius of the umbra at the indicated point by considering the change along the line from the point where the umbra ends to the radius at the indicated point. We use

$$\text{Vertical change} = \text{Slope} \times \text{Horizontal change.}$$

Now the vertical change is the radius of the umbra we are asked to find, and the horizontal change is $860,000 - 239,000 = 621,000$. Using the slope calculated above, we find

$$\text{Radius of the umbra} = \frac{3960}{860,000} \times 621,000 = 2859.49 \text{ miles.}$$

Thus the radius of the umbra at the indicated point is about 2859 miles. Since the radius of the moon is smaller than this, the moon can fit inside Earth's umbra. When this happens there is a total lunar eclipse.

19. **The umbra of the moon**: We take the positive horizontal axis to pass from the apex (at the center of Earth) to the center of the sun. We find the slope of the line from the apex to the surface of the sun by considering the change from the apex to the surface of the moon. (This can be visualized by replacing Earth by the moon in Figure 3.40.) We find that the slope is

$$\text{Slope} = \frac{\text{Vertical change}}{\text{Horizontal change}} = \frac{\text{Radius of moon}}{\text{Distance from Earth to moon}} = \frac{1100}{239,000} \text{ mile per mile.}$$

Now we find the radius of the sun by considering the change from the apex to the surface of the sun. We use

$$\text{Vertical change} = \text{Slope} \times \text{Horizontal change.}$$

Now the vertical change is the desired radius of the sun, and the horizontal change is the distance from Earth to the sun. Using the slope calculated above, we find

$$\text{Radius of the sun} = \frac{1100}{239,000} \times 93,498,600 = 430,328.28 \text{ miles.}$$

Thus the radius of the sun is about 430,328 miles.

3.2 LINEAR FUNCTIONS

E-1. **Parallel lines**: We know from the preceding section that parallel lines have the same slope. The line $y = 3x - 2$ has slope 3, so we want to find the equation of the line with slope 3 that passes through the point $(3, 3)$. We use the point-slope form:

$$
\begin{aligned}
y - 3 &= 3(x - 3) \\
y - 3 &= 3x - 9 \\
y &= 3x - 6.
\end{aligned}
$$

E-3. **Finding equations of lines**:

(a) We use the point-slope form:

$$
\begin{aligned}
y - 1 &= 3(x - 2) \\
y - 1 &= 3x - 6 \\
y &= 3x - 5.
\end{aligned}
$$

(b) We use the point-slope form:

$$
\begin{aligned}
y - 1 &= -4(x - 1) \\
y - 1 &= -4x + 4 \\
y &= -4x + 5.
\end{aligned}
$$

(c) We use the two-point form. We have

$$
\begin{aligned}
y - 2 &= \left(\frac{10 - 2}{5 - 1} \right) (x - 1) \\
y - 2 &= 2x - 2 \\
y &= 2x.
\end{aligned}
$$

(d) We use the two-point form. We have

$$
\begin{aligned}
y - 1 &= \left(\frac{2 - 1}{-2 - 3} \right) (x - 3) \\
y - 1 &= -\frac{1}{5}x + \frac{3}{5} \\
y &= -\frac{1}{5}x + \frac{8}{5}.
\end{aligned}
$$

S-1. **Slope from two values**: We have that the slope m is:

$$
m = \frac{\text{Change in function}}{\text{Change in variable}} = \frac{19 - 7}{5 - 2} = \frac{12}{3} = 4.
$$

S-3. Function value from slope and run: We have that the slope is $m = 2.7$:

$$2.7 = \frac{\text{Change in function}}{\text{Change in variable}} = \frac{f(5) - f(3)}{5 - 3} = \frac{f(5) - 7}{5 - 3} = \frac{f(5) - 7}{2}.$$

Thus $f(5) - 7 = 2 \times 2.7 = 5.4$, and so $f(5) = 5.4 + 7 = 12.4$.

S-5. Run from slope and rise: We have that the slope is $m = -3.4$:

$$-3.4 = \frac{\text{Change in function}}{\text{Change in variable}} = \frac{f(x) - f(1)}{x - 1} = \frac{0 - 6}{x - 1} = \frac{-6}{x - 1}.$$

Thus $-6 = -3.4 \times (x - 1)$, so $x - 1 = \frac{-6}{-3.4} = 1.76$, and therefore $x = 1.76 + 1 = 2.76$.

S-7. Linear equation from slope and point: Since f is a linear function with slope 4, $f = 4x + b$ for some b. Now $f(3) = 5$, so $5 = 4 \times 3 + b$, and therefore $b = 5 - 4 \times 3 = -7$. Thus $f = 4x - 7$.

S-9. Linear equation from two points: We first compute the slope:

$$m = \frac{\text{Change in function}}{\text{Change in variable}} = \frac{f(9) - f(4)}{9 - 4} = \frac{2 - 8}{9 - 4} = \frac{-6}{5} = -1.2.$$

Since f is a linear function with slope -1.2, $f = -1.2x + b$ for some b. Now $f(4) = 8$, so $8 = -1.2 \times 4 + b$, and therefore $b = 8 + 1.2 \times 4 = 12.8$. Thus $f = -1.2x + 12.8$.

S-11. Linear equation from slope and points: We are given the slope $m = -1.7$ and the initial value $b = -3.7$ (because $f(0) = -3.7$). Thus $f = -1.7x - 3.7$.

S-13. Linear equation from two points: We first compute the slope:

$$m = \frac{\text{Change in function}}{\text{Change in variable}} = \frac{f(0) - f(-2)}{0 - (-2)} = \frac{3 - 2}{2} = \frac{1}{2} = 0.5.$$

Now we need the initial value b. Since $f(0) = 3$, that value is $b = 3$. Thus $f = 0.5x + 3$.

1. **Getting Celsius from Fahrenheit**:

   (a) Water freezes at 0 degrees Celsius, which is 32 degrees Fahrenheit. So $C = 0$ when $F = 32$. Water boils at 100 degrees Celsius, which is 212 degrees Fahrenheit. So $C = 100$ when $F = 212$. These two bits of information allow us to get the slope of the function:

   $$\text{Slope} = \frac{\text{Change in } C}{\text{Change in } F} = \frac{100 - 0}{212 - 32} = 0.56.$$

   Thus $C = 0.56F + b$, and we need to find b. When $C = 0$, then $F = 32$. This gives $0 = 0.56 \times 32 + b$. Solving for b, we get $b = -17.92$. Thus $C = 0.56F - 17.92$.

   There are some variations that yield slightly different formulas. The differences are due to the rounding of $\frac{5}{9}$ as 0.56. If the fraction is left as a fraction, then the formula will be $C = \dfrac{5}{9}F - \dfrac{160}{9}$, which, to two decimal places, is $C = 0.56F - 17.78$. If the second data point, $C = 100$ when $F = 212$, is used to find b from $C = 0.56F + b$, the resulting calculation will yield $b = -18.72$, giving the formula $C = 0.56F - 18.72$.

(b) The slope we found in Part (a) was 0.56. This means that for every one degree increase in the Fahrenheit temperature, the Celsius temperature will increase by 0.56 degree.

(c) We need to solve $F = 1.8C + 32$ for C. We have

$$
\begin{aligned}
F &= 1.8C + 32 \\
F - 32 &= 1.8C \qquad \text{\textbf{Subtract 32 from each side.}} \\
\frac{1}{1.8} \times (F - 32) &= C \qquad \text{\textbf{Divide both sides by 1.8.}} \\
\frac{1}{1.8}F - \frac{1}{1.8} \times 32 &= C \\
0.56F - 17.78 &= C
\end{aligned}
$$

This is the same as the formula we found in Part (a).

3. **Digitized pictures on a disk drive**:

(a) Each additional picture stored increases the total storage space used by 2 megabytes. This means that the change in total storage space used is always the same, 2 megabytes, for a change of 1 in the number of pictures that are stored. Thus the storage space used S is a linear function of the number of pictures stored n.

(b) From Part (a), S is a linear function of n with slope 2, since the slope represents the additional storage space used with the addition of 1 picture. The initial value of S is the amount of storage space used if $n = 0$, that is, if no pictures are stored. Since the formatting information, operating system, and applications software use 6000 megabytes, that is the initial value. Thus a formula for $S(n)$ is $S = 2n + 6000$.

(c) The total amount of storage space used on the disk drive if there are 350 pictures stored on the drive is expressed in functional notation as $S(350)$. (Of course, this depends on the choice of function name we made in Part (b).) Its value is $S(350) = 2 \times 350 + 6000 = 6700$ megabytes.

(d) If there are 769,000 megabytes free and the drive holds 800,000 megabytes, then there are $800{,}000 - 769{,}000 = 31{,}000$ megabytes used. To find out how many pictures are stored, we need to solve this equation for n:

$$2n + 6000 = 31{,}000$$

$$2n = 31{,}000 - 6000 = 25{,}000 \qquad \textbf{Subtract 6000 from both sides.}$$

$$n = \frac{25{,}000}{2} = 12{,}500 \text{ pictures} \qquad \textbf{Divide both sides by 2.}$$

Thus there are 12,500 digitized pictures stored. There are 769,000 megabytes of room left, and that space can hold $\dfrac{769{,}000}{2} = 384{,}500$ more pictures before the disk drive is full.

5. **Total cost:**

(a) Because the variable cost is a constant $20 per widget, for each additional widget produced per month the monthly cost increases by the same amount, $20. This means that C always increases by 20 when N increases by 1. Thus C has a constant rate of change and so is a linear function of N. The slope is the constant rate of change, namely 20 dollars per widget. The initial value is the monthly cost when no widgets are manufactured, and this is the amount of the fixed costs, namely $1500. Since the slope is 20 and the initial value is 1500, the formula is $C = 20N + 1500$.

(b) Let A represent the monthly costs (in dollars) for this other manufacturer and N the number of widgets produced in a month. As in Part (a), the slope of this linear function is given by the variable cost, which in this case is 12 dollars per widget. Thus $A = 12N + b$ for some constant b. In fact, the constant b is the initial value of the function, and, as in Part (a), this represents the fixed costs. We find b using the fact that $A = 3100$ when $N = 150$: We have $3100 = 12 \times 150 + b$, so $b = 3100 - 12 \times 150 = 1300$. Thus the amount of fixed costs is $1300.

Another way of finding this is to start with the total cost of $3100 at a production level of $N = 150$ and subtract $12 for each widget produced to account for the variable cost; the resulting amount of $3100 - 12 \times 150 = 1300$ dollars is the amount of fixed costs.

(c) Let B represent the monthly costs (in dollars) for this manufacturer and N the number of widgets produced in a month. As in Parts (a) and (b), the slope of this linear function is the variable cost, and the initial value of the function represents the fixed costs. We know that $B = 2700$ when $N = 100$ and that $B = 3500$ when

$N = 150$. Thus the slope of B is given by

$$\text{Slope} = \frac{\text{Change in } B}{\text{Change in } N} = \frac{3500 - 2700}{150 - 100} = \frac{800}{50} = 16 \text{ dollars per widget.}$$

Hence the variable cost is \$16 per widget. We now know that $B = 16N + b$, where b is the initial value (representing the fixed costs). We find b using the fact that $B = 2700$ when $N = 100$: We have $2700 = 16 \times 100 + b$, so $b = 2700 - 16 \times 100 = 1100$. Thus the amount of fixed costs is \$1100.

7. **Slowing down in a curve:**

(a) Because S decreases by 0.746 when D increases by 1, the slope of S is -0.746 mile per hour per degree. The initial value of S is 46.26 miles per hour because that is the speed on a straight road. Thus the formula is $S = -0.746D + 46.26$. Of course, this can also be written as $S = 46.26 - 0.746D$.

(b) The speed for a road with a curvature of 10 degrees is expressed as $S(10)$ in functional notation. The value is $S(10) = -0.746 \times 10 + 46.26 = 38.8$ miles per hour.

9. **Currency conversion:**

(a) We know we can get $P = 31$ pounds for $D = 83$ dollars, and we can get $P = 0$ pounds for $D = 0$ dollars. Thus

$$\text{Slope} = \frac{\text{Change in pounds}}{\text{Change in dollars}} = \frac{31}{83} = 0.37 \text{ pound per dollar.}$$

This means each dollar is worth 0.37 pound.

(b) Since we get 0.37 pound per dollar, for 130 dollars we get $0.37 \times 130 = 48.10$ pounds. So the tourist received 48.10 pounds for the 130 dollars.

(c) We want to know how many dollars we get for 12.32 pounds. That is, we want to solve $0.37D = 12.32$ for D. We need only divide each side by 0.37 to get $D = 33.30$ dollars. The tourist received \$33.30.

11. **Adult male height and weight:**

(a) According to this rule of thumb, if a man is 1 inch taller than another, then we expect him to be heavier by 5 pounds. This says that the rate of change in adult male weight is constant, namely 5 pounds per inch, and thus weight is a linear function of height. The slope is the rate of change, 5 pounds per inch.

(b) Let h denote the height in inches and w the weight in pounds. Part (a) tells us that w is a linear function of h with slope 5. Thus $w = 5h + b$. To get the value of b, we note that when the weight is $w = 170$ pounds, the height is $h = 70$ inches. Thus $170 = 5 \times 70 + b$, which gives $b = 170 - 5 \times 70 = -180$. We get $w = 5h - 180$.

(c) If the weight is $w = 152$ pounds, then $5h - 180 = 152$. We want to solve this for h to find the height:

$$
\begin{aligned}
5h - 180 &= 152 \\
5h &= 152 + 180 \qquad \textbf{Add } 180 \textbf{ to each side.} \\
h &= \frac{152 + 180}{5} \qquad \textbf{Divide both sides by } 5. \\
h &= 66.4.
\end{aligned}
$$

We would expect the man to be 66.4 inches tall.

(d) According to the rule of thumb, if a man is 75 inches tall then his weight is $w = 5 \times 75 - 180 = 195$ pounds. The atypical man with a height of 75 inches and a weight of 190 pounds would therefore be light for his height.

13. **Lean body weight in females**: We noted in Exercise 12 that lean body weight in young adult males increases by 1.08 pounds for every pound of increase in total weight, assuming that the abdominal circumference remains the same. For young adult females, the formula shows that the slope of lean body weight as a function of total weight alone (for fixed values of R, A, H, and F) is 0.73 pound per pound. This means that their lean body weight increases by only 0.73 pound for every pound of increase in their total weight, assuming that all other factors remain the same.

15. **Vertical reach of fire hoses**:

(a) Because the vertical factor V always increases by 5 when d increases by $\frac{1}{8}$ inch, the vertical factor has a constant rate of change and hence is linear.

(b) First we find the slope of V. Because V always increases by 5 when d increases by $\frac{1}{8}$, the slope is given by

$$
\text{Slope} = \frac{\text{Change in } V}{\text{Change in } d} = \frac{5}{\frac{1}{8}} = 40.
$$

We now know that $V = 40d + b$, where b is a constant. We find b using the fact that $V = 85$ when $d = 0.5$: We have $85 = 40 \times 0.5 + b$, so $b = 85 - 40 \times 0.5 = 65$. Thus the formula is $V = 40d + 65$.

(c) We are given that $p = 50$ and $d = 1.75$. Using the formula for V from Part (b), we have $V = 40 \times 1.75 + 65$, and thus

$$
S = \sqrt{Vp} = \sqrt{(40 \times 1.75 + 65)50} = 82.16 \text{ feet.}
$$

Hence the vertical stream will travel 82.16 feet high.

(d) We are given that $d = 1.25$ and $p = 70$. Using the formula for V from Part (b), we have $V = 40 \times 1.25 + 65$, and thus

$$S = \sqrt{Vp} = \sqrt{(40 \times 1.25 + 65)70} = 89.72 \text{ feet}.$$

Hence the horizontal stream will travel 89.72 feet high. This is greater than the height of 60 feet, so the stream can reach the fire.

17. **More on budget constraints**:

(a) If you buy a pounds of apples and each pound costs \$0.50, you will have spent $0.50 \times a = 0.50a$ dollars on apples.

(b) If you buy g pounds of grapes and each pound costs \$1, you will have spent $1 \times g = g$ dollars on grapes.

(c) Since you will spend a total of \$5 on some combination of grapes and apples, then

Money spent on apples $+$ Money spent on grapes $= \$5$,

so $0.50a + g = 5$.

(d) To solve $0.50a + g = 5$ for g, subtract $0.50a$ from each side to get $g = -0.50a + 5$, which is the same as the answer for Part (d) from Exercise 16.

19. **Sleeping longer**: If we let M be the number of hours the man sleeps and t the number of days since his observation, then M is a linear function of t, with slope $\frac{1}{4}$ hour per day (since 15 minutes is $\frac{1}{4}$ hour) and initial value 8 hours. Thus, $M = \frac{1}{4}t + 8$. Since we want to know when the man sleeps 24 hours, we want to find t so that $M(t) = 24$. Thus we want to solve the equation $\frac{1}{4}t + 8 = 24$. We find that $\frac{1}{4}t = 16$, so $t = 64$. Thus, 64 days after his observation the man will sleep 24 hours.

To answer this without using the language of linear functions, note that the exercise is asking how many $\frac{1}{4}$-hour segments we need to add to 8 to get 24, i.e., how many $\frac{1}{4}$-hour segments there are in 16 hours. Clearly there are 64.

21. **Life on other planets**:

(a) To calculate the most pessimistic value for N we take the pessimistic value for each of the variables other than S (because N is an increasing function of each variable). Thus the most pessimistic value for N is

$$N = (2 \times 10^{11}) \times 0.01 \times 0.01 \times 0.01 \times 0.01 \times 10^{-8} = 2 \times 10^{-5}.$$

Hence the most pessimistic value for N is 2×10^{-5}, or 0.00002. To calculate the most optimistic value for N we take the optimistic value for each of the variables other than S. Thus the most optimistic value for N is

$$N = (2 \times 10^{11}) \times 0.5 \times 1 \times 1 \times 1 \times 10^{-4} = 10{,}000{,}000.$$

Hence the most optimistic value for N is 10,000,000.

(b) Keeping L as a variable and using pessimistic values for each of the other variables (except for S) gives

$$N = (2 \times 10^{11}) \times 0.01 \times 0.01 \times 0.01 \times 0.01 \times L = 2000L.$$

Now we want to find what value of the variable L gives the function value $N = 1$. Thus we need to solve the equation $2000L = 1$. The result is $L = \dfrac{1}{2000} = 5 \times 10^{-4}$. Thus a value of $L = 5 \times 10^{-4}$, or $L = 0.0005$, will give 1 communicating civilization per galaxy with pessimistic values for the other variables.

(c) Because there is such a difference, even in orders of magnitude, between the pessimistic and optimistic values for the variables and for the function N, the formula does not seem helpful in determining if there is life on other planets.

3.3 MODELING DATA WITH LINEAR FUNCTIONS

E-1. **Testing for linearity**:

(a) Here are the average rates of change for the data table:

Interval	2 to 5	5 to 7	7 to 8	8 to 11
Average rate of change	3	3	3	3

The average rates of change are all 3, so the data are linear, and the slope is 3. Because we know the line passes through $(2, 5)$ we can use the point-slope form to get the equation of the line:

$$\begin{aligned} y - 5 &= 3(x - 2) \\ y - 5 &= 3x - 6 \\ y &= 3x - 1. \end{aligned}$$

(b) Here are the average rates of change for the data table:

Interval	3 to 4	4 to 6	6 to 9	9 to 10
Average rate of change	2	2	$\dfrac{8}{3}$	2

The average rates of change are not all the same, so the data are not linear.

(c) Here are the average rates of change for the data table:

Interval	1 to 3	3 to 7	7 to 9	9 to 12
Average rate of change	4	4	4	4

The average rates of change are all 4, so the data are linear, and the slope is 4. Because we know the line passes through $(1, -1)$ we can use the point-slope form to get the equation of the line:

$$\begin{aligned} y - (-1) &= 4(x - 1) \\ y + 1 &= 4x - 4 \\ y &= 4x - 5. \end{aligned}$$

E-3. **Finding data points**: Using the data points $(2, 5)$ and $(8, 23)$, we find that the slope is $\dfrac{23 - 5}{8 - 2} = 3$. Once we have the slope there are different ways to fill in the missing entries. We will find the equation of the linear model and use this to fill in the blanks. Because we know the slope, 3, and data point $(2, 5)$ we can use the point-slope form to get the equation of the line:

$$\begin{aligned} y - 5 &= 3(x - 2) \\ y - 5 &= 3x - 6 \\ y &= 3x - 1. \end{aligned}$$

Since $y = 3x - 1$, for $x = 5$ we have $y = 3 \times 5 - 1 = 14$. If $y = 17$ then $3x - 1 = 17$, and solving for x gives $x = 6$. If $y = 29$ then $3x - 1 = 29$, and solving for x gives $x = 10$. Thus the completed table is

x	2	5	6	8	10
y	5	**14**	17	23	29

S-1. **Testing data for linearity**: Calculating differences, we see that y changes by 5 for each change in x of 2. Thus these data exhibit a constant rate of change and so are linear.

S-3. **Making a linear model**: The data in Exercise S-1 show a constant rate of change of 5 in y per change of 2 in x and therefore the slope is $m = \dfrac{\text{Change in } y}{\text{Change in } x} = \dfrac{5}{2} = 2.5$. We know now that $y = 2.5x + b$ for some b. Since $y = 12$ when $x = 2$, $12 = 2.5 \times 2 + b$, so $b = 12 - 2.5 \times 2 = 7$. The equation is $y = 2.5x + 7$.

S-5. **Graphing discrete data**: The figure below shows a plot of the data in Exercise S-2.

S-7. **Entering and graphing data**: The entered data are shown in the figure on the left below, and the plot of the data is shown in the figure on the right below.

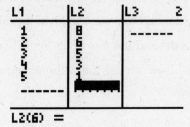 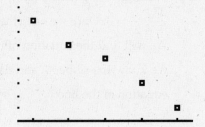

S-9. **Editing data**: In the figure on the left below, we have entered the squares of the data points in the table from Exercise S-7. The plot of the squares is shown in the figure on the right below.

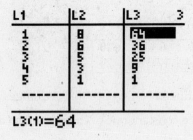 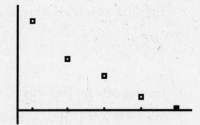

S-11. **Entering and graphing data**: The entered data are shown in the figure on the left below, and the plot of the data is shown in the figure on the right below.

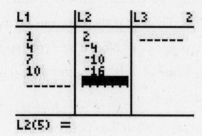
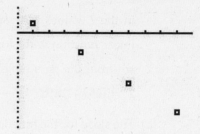

1. **Gasoline prices**: Let t be the time in days since January 1, 2004, and let P be the price in cents per gallon.

   (a) The table below shows the differences in successive time periods.

Change in t	0 to 14	14 to 28	28 to 42	42 to 56
Change in P	6.34	6.34	6.34	6.34

   The table of differences shows a constant change of 6.34 over each 14-day period. This shows that the data can be modeled using a linear function.

   (b) In the data display in the left-hand picture below, the window size is automatically selected by the calculator. The horizontal span is -5.6 to 61.6, and the vertical span is 147.69 to 181.67. The horizontal axis is time in days, and the vertical axis is the price in cents per gallon.

   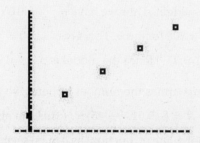
   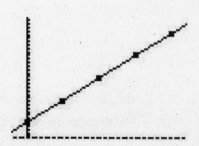

   (c) The slope of the line is $m = \dfrac{6.34}{14} = 0.4529$ cents per gallon per day. The initial value is 152 cents per gallon according to the table. Thus the equation is given by
   $$P = 0.4529t + 152.$$

   (d) Using the same spans as in Part (b), we obtain the right-hand picture above.

3. **Tuition at American public universities:**

(a) Let d be the number of years since 1994 and T the average tuition (in dollars) for public universities. The difference in the d values is always 1, and the difference in the T values is always 168. This shows that the data can be modeled by a linear function with slope 168 dollars per year. Now $d = 0$ corresponds to 1994, so (from the data table) the initial value of T is 2816, and thus the equation is $T = 168d + 2816$.

(b) The slope for the function from Part (a) is 168 dollars per year.

(c) The slope for the function from Exercise 2 is 866 dollars per year.

(d) Part (b) tells us that the tuition for public institutions increases at a rate of $168 per year, while Part (c) tells us that the tuition for private institutions increases at a rate of $866 per year. The average tuition at private universities is increasing at a greater rate.

(e) The percentage increase for private universities from 1997 to 1998 is $\dfrac{866}{16,419} = 0.053 = 5.3\%$, and the percentage increase for public universities from 1997 to 1998 is $\dfrac{168}{3320} = 0.051 = 5.1\%$. The average tuition for private universities had the larger percentage increase from 1997 to 1998.

5. **Total revenue and profit:**

(a) The difference in the N values is always 50, and the difference in the p values is always -0.50. This shows that the data can be modeled by a linear function with slope $\dfrac{-0.50}{50} = -0.01$ dollar per widget. Thus we have $p = -0.01N + b$, and we use the fact that $p = 43.00$ when $N = 200$ to find b. This gives $43.00 = -0.01 \times 200 + b$, and thus $b = 43.00 + 0.01 \times 200 = 45$. Hence the model is $p = -0.01N + 45$.

(b) Now the total revenue is the price times the number of items, so $R = pN$. Using the formula from Part (a) gives $R = (-0.01N + 45)N$. (This can also be written as $R = -0.01N^2 + 45N$.) This is not a linear function, as can be seen by examining a table of values (or a graph) or by observing that the coefficient of N is not constant.

(c) We have $P = R - C$. From Part (a) of Exercise 4 we know that C has slope 35 and initial value 900, so its formula is $C = 35N + 900$. Using the formula for R from Part (b) of this exercise, we have $P = (-0.01N + 45)N - (35N + 900)$. (This can also be written as $P = -0.01N^2 + 10N - 900$.) This is not a linear function, as can be seen by examining a table of values (or a graph) or by observing that the coefficient of N is not constant.

7. **The Kelvin temperature scale:**

   (a) The change in F is always 36 degrees Fahrenheit for each 20 degree change in K (the temperature in kelvins). This means that F is a linear function of K.

   (b) The slope is calculated as

   $$\frac{\text{Change in } F}{\text{Change in } K} = \frac{36}{20} = 1.8.$$

   Thus the slope is 1.8 degrees Fahrenheit per kelvin.

   (c) The slope of the function is 1.8, so $F = 1.8K + b$ for some b. To find b, we use one point on the line; for example, when $K = 200$, then $F = -99.67$. These values give $-99.67 = 1.8 \times 200 + b$, so $b = -459.67$. Thus $F = 1.8K - 459.67$.

   (d) Now $F = 98.6$, and we need to find K. Thus we have to solve the equation $1.8K - 459.67 = 98.6$:

   $$
   \begin{aligned}
   1.8K - 459.67 &= 98.6 \\
   1.8K &= 558.27 \quad \textbf{Add 459.67 to both sides.} \\
   K &= 310.15 \quad \textbf{Divide both sides by 1.8.}
   \end{aligned}
   $$

   Thus, body temperature on the Kelvin scale is 310.15 kelvins.

   (e) Since the slope of F is 1.8, when temperature increases by one kelvin, the Fahrenheit temperature will increase by 1.8 degrees.

   If we solve $F = 1.8K - 459.67$ for K we get $K = \dfrac{1}{1.8}F + \dfrac{459.67}{1.8}$, or $K = 0.56F + 255.37$. So a one degree increase in Fahrenheit will increase the temperature by 0.56 kelvin.

   (f) If $K = 0$ then $F = 1.8 \times 0 - 459.67 = -459.67$ degrees Fahrenheit.

9. **Market supply:**

   (a) By subtracting entries, we see that for each increase of 0.5 in the quantity S, there is an increase of 1.05 in price P. Thus P is a linear function of S with slope $\dfrac{1.05}{0.5} = 2.1$. Since the slope is 2.1, we have $P = 2.1S + b$. Now b can be determined using the first data point: $S = 1$ when $P = 1.35$. Substituting, we have $1.35 = 2.1 \times 1.0 + b$, so $b = -0.75$, and therefore $P = 2.1S - 0.75$.

(b) The given table suggests a horizontal span from 0 to 3 billion bushels of wheat and a vertical span from 0 to 5 dollars per bushel. The horizontal axis is billions of bushels produced, and the vertical axis is price per bushel .

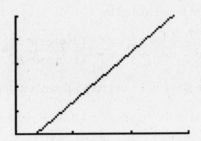

(c) If the price increases, then the wheat suppliers will want to produce more wheat. This means that the market supply curve will be going up to the right, so it is increasing.

(d) If the price P is \$3.90 per bushel, then S can be determined using the formula from Part (a): $2.1S - 0.75 = 3.90$. Solving (by adding 0.75 to each side, then dividing by 2.1) yields $S = 2.21$. Thus suppliers would be willing to produce 2.21 billion bushels in a year for a price of \$3.90 per bushel.

11. **Sports car:**

(a) Let t be the time since the car was at rest, in seconds, and V the velocity, in miles per hour. The differences are in the table below:

Change in t	0.5	0.5	0.5
Change in V	5.9	5.9	5.9

The ratio of change in V to change in t is always $\dfrac{5.9}{0.5} = 11.8$, so V can be modeled by a linear function of t with slope 11.8.

(b) The slope is 11.8 miles per hour per second. In practical terms, this means that, after every second, the car will be going 11.8 miles per hour faster. This says that the acceleration is constant.

(c) From Part (b), we know that $V = 11.8t + b$ for some b. Using the first data point, $t = 2.0$ and $V = 27.9$, we have $27.9 = 11.8 \times 2.0 + b$. Thus $b = 4.3$, and the final formula is $V = 11.8t + 4.3$.

(d) According to the formula in Part (c), the initial velocity is $V(0) = 4.3$ miles per hour. On the other hand, the exercise states that when $t = 0$, the car was at rest, so we would expect the initial velocity to be 0, not 4.3. In practical terms, the formula is valid only after the car has been moving a while since the acceleration is not constant the whole time: acceleration is large initially and decreases afterwards.

(e) We need to find when V is 60. Using the formula, we have $11.8t + 4.3 = 60$, so

$$t = \frac{60 - 4.3}{11.8} = 4.72 \text{ seconds.}$$

This car goes from 0 to 60 mph in 4.72 seconds.

13. **Later high school graduates**: Let N be the number graduating (in millions) and t the number of years since 1995.

(a) The difference in the t values is always 2, and the difference in the N values is always 0.16. This shows that the data can be modeled by a linear function with slope $\dfrac{0.16}{2} = 0.08$ million graduating per year. The slope means that every year the number graduating increases by 0.08 million (or 80,000).

(b) Now $t = 0$ corresponds to 1995, so (from the data table) the initial value of N is 2.59, and thus the equation is $N = 0.08t + 2.59$.

(c) The number graduating from high school in 2002 is expressed in functional notation as $N(7)$ since 2002 is 7 years since 1995. Its value is $N(7) = 0.08 \times 7 + 2.59 = 3.15$ million graduating.

(d) Since 1994 is 1 year before 1995, we use $t = -1$ in the formula from Part (b). The value is $N(-1) = 0.08 \times (-1) + 2.59 = 2.51$ million graduating. This is much closer to the actual number than the value of 2.02 million graduating given in Part (d) of Exercise 12. The trend in high school graduations was a steady decline from 1985 until the early 1990s followed by a steady increase starting in 1994 or slightly earlier.

15. **Sound speed in oceans**:

(a) Here is a table of differences:

Change in S	0.6	0.6	0.6	0.6
Change in c	0.72	0.72	0.82	0.62

From this table it is clear that the fourth entry for speed (at the salinity of $S = 36.8$) is in error—it is too high by 0.1. When $S = 36.8$, c should be $1517.62 - 0.1 = 1517.52$. With this choice, all entries are consistent with a linear model.

(b) The linear model has slope

$$\frac{\text{Change in } c}{\text{Change in } S} = \frac{0.72}{0.6} = 1.2 \text{ meters per second per parts per thousand.}$$

The slope means that, at the given depth and temperature, for an increase in salinity of 1 part per thousand the speed of sound increases by 1.2 meters per second.

(c) We use the slope, 1.2, and the last entry in the table. Because

$$\text{Change in } c = \text{Slope} \times \text{Change in } S,$$

we have

$$c(39) - c(37.4) = 1.2 \times (39 - 37.4),$$

and thus

$$c(39) = c(37.4) + 1.2 \times (39 - 37.4) = 1518.24 + 1.2 \times (39 - 37.4) = 1520.16.$$

This means that the speed of sound is 1520.16 meters per second when the salinity is 39 parts per thousand. *Note*: The value of $c(39)$ can also be found by first using the slope and an entry in the table to find the formula for c, namely $c = 1.2S + 1473.36$.

17. **Measuring the circumference of the Earth**: Because the rate of travel is 100 stades per day, a trip requiring 50 days is $100 \times 50 = 5000$ stades long. This is estimated to be $\frac{1}{50}$ of the Earth's circumference, so Eratosthenes' measure of the circumference of the Earth is $5000 \times 50 = 250,000$ stades. This is $250,000 \times 0.104 = 26,000$ miles.

3.4 LINEAR REGRESSION

E-1 **Fitting lines**: Answers will vary according to the choice of the estimate line for the federal education expenditures. For example, I might choose a line E with the same slope, 1.275, as Est, but which passes through the third data point $(2, 30.4)$. Such a line has equation $E - 30.4 = 1.275(x - 2)$, so $E = 1.275x + 27.85$. The table below shows the calculations to find the squares of the errors for E.

Year	Data	Estimate	Error of estimate	Square of error of estimate
x	y	E	$y - E$	$(y - E)^2$
0	27.0	27.850	-0.850	0.7225
1	29.0	29.125	-0.125	0.0156
2	30.4	30.400	0	0
3	30.9	31.675	-0.775	0.6006
4	32.1	32.950	-0.850	0.7225

The sum of the squares of the errors is the sum of the last column, which is 2.0612. Clearly this is not better fit than Est.

E-3 Calculating the error: For each data set in Exercise S-2, we form a table that shows the calculations yielding the sum-of-squares error for the regression line Reg.

(a) Here we use Reg $= 3.2x + 0.4$.

x	y	Regression Reg	Error of regression $y - \text{Reg}$	Square of error of regression $(y - \text{Reg})^2$
0	1	0.4	0.6	0.36
2	6	6.8	−0.8	0.64
4	13	13.2	−0.2	0.04
6	20	19.6	0.4	0.16

Thus the sum-of-squares error for the regression line is $0.36+0.64+0.04+0.16 = 1.2$.

(b) Here we use Reg $= 1.84x - 0.57$.

x	y	Regression Reg	Error of regression $y - \text{Reg}$	Square of error of regression $(y - \text{Reg})^2$
0	−1	−0.58	−0.42	0.1764
3	6	4.95	1.05	1.1025
5	8	8.63	−0.63	0.3969
9	16	15.99	0.01	0.0001

Thus the sum-of-squares error for the regression line is $0.1764 + 1.1025 + 0.3969 + 0.0001 = 1.6759$, or about 1.68.

(c) Here we use Reg $= 3.38x + 2.46$.

x	y	Regression Reg	Error of regression $y - \text{Reg}$	Square of error of regression $(y - \text{Reg})^2$
1	6	5.84	0.16	0.0256
2	9	9.22	−0.22	0.0484
6	23	22.74	0.26	0.0676
7	26	26.12	−0.12	0.0144

Thus the sum-of-squares error for the regression line is $0.0256 + 0.0484 + 0.0676 + 0.0144 = 0.156$.

E-5 The meaning of errors:

(a) If the sum of the squares of the errors of an estimate is 0, then, since each square is non-negative, this would mean that each error is 0. In this case, then, the line would fit the data points exactly.

(b) For each data set and regression line found in Exercise E-2, the errors are found as the fourth column in the tables in Exercise E-3. The sum of each fourth column is zero, allowing for rounding. (In general, it is always true that the sum of the errors of the regression line is 0.)

S-1. **Slope of regression line**: If the slope of the regression line is positive, then the line is increasing, so we expect overall the data values to be increasing.

S-3. **Meaning of slope of regression line**: If the slope of the regression line for federal agricultural spending is larger than the slope of the regression line for federal spending on research and development, then this means that federal spending on agriculture is increasing faster than federal spending on research and development.

S-5. **Plotting data and regression lines**: The figure below on the left is a plot of the data. The regression line is $y = -0.26x + 2.54$. The figure on the right below is the plot of the data with the graph of the regression line added.

S-7. **Plotting data and regression lines**: The figure below on the left is a plot of the data. The regression line is $y = 1.05x + 2.06$. The figure on the right below is the plot of the data with the graph of the regression line added.

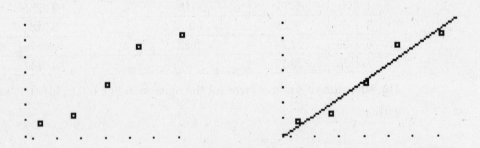

S-9. **Plotting data and regression lines**: The figure below on the left is a plot of the data. The regression line is $y = -0.32x - 1.59$. The figure on the right below is the plot of the data with the graph of the regression line added.

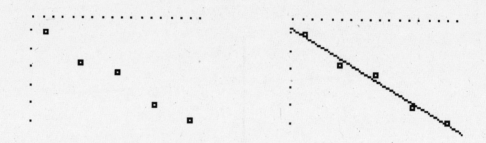

S-11. **Meaning of slope of regression line**: If the slope of the regression line for average life expectancy of children as a function of year of birth is 0.16, then for each increase of 1 in the year of birth the life expectancy increases by 0.16 year, or about 2 months.

1. **Is a linear model appropriate?**

    (a) The plotted points are shown below.

    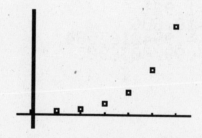

    The plot of the data appears to be fit better by a curve rather than by a straight line.

    (b) We take the variable to be years since 1996. The plotted points are shown below.

    It looks like the points almost fall on a straight line. It appears reasonable to approximate this data with a straight line.

3. **Tourism**: Let t be the number of years since 1994 and T the number of tourists, in millions.

   (a) A plot of the data points is shown below.

   (b) From the left-hand figure below, the equation of the regression line is $T = 1.92t + 18.61$. Adding this line to the plot above gives the right-hand figure below. The horizontal axis is years since 1994, and the vertical axis is millions of tourists.

   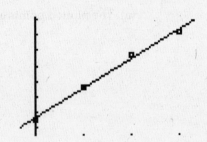

   ```
   LinReg
    y=ax+b
    a=1.921
    b=18.606
    r²=.9940163584
    r=.9970036903
   ```

   (c) The slope of the regression line is 1.92, and this means that every year the number of tourists who visited the United States from overseas increases by about 1.92 million.

   (d) The number of tourists who visited the United States in 2001 is $T(7)$ in functional notation, since $t = 7$ corresponds to seven years after 1994, that is, the year 2001. Using the regression line, we calculate $T(7) = 1.92 \times 7 + 18.61 = 32.05$ million tourists. This is a good bit higher than the actual number. Surely one factor was the coordinated terrorist attack of September 11, 2001. The general economic slowdown in that time period may have been another factor.

5. **Long jump**: Let t be the number of years since 1900 and L the length of the winning long jump, in meters.

   (a) Using the calculator, we find the equation of the regression line: $L = 0.034t + 7.197$.

(b) In practical terms the meaning of the slope, 0.034 meter per year, of the regression line is that each year the length of the winning long jump increased by an average of 0.034 meter, or about 1.34 inches.

(c) The figure below shows a plot the data points together with a graph of the regression line.

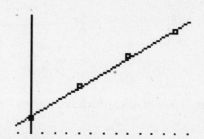

(d) The regression line formula has positive slope and so increases without bound. Surely there is some length that cannot be attained by a long jump, and therefore the regression line is not necessarily a good model of the winning length over a long period of time.

(e) The regression line model gives the value $L(20) = 0.034 \times 20 + 7.197 = 7.877$, or about 7.88, meters, which is over two feet longer than the length of the winning long jump in the 1920 Olympic Games. This result is consistent with our answer to Part (d) above.

7. **The effect of sampling error on linear regression:**

(a) The plotted points are shown below.

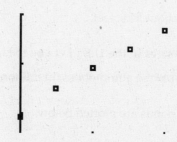

(b) From the left-hand figure below, the equation of the regression line is $D = 0.88t + 51.98$. In practical terms, the slope of 0.88 foot per hour means that the flood waters are rising by 0.88 foot each hour.

(c) Adding the regression line to the plot gives the right-hand figure below. The horizontal axis is time since flooding began, and the vertical axis is depth.

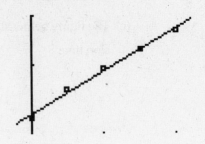

(d) Adding $D = 0.8t + 52$ (thick line) gives the following graph.

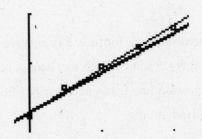

So the data does give a close approximation of the depth function, since the two lines are very close. The regression model shows a water level which is a bit too high.

(e) From the depth function $D = 0.8t + 52$, the actual depth of the water after 3 hours is $0.8 \times 3 + 52 = 54.4$ feet.

(f) The regression line equation predicts the depth after 3 hours to be $0.88 \times 3 + 51.98 = 54.62$, or about 54.6, feet.

9. **Japanese auto sales in the U.S.:** Let t be the number of years since 1992 and J the total U.S. sales of Japanese automobiles, in millions.

(a) The data points are plotted below.

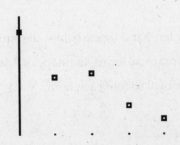

It may be reasonable to model this with a linear function, but there is room for argument here. There is enough deviation from a straight line to cast some doubt on whether the use of a linear model would be appropriate. The collection of additional data would be wise.

(b) From the left-hand figure below, the equation of the regression line is $J = -0.044t + 2.46$. The slope is -0.044, which means that each year the total US sales of Japanese cars decreases by 0.044 million cars, or about 44,000 cars.

(c) The regression line has been added in the right-hand figure below. The horizontal axis is years since 1992, and the vertical axis is millions of auto sales.

To the eyes of the authors, this picture makes the use of the regression line appear more appropriate than does the appearance of the plotted data alone. Gathering more data still seems advisable.

(d) Now 1997 corresponds to $t = 5$, and 1998 corresponds to $t = 6$. From the equation of the regression line, we expect $J = -0.044 \times 5 + 2.46 = 2.24$ million cars to be sold in 1997 and $J = -0.044 \times 6 + 2.46 = 2.196$, or about 2.20, million cars to be sold in 1998. The regression line estimates are higher than the Department of Commerce's figures in both cases.

11. **Antimasonic voting**: Using the notation of M as the percentage antimasonic voting and C as the number of church buildings, the statement to be analyzed is "M depends directly on C," that is, M is a linear function of C. The data plot is shown below.

From the left-hand figure below, the equation for the regression line is $M = 5.89C + 45.55$; moreover, plotting that equation with the data points gives the right-hand figure. The horizontal axis is number of churches and the vertical axis is percent antimasonic voting.

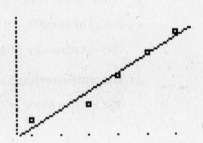

The figure shows that the regression line fits the data well. So the data does support the premise of the stated direct dependence.

13. **Expansion of steam**:

  (a) The linear regression model of volume V as a function of T is $V = 12.57T + 7501.00$.

  (b) If one fire is 100 degrees hotter than another, then T increases by 100, so V increases by $12.57 \times 100 = 1257$; that is, the volume of steam produced by 50 gallons of water in the hotter fire is 1257 cubic feet more.

  (c) The volume $V(420) = 12.57 \times 420 + 7501 = 12{,}780.4$ cubic feet. In practical terms this answer means that 50 gallons of water applied to a fire of 420 degrees Fahrenheit will produce about 12,780 cubic feet of steam.

  (d) If 50 gallons of water expanded to 14,200 cubic feet of steam, then $V = 14{,}200$, and so $14{,}200 = 12.57T + 7501$. Solving for T, we find that $12.57T = 14{,}200 - 7501 = 6699$. Thus $T = \dfrac{6699}{12.57} = 532.94$, and therefore the temperature of the fire was 532.94 degrees Fahrenheit.

15. **Whole crop weight versus rice weight**:

  (a) Using regression, we find a linear model $B = 0.29W + 0.01$ for B as a function of W.

  (b) Comparing the predicted rice weight from the linear model of Part (a) with the data, we see that the third sample, with $W = 13.7$ and $B = 3.7$ has a significantly lower rice weight than expected from the whole crop weight. This can also be seen by plotting the data with the regression line.

W Data	B Data	Predicted $B = 0.29W + 0.01$
6	1.8	1.75
11.1	3.2	3.23
13.7	3.7	3.98
14.9	4.3	4.33
17.6	5.2	5.11

(c) An increase in whole crop weight W of one ton per hectare is expected to produce an increase in rice weight B of 0.29 ton per hectare.

17. **Energy cost of running**:

(a) From the figure below, the equation of regression line for E in terms of v is $E = 0.34v + 0.37$.

```
LinReg
 y=ax+b
 a=.3418326693
 b=.3717131474
```

(b) The cost of transport is defined to be the slope of the regression line in Part (a), so the cost of transport is 0.34, in units of (unit of E) per (unit of v), which is milliliter of oxygen per gram per hour per kilometer per hour.

(c) If a rhea weighs 22,000 grams, then $W = 22,000$. The formula for C gives $C = 8.5 \times 22,000^{-0.40} = 0.16$. The figure found for the cost of transport in Part (b) is much higher than what the general formula would predict. Based on this, the rhea is a much less efficient runner that a typical animal of its size.

(d) If the rhea is at rest, then $v = 0$. Using $v = 0$ in the regression equation from Part (a) gives $E = 0.37$ as an estimate for the oxygen consumption for a rhea at rest. Surely this estimate is higher than the actual oxygen consumption level at rest, since we expect animals to be very inefficient in running at very low speeds.

19. **Laboratory experiment**: Answers will vary. Please refer to the website for details and instructions.

21. **Research project**: Answers will vary. Please refer to the website for details and instructions.

3.5 SYSTEMS OF EQUATIONS

E-1. **Using elimination to solve systems of equations:**

(a) We start with

$$x + y = 5$$
$$x - y = 1.$$

If we add the equations together, since the y's cancel out, keep the first equation and replace the second, we have

$$x + y = 5$$
$$2x = 6.$$

Thus $x = 3$. Plugging $x = 3$ into the first equation, we have $3 + y = 5$ and so $y = 2$.

(b) We start with

$$2x - y = 0$$
$$3x + 2y = 14.$$

If we multiply the first equation by 3 and the second equation by 2, then we have

$$6x - 3y = 0$$
$$6x + 4y = 28.$$

Subtracting the first from the second and, as before, keeping the first, but replacing the second by the difference, we have

$$6x - 3y = 0$$
$$7y = 28.$$

Thus $y = 4$. Plugging $y = 4$ into the first equation, we have $6x - 3 \times 4 = 0$, so $6x = 12$, and therefore $x = 2$.

(c) We start with

$$3x + 2y = 6$$
$$4x - 3y = 8.$$

If we multiply the first equation by 4 and the second equation by 3, then we have

$$12x + 8y = 24$$
$$12x - 9y = 24.$$

Subtracting first from the second and, as before, keeping the first, but replacing the second by the difference, we have

$$12x + 8y = 24$$
$$-17y = 0.$$

Thus $y = 0$. Plugging $y = 0$ into the first equation, we have $12x = 24$, and so $x = 2$.

E-3. Solving using the augmented matrix:

(a) We write this system of equations as an augmented matrix:

$$\begin{pmatrix} 1 & 1 & 1 & 3 \\ 2 & -1 & 2 & 3 \\ 3 & 3 & -1 & 9 \end{pmatrix}$$

Eliminating the entries under the upper left-hand corner, we use the indicated row operations:

$$\begin{pmatrix} 1 & 1 & 1 & 3 \\ 0 & -3 & 0 & -3 \\ 0 & 0 & -4 & 0 \end{pmatrix} \begin{matrix} \\ R_2 - 2R_1 \\ R_3 - 3R_1 \end{matrix}$$

This matrix corresponds to the system of equations

$$x + y + z = 3$$
$$-3y = -3$$
$$-4z = 0,$$

which can be solved by using the third equation to solve for z, the second to solve for y, and the first to solve for x: $z = 0$, $y = \frac{-3}{-3} = 1$, and $x = 3 - y - z = 3 - 1 - 0 = 2$.

(b) We write this system of equations as an augmented matrix:

$$\begin{pmatrix} 1 & -1 & -1 & -2 \\ 3 & -1 & 3 & 8 \\ 2 & 2 & 1 & 6 \end{pmatrix}$$

Eliminating the entries under the upper left-hand corner, we use the indicated row operations:

$$\begin{pmatrix} 1 & -1 & -1 & -2 \\ 0 & 2 & 6 & 14 \\ 0 & 4 & 3 & 10 \end{pmatrix} \begin{matrix} \\ R_2 - 3R_1 \\ R_3 - 2R_1 \end{matrix}$$

Eliminating the entry under the 2 in the second row:

$$\begin{pmatrix} 1 & -1 & -1 & -2 \\ 0 & 2 & 6 & 14 \\ 0 & 0 & -9 & -18 \end{pmatrix} \begin{matrix} \\ \\ R_3 - 2R_2 \end{matrix}$$

This matrix corresponds to the system of equations

$$x - y - z = -2$$

$$2y + 6z = 14$$

$$-9z = -18,$$

which can be solved exactly as in (a) above, namely, $z = \frac{-18}{-9} = 2$, $y = \frac{14-6z}{2} = \frac{14-6\times 2}{2} = 1$, and $x = -2 + y + z = -2 + 1 + 2 = 1$

(c) We write this system of equations as an augmented matrix:

$$\begin{pmatrix} 1 & 1 & 1 & 1 & 8 \\ 2 & -1 & 1 & -1 & -1 \\ 3 & 1 & -1 & 2 & 9 \\ 1 & 2 & 2 & 1 & 12 \end{pmatrix}$$

Eliminating the entries under the upper left-hand corner, we use the indicated row operations:

$$\begin{pmatrix} 1 & 1 & 1 & 1 & 8 \\ 0 & -3 & -1 & -3 & -17 \\ 0 & -2 & -4 & -1 & -15 \\ 0 & 1 & 1 & 0 & 4 \end{pmatrix} \begin{matrix} \\ R_2 - 2R_1 \\ R_3 - 3R_1 \\ R_4 - R_1 \end{matrix}$$

We multiply to simplify the elimination in the second column:

$$\begin{pmatrix} 1 & 1 & 1 & 1 & 8 \\ 0 & -6 & -2 & -6 & -34 \\ 0 & -6 & -12 & -3 & -45 \\ 0 & 6 & 6 & 0 & 24 \end{pmatrix} \begin{matrix} \\ 2R_2 \\ 3R_3 \\ 6R_4 \end{matrix}$$

And then we eliminate below the -6 in the second row:

$$\begin{pmatrix} 1 & 1 & 1 & 1 & 8 \\ 0 & -6 & -2 & -6 & -34 \\ 0 & 0 & -10 & 3 & -11 \\ 0 & 0 & 4 & -6 & -10 \end{pmatrix} \begin{matrix} \\ \\ R_3 - R_2 \\ R_4 + R_2 \end{matrix}$$

Finally we eliminate below the -10 in the third row:

$$\begin{pmatrix} 1 & 1 & 1 & 1 & 8 \\ 0 & -6 & -2 & -6 & -34 \\ 0 & 0 & -10 & 3 & -11 \\ 0 & 0 & 0 & -24 & -72 \end{pmatrix} \begin{matrix} \\ \\ \\ 5R_4 + 2R_3 \end{matrix}$$

This matrix corresponds to the system of equations

$$x + y + z + w = 8$$

$$-6y - 2z - 6w = -34$$

$$-20z + 6w = -22$$

$$-24w = -72,$$

which can be solved exactly as in (b) above, yielding $w = 3$, $z = 2$, $y = 2$, and $x = 1$.

E-5. **A system of equations with no solution**: We start with

$$x + y + z = 1$$
$$x + y - z = 2$$
$$x + y = 5,$$

and form its augmented matrix

$$\begin{pmatrix} 1 & 1 & 1 & 1 \\ 1 & 1 & -1 & 2 \\ 1 & 1 & 0 & 5 \end{pmatrix}$$

Eliminating the terms below the upper left-hand corner, we obtain

$$\begin{pmatrix} 1 & 1 & 1 & 1 \\ 0 & 0 & -2 & 1 \\ 0 & 0 & -1 & 4 \end{pmatrix} \begin{matrix} \\ R_2 - R_1 \\ R_3 - R_1 \end{matrix}$$

Eliminating the term below the -2 in the second row, we obtain

$$\begin{pmatrix} 1 & 1 & 1 & 1 \\ 0 & 0 & -2 & 1 \\ 0 & 0 & 0 & 7 \end{pmatrix} \begin{matrix} \\ \\ 2R_3, \text{ then } R_3 - R_2 \end{matrix}$$

The third row of this matrix corresponds to the equation $0x + 0y + 0z = 7$, which is impossible, so the original system has no solutions.

E-7. **Cramer's rule**:

(a) We have

$$x + 2y = 5$$
$$x - y = -1$$

so by Cramer's rule

$$x = \frac{\begin{vmatrix} 5 & 2 \\ -1 & -1 \end{vmatrix}}{\begin{vmatrix} 1 & 2 \\ 1 & -1 \end{vmatrix}} = \frac{5 \times (-1) - 2 \times (-1)}{1 \times (-1) - 2 \times 1} = \frac{-3}{-3} = 1$$

and

$$y = \frac{\begin{vmatrix} 1 & 5 \\ 1 & -1 \end{vmatrix}}{\begin{vmatrix} 1 & 2 \\ 1 & -1 \end{vmatrix}} = \frac{1 \times (-1) - 5 \times 1}{1 \times (-1) - 2 \times 1} = \frac{-6}{-3} = 2.$$

(b) We have

$$x + y = 3$$
$$x - y = 0$$

so by Cramer's rule

$$x = \frac{\begin{vmatrix} 3 & 1 \\ 0 & -1 \end{vmatrix}}{\begin{vmatrix} 1 & 1 \\ 1 & -1 \end{vmatrix}} = \frac{3 \times (-1) - 1 \times 0}{1 \times (-1) - 1 \times 1} = \frac{-3}{-2} = 1.5$$

and

$$y = \frac{\begin{vmatrix} 1 & 3 \\ 1 & 0 \end{vmatrix}}{\begin{vmatrix} 1 & 1 \\ 1 & -1 \end{vmatrix}} = \frac{1 \times 0 - 3 \times 1}{1 \times (-1) - 1 \times 1} = \frac{-3}{-2} = 1.5.$$

(c) We have

$$5x + 3y = 3$$
$$2x + 3y = 4$$

so by Cramer's rule

$$x = \frac{\begin{vmatrix} 3 & 3 \\ 4 & 3 \end{vmatrix}}{\begin{vmatrix} 5 & 3 \\ 2 & 3 \end{vmatrix}} = \frac{3 \times 3 - 3 \times 4}{5 \times 3 - 3 \times 2} = \frac{-3}{9} = -\frac{1}{3}$$

and

$$y = \frac{\begin{vmatrix} 5 & 3 \\ 2 & 4 \end{vmatrix}}{\begin{vmatrix} 5 & 3 \\ 2 & 3 \end{vmatrix}} = \frac{5 \times 4 - 3 \times 2}{5 \times 3 - 3 \times 2} = \frac{14}{9}.$$

(d) We have

$$6x - y = 5$$
$$3x + 4y = 2$$

so by Cramer's rule

$$x = \frac{\begin{vmatrix} 5 & -1 \\ 2 & 4 \end{vmatrix}}{\begin{vmatrix} 6 & -1 \\ 3 & 4 \end{vmatrix}} = \frac{5 \times 4 - (-1) \times 2}{6 \times 4 - (-1) \times 3} = \frac{22}{27}$$

and

$$y = \frac{\begin{vmatrix} 6 & 5 \\ 3 & 2 \end{vmatrix}}{\begin{vmatrix} 6 & -1 \\ 3 & 4 \end{vmatrix}} = \frac{6 \times 2 - 5 \times 3}{6 \times 4 - (-1) \times 3} = \frac{-3}{27} = -\frac{1}{9}.$$

S-1. **An explanation**: Each graph consists of pairs of points that make the corresponding equation true. The solution of the system is the point that makes both true, and that is the common intersection point.

S-3. **Crossing graphs**: We start with

$$3x + 4y = 6$$
$$2x - 6y = 5$$

and solve each equation for y. For the first equation we have $3x + 4y = 6$, so $4y = 6 - 3x$ and therefore $y = \dfrac{6 - 3x}{4}$. For the second equation we have $2x - 6y = 5$, so $-6y = 5 - 2x$ and therefore $y = \dfrac{5 - 2x}{-6}$. The figure below shows the graph of both $y = \dfrac{6 - 3x}{4}$ and $y = \dfrac{5 - 2x}{-6}$ using a horizontal span of 0 to 5 and a vertical span of -3 to 3. The intersection point, which is the solution to the system, is at $x = 2.15$, $y = -0.12$.

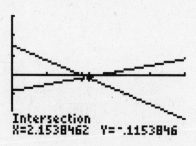

```
Intersection
X=2.1538462  Y=-.1153846
```

S-5. **Crossing graphs**: We start with

$$0.7x + 5.3y = 6.6$$
$$5.2x + 2.2y = 1.7$$

and solve each equation for y. For the first equation we have $0.7x + 5.3y = 6.6$, so $5.3y = 6.6 - 0.7x$ and therefore $y = \dfrac{6.6 - 0.7x}{5.3}$. For the second equation we have $5.2x + 2.2y = 1.7$, so $2.2y = 1.7 - 5.2x$ and therefore $y = \dfrac{1.7 - 5.2x}{2.2}$. The figure on the next page shows the graph of both $y = \dfrac{6.6 - 0.7x}{5.3}$ and $y = \dfrac{1.7 - 5.2x}{2.2}$ using a horizontal span of -2 to 2 and a vertical span of 0 to 5. The intersection point, which is the solution to the system, is at $x = -0.21$, $y = 1.27$.

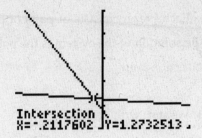

Intersection
X=⁻.2117602 Y=1.2732513 .

S-7. **Hand calculation**: To solve Exercise S-3 by hand calculation, we take the first equation $3x + 4y = 6$ and solve for y. As shown above for Exercise S-3, this yields $y = \dfrac{6 - 3x}{4}$. Now we take the second equation $2x - 6y = 5$, substitute the expression above for y and then solve for x:

$$
\begin{aligned}
2x - 6y &= 5 \\
2x - 6\left(\frac{6 - 3x}{4}\right) &= 5 \\
2x - \frac{6}{4}(6 - 3x) &= 5 \\
2x - 6 \times \frac{6}{4} + 3 \times \frac{6}{4}x &= 5 \\
\left(2 + \frac{18}{4}\right)x &= 5 + \frac{36}{4} = 5 + 9 = 14 \\
x &= \frac{14}{2 + (9/2)} = 2.15.
\end{aligned}
$$

Now $y = \dfrac{6 - 3x}{4} = \dfrac{6 - 3 \times \left(\dfrac{14}{2 + (9/2)}\right)}{4} = -0.12.$

S-9. **Hand calculation**: To solve Exercise S-5 by hand calculation, we take the first equation $0.7x + 5.3y = 6.6$ and solve for y. As shown above for Exercise S-5, this yields $y = \dfrac{6.6 - 0.7x}{5.3}$. Now we take the second equation $5.2x + 2.2y = 1.7$, substitute the expression above for y and then solve for x:

$$
\begin{aligned}
5.2x + 2.2y &= 1.7 \\
5.2x + 2.2\left(\frac{6.6 - 0.7x}{5.3}\right) &= 1.7 \\
5.2x + \frac{2.2}{5.3}(6.6 - 0.7x) &= 1.7 \\
5.2x + 6.6\frac{2.2}{5.3} - 0.7\frac{2.2}{5.3}x &= 1.7 \\
\left(5.2 - 0.7\frac{2.2}{5.3}\right)x &= 1.7 - 6.6\frac{2.2}{5.3} \\
x &= \frac{1.7 - 6.6\dfrac{2.2}{5.3}}{5.2 - 0.7\dfrac{2.2}{5.3}} = -0.21.
\end{aligned}
$$

Now $y = \dfrac{6.6 - 0.7x}{5.3} = \dfrac{6.6 - 0.7 \times (-0.21)}{5.3} = 1.27.$

S-11. **Crossing graphs**: We start with

$$3x - y = 5$$

$$2x + y = 0$$

and solve each equation for y. For the first equation we have $3x - y = 5$, so $y = 3x - 5$. For the second equation we have $2x + y = 0$, so $y = -2x$. The figure below shows the graph of both $y = 3x - 5$ and $y = -2x$ using a horizontal span of 0 to 2 and a vertical span of -5 to 1. The intersection point, which is the solution to the system, is at $x = 1$, $y = -2$.

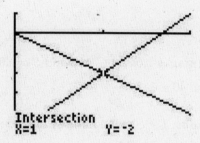

Intersection
X=1   Y=-2

1. **A party**: Let C be the number of bags of chips we need to buy and S be the number of sodas we need to buy. We have \$36 to spend, and so

$$\text{Cost of chips} \ + \ \text{Cost of sodas} \ = \ \text{Available money}$$

$$2C + 0.5S \ = \ 36.$$

Additionally, we know that we will buy 5 times as many sodas as bags of chips:

$$\text{Sodas} \ = \ 5 \times \text{Chips}$$

$$S \ = \ 5C.$$

Thus, we need to solve the system of equations

$$2C + 0.5S \ = \ 36$$

$$S \ = \ 5C.$$

The first step is to solve each equation for one of the variables. Since the second equation is already solved for S, we solve the first for S as well:

$$S \ = \ \frac{36 - 2C}{0.5}$$

$$S \ = \ 5C.$$

We will certainly buy fewer than 20 bags of chips, and so we use a horizontal span of 0 to 20. The following table of values leads us to choose a vertical span of 0 to 90 sodas. In the right-hand figure below, the thick line is the graph of $S = 5C$. The horizontal axis represents the number of bags of chips, and the vertical axis is the number of sodas.

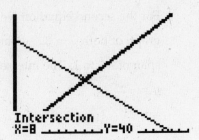

We see that the graphs cross at $C = 8$ and $S = 40$. This means that we should buy 8 bags of chips and 40 sodas.

3. **An order for bulbs**: Let c be the number of crocus bulbs, and let d be the number of daffodil bulbs. We get our first equation from the fact that the total number of bulbs is 55:

$$\text{Crocus} + \text{Daffodil} = \text{Total bulbs}$$
$$c + d = 55.$$

The second equation we need comes from budget constraints:

$$\text{Cost of crocuses} + \text{Cost of daffodils} = \text{Total cost}$$
$$0.35c + 0.75d = 25.65.$$

Thus we need to solve the system of equations

$$c + d = 55$$
$$0.35c + 0.75d = 25.65.$$

Solving each of these equations for d, we get

$$d = 55 - c$$
$$d = \frac{25.65 - 0.35c}{0.75}.$$

Since there are certainly no more than 55 crocus bulbs, we use a horizontal span of 0 to 55. From the table below, we choose a vertical span of 0 to 35. In the graph below, the thick line corresponds to $d = \dfrac{25.65 - 0.35c}{0.75}$. The horizontal axis is the number of crocus bulbs, and the vertical axis is the number of daffodil bulbs.

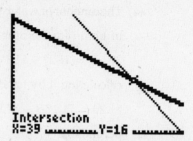

X	Y₁	Y₂
0	55	34.2
10	45	29.533
20	35	24.867
30	25	20.2
40	15	15.533
50	5	10.867
60	-5	6.2

X=0

Intersection
X=39 ⎯⎯⎯⎯ Y=16 ⎯⎯⎯⎯

We see that the graphs intersect at the point $c = 39$ and $d = 16$. So we should buy 39 crocus bulbs and 16 daffodil bulbs.

5. **Population growth**: Let f be the number of foxes and r the number of rabbits. Since the rates of change for both foxes and rabbits are constant (33 per year and 53 per year respectively), both are linear functions of time. After t years, there will be $f = 255 + 33t$ foxes and $r = 104 + 53t$ rabbits. We want to know when these values will be the same. That is, we need to find out when $f = r$, or when $255 + 33t = 104 + 53t$. This is a linear equation which we can solve by hand calculation or by graphing. Here is the hand calculation:

$$
\begin{aligned}
255 + 33t &= 104 + 53t & \\
151 + 33t &= 53t & \textbf{Subtract 104 from each side.} \\
151 &= 20t & \textbf{Subtract } 33t \textbf{ from both sides.} \\
7.55 &= t & \textbf{Divide each side by 20.}
\end{aligned}
$$

Hence, the number of foxes will be the same as the number of rabbits in 7.55 years. At that time, there will be $f = 255 + 33 \times 7.55 = 504.15$ foxes (this is better reported as 504 foxes) and the same number of rabbits.

7. **Competition between populations**: The equilibrium point occurs when the per capita growth rate of each population is 0. Thus we have two equations:

$$
\begin{aligned}
3(1 - m - n) &= 0 \\
2(1 - 0.7m - 1.1n) &= 0.
\end{aligned}
$$

This can solved by hand directly. The first equation is the same as $1 - m - n = 0$, and so $m = 1 - n$. Substituting this into the second equation, we get $2(1 - 0.7(1 - n) - 1.1n) = 0$.

Dividing by 2 and expanding, we get $1 - 0.7 + 0.7n - 1.1n = 0$, so $0.3 - 0.4n = 0$. Thus $0.4n = 0.3$, and so $n = \dfrac{0.3}{0.4} = 0.75$ thousand animals. We can substitute this value to find m. Now $m = 1 - n = 1 - 0.75 = 0.25$ thousand animals.

9. **Boron uptake**:

   (a) The amount of water-soluble boron B available resulting in the same plant content of boron for Decatur silty clay and Hartsells fine sandy loam is where $33.78 + 37.5B = 31.22 + 71.17B$. You can solve this problem graphically, using a table, or calculating it by hand. By hand, we move terms to the same side to get $33.78 - 31.22 = 71.17B - 37.5B = (71.17 - 37.5)B$. Thus $B = \dfrac{33.78 - 31.22}{71.17 - 37.5} = 0.08$ part per million.

   (b) The value of boron B from Part (a) gives the same plant content C for either soil type. For the Hartsells fine sandy loam, C increases by 71.17 parts per million with each increase of 1 in B, whereas for the Decatur silty clay, C only increases by 37.5 parts per million with each increase of 1 in B. Thus if B is larger than 0.08 part per million, the Hartsells fine sandy loam will have the larger plant content of boron.

11. **Fahrenheit and Celsius**: When Fahrenheit is exactly twice as much as Celsius, we have $F = 2C$. That means we need to solve the system of equations

$$F = \frac{9}{5}C + 32$$
$$F = 2C.$$

Since both equations are already solved for F, we need only graph both functions and find the intersection. The table of values below leads us to choose a horizontal span of 150 to 200 and a vertical span of 300 to 400. In the graph below, the thick line corresponds to $F = 2C$. The horizontal axis is Celsius temperature, and the vertical axis is Fahrenheit temperature.

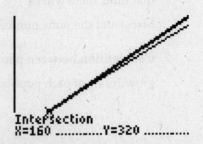

X	Y1	Y2
0	32	0
50	122	100
100	212	200
150	302	300
200	392	400
250	482	500
300	572	600

X=0

Intersection
X=160 ....Y=320 ....

We see that the graphs intersect when $C = 160$ and $F = 320$. Thus the temperature we seek is 160 degrees Celsius and 320 degrees Fahrenheit.

13. **Parabolic mirrors**:

(a) The point $(2, 4)$ is (a, a^2) where $a = 2$, so the light reflected from $(2, 4)$ follows the line $4 \times 2y + (1 - 4 \times 4)x = 2$, that is, $8y - 15x = 2$. The point $(3, 9)$ is (a, a^2) where $a = 3$, so the light reflected from $(3, 9)$ follows the line $4 \times 3y + (1 - 4 \times 9)x = 3$, that is, $12y - 35x = 3$. The light rays meet at the point where these two equations are both satisfied. Using crossing graphs, for example, we first solve each equation for y. The first equation is $8y - 15x = 2$, so $8y = 2 + 15x$ and therefore $y = \dfrac{2 + 15x}{8}$, while the second equation is $12y - 35x = 3$, so $12y = 3 + 35x$ and therefore $y = \dfrac{3 + 35x}{12}$. Graphing, we find a common solution at $x = 0$, $y = 0.25$, so the light rays meet at the point $(0, 0.25)$, or $(0, \frac{1}{4})$.

(b) To show that all the reflected light rays pass through the point $(0, 0.25)$, we need only to show that $x = 0$, $y = 0.25$ satisfies the equation $4ay + (1 - 4a^2)x = a$. But plugging in these values, we have $4ay + (1 - 4a^2)x = 4a \times 0.25 + (1 - 4a^2) \times 0 = 4a \times 0.25 + 0 = a \times 1 = a$, so each reflected line does pass through the focal point $(0, 0.25)$.

15. **Another interesting system of equations**: Solving both equations for y, we see that we need to graph the two functions

$$y = \frac{3 - x}{2}$$
$$y = \frac{6 - 2x}{4}.$$

If we do this, we find that both graphs produce the same line, showing that the two functions are in fact identical. If you divide the equation $-2x - 4y = -6$ by -2, you get the equation $x + 2y = 3$, which is exactly the first equation. Since the two equations are the same, every point on the line $x + 2y = 3$ is also on the other line, and hence every such point is a solution of the system of equations.

17. **An application of three equations in three unknowns**: Let N be the number of nickels, D the number of dimes, and Q the number of quarters. Since there are 21 coins in the bag, $N + D + Q = 21$. Since there is one more dime than nickel in the bag, $D = N + 1$. Since there is \$3.35 in the bag, $0.05N + 0.10D + 0.25Q = 3.35$.

This gives three linear equations in the variables N, D, and Q:

$$N + D + Q = 21$$
$$D = N + 1$$
$$0.05N + 0.10D + 0.25Q = 3.35.$$

The second equation is already solved for D, so we will substitute that into the other two equations. The first equation becomes $N + (N + 1) + Q = 21$, or $2N + Q + 1 = 21$. Solving for Q, we get $Q = 20 - 2N$.

The third equation becomes $0.05N + 0.10(N + 1) + 0.25Q = 3.35$, or $0.05N + 0.10N + 0.10 + 0.25Q = 3.35$. Solving for Q gives

$$Q = \frac{3.25 - 0.15N}{0.25}.$$

Graphing both of these equations using a horizontal span from 0 to 10 nickels and a vertical span from 0 to 30 quarters yields the following picture.

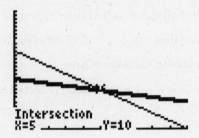

The graphs intersect when $N = 5$ and $Q = 10$.

We know that $D = N + 1$. Since $N = 5$, $D = 5 + 1 = 6$. So the solution to the equation is $N = 5$, $D = 6$, and $Q = 10$: there are 5 nickels, 6 dimes, and 10 quarters.

Chapter 3 Review Exercises

1. **Drainage pipe slope**: We have Slope $= \dfrac{\text{Rise}}{\text{Run}}$. In this case, the rise is -0.5 foot (since 6 inches make 0.5 foot) for a run of 8 feet. Thus the slope is $\dfrac{-0.5}{8} = -0.0625$ foot per foot. If we measure the rise in inches then we find $\dfrac{-6}{8} = -0.75$ inch per foot for the slope.

2. **Height from slope and horizontal distance**: We have

$$\text{Vertical change} = \text{Slope} \times \text{Horizontal change}.$$

In this case, the slope of the ladder is 3.7, while the horizontal distance is 4.4 feet, so the vertical height is $3.7 \times 4.4 = 16.28$ feet.

3. **A ramp into a van:**

   (a) We have Slope $= \dfrac{\text{Rise}}{\text{Run}}$. In this case, the rise is 18 inches, the distance from the van door to the ground, and the run is 4 feet, since that is the horizontal distance from the ramp to the van. Thus the slope is $\dfrac{18}{4} = 4.5$ inches per foot. If we measure the rise in feet then we find $\dfrac{1.5}{4} = 0.375$ foot per foot for the slope.

   (b) We have
   $$\text{Rise} = \text{Slope} \times \text{Run}.$$
   In this case we take the slope to be 1.2 inches per foot. The rise is still 18 inches, so we have $18 = 1.2 \times \text{Run}$. Thus the run is $\dfrac{18}{1.2} = 15$ feet. A larger slope will require a greater distance, so the ramp should rest at least 15 feet from the van.

4. **A reception tent:**

   (a) We have Slope $= \dfrac{\text{Rise}}{\text{Run}}$. The rise from the top of an outside pole to the top of the center pole is $25 - 10 = 15$ feet, and the corresponding run is 30 feet. Thus the slope is $\dfrac{15}{30} = 0.5$ foot per foot.

   (b) We have
   $$\text{Rise} = \text{Slope} \times \text{Run}.$$
   We know from Part (a) that the slope is 0.5 foot per foot. The run is 5 feet, so the rise is $0.5 \times 5 = 2.5$ feet. Thus the height has increased by 2.5 feet from the outside poles. We add the height of an outside pole to get that the height at this point is $2.5 + 10 = 12.5$ feet.

   (c) We have
   $$\text{Rise} = \text{Slope} \times \text{Run}.$$
   We know from Part (a) that the slope is 0.5 foot per foot. The rise from the ground to the top of an outer pole is 10 feet, so we have $10 = 0.5 \times \text{Run}$. Thus the run is $\dfrac{10}{0.5} = 20$ feet. Hence the ropes are attached to the ground 20 feet from the outside poles.

5. **Function value from slope and run:** We have that the slope is $m = -2.2$, so
   $$-2.2 = \frac{\text{Change in function}}{\text{Change in variable}} = \frac{g(6.8) - g(3.7)}{6.8 - 3.7} = \frac{g(6.8) - 5.1}{6.8 - 3.7} = \frac{g(6.8) - 5.1}{3.1}.$$
   Thus $g(6.8) - 5.1 = 3.1 \times (-2.2)$, and so $g(6.8) = 5.1 - 3.1 \times (-2.2) = -1.72$.

6. **Lanes on a curved track:**

   (a) To find the inner radius of the first lane, we put $n = 1$ in the formula and find $R(1) = 100/\pi$. Thus the radius is about 31.83 meters.

   (b) The width of a lane is the change in R when we increase n by 1, and that is the slope of the linear function R. From the formula we see that the slope is 1.22. Thus the width of a lane is 1.22 meters. This can also be found by subtracting successive values of the radius; for example, $R(2) - R(1) = 1.22$.

   (c) We first find for what value of n we have $R(n) = 35$. This means we must solve the linear equation $100/\pi + 1.22(n - 1) = 35$ for n. We have

   $$
   \begin{aligned}
   100/\pi + 1.22(n - 1) &= 35 \\
   1.22(n - 1) &= 35 - 100/\pi \\
   n - 1 &= \frac{35 - 100/\pi}{1.22} \\
   n &= \frac{35 - 100/\pi}{1.22} + 1 = 3.60.
   \end{aligned}
   $$

   Since the number of lanes is a whole number, if you wish to run in a lane with a radius of at least 35 meters you should pick lane 4 or a higher-numbered lane.

7. **Linear equation from two points:** We first compute the slope:

   $$
   m = \frac{\text{Change in function}}{\text{Change in variable}} = \frac{g(1.1) - g(-2)}{1.1 - (-2)} = \frac{3.3 - 7.2}{3.1} = \frac{-3.9}{3.1} = -1.258.
   $$

   (We keep the extra digit for accuracy in the rest of the calculations.) Since g is a linear function with slope -1.258, we have $g = -1.258x + b$ for some b. Now $g(1.1) = 3.3$, so $3.3 = -1.258 \times 1.1 + b$, and therefore $b = 3.3 + 1.258 \times 1.1 = 4.68$. Thus if we now round the slope to 2 decimal places we get $g = -1.26x + 4.68$.

8. **Working on a commission:**

   (a) Each additional dollar of total sales increases the income by 5% of $1, or 0.05 dollar. This means that the change in I is always the same, 0.05 dollar, for a change of 1 in S. Thus I is a linear function of S.

   (b) If he sells $1600 in a month then he earns $1000 plus 5% of $1600, or $1000 + 0.05 \times 1600 = 1080$ dollars.

   (c) From Part (a), I is a linear function of S with slope 0.05, since the slope represents the additional income for 1 additional dollar in sales. Since I is a linear function with slope 0.05, we have $I = 0.05S + b$ for some b. Now $I(1600) = 1080$ by Part (b), so $1080 = 0.05 \times 1600 + b$, and therefore $b = 1080 - 0.05 \times 1600 = 1000$. Thus $I = 0.05S + 1000$.

(d) We want to find the value of S so that $I = 1350$. That means we need to solve the equation $0.05S + 1000 = 1350$ for S. We find:

$$0.05S + 1000 = 1350$$
$$0.05S = 1350 - 1000$$
$$S = \frac{1350 - 1000}{0.05} = 7000.$$

Thus his monthly sales must be $7000.

9. **Testing data for linearity**:

(a) There is a constant change of 0.3 in x and a constant change of -0.06 in f. Thus these data exhibit a constant rate of change and so are linear.

(b) From Part (a) we know that the data show a constant rate of change of -0.6 in f per change of 0.3 in x. Hence the slope is $m = \dfrac{\text{Change in } f}{\text{Change in } x} = \dfrac{-0.6}{0.3} = -2.$ We know now that $f = -2x + b$ for some b. Since $f = 8$ when $x = 3$, we have $8 = -2 \times 3 + b$, so $b = 8 + 2 \times 3 = 14$. The formula is $f = -2x + 14$.

10. **An epidemic**:

(a) There is a constant change of 5 in d and a constant change of 6 in T. Thus these data exhibit a constant rate of change and so are linear.

(b) The slope represents the number of new cases per day.

(c) The slope is $m = \dfrac{\text{Change in } T}{\text{Change in } d} = \dfrac{6}{5} = 1.2.$

(d) Now T is a linear function with slope 1.2, and the initial value is 35 (from the first entry in the table). Thus the formula is $T = 1.2d + 35$.

(e) To predict the total number of diagnosed flu cases after 17 days we put $d = 17$ in the formula from Part (d): $T = 1.2 \times 17 + 35 = 55.4$. Thus we expect 55 flu cases to be diagnosed after 17 days.

11. **Testing and plotting linear data**:

(a) There is a constant change of 2 in x and a constant change of 0.24 in f. Thus these data exhibit a constant rate of change and so are linear.

(b) The slope is $m = \dfrac{\text{Change in } f}{\text{Change in } x} = \dfrac{0.24}{2} = 0.12$. We know now that $f = 0.12x + b$ for some b. Since $f = -1.12$ when $x = -1$, we have $-1.12 = 0.12 \times (-1) + b$, so $b = -1.12 + 0.12 = -1$. The formula is $f = 0.12x - 1$.

(c) The figure below shows the plot of the data together with the graph of $f = 0.12x - 1$.

12. **Marginal tax rate:**

(a) There is a constant change of $50 in the taxable income and a constant change of $14 in the tax due. Thus these data exhibit a constant rate of change and so are linear.

(b) From Part (a) the rate of change in tax due as a function of taxable income is $\dfrac{14}{50} = 0.28$ dollar per dollar. Thus the additional tax due on each dollar is $0.28.

(c) From the table the tax due on a taxable income of $97,000 is $21,913. If the taxable income increases by $1000 to $98,000, by Part (b) the tax due will increase by $0.28 \times 1000 = 280$ dollars. Thus the tax due on a taxable income of $98,000 is $21,913 + 280 = 22,913$ dollars.

(d) Let T denote the tax due, in dollars, on an income of A dollars over $97,000. The slope of T is 0.28 by Part (b). The initial value of T is the tax due on 0 dollars over $97,000, that is, on $97,000. That tax is $21,913. Thus the formula is $T = 0.28A + 21,913$.

13. **Plotting data and regression lines:**

(a) The figure below on the left is a plot of the data.

(b) The regression line is $y = -16.3x + 37.19$.

(c) The figure on the right below is the plot of the data with the regression line added.

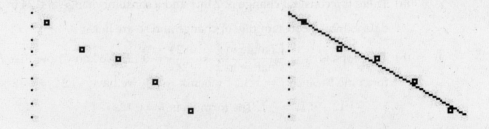

14. **Meaning of slope of regression line**: If the slope of the regression line for number of births per thousand to unmarried 18- to 19-year-old women as a function of year is -1.13, then for each year that passes the number of births to unmarried 18- to 19-year old women decreases by about 1.16 births per thousand.

15. **Life expectancy**:

    (a) Let E be life expectancy in years and t the time in years since 1996. Using regression gives the model $E = 0.2t + 76.2$.

    (b) The figure below is the plot of the data with the regression line added.

    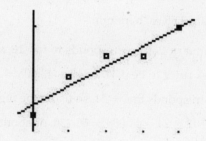

    (c) The slope of the regression line is 0.2, and this means that for each increase of 1 in the year of birth the life expectancy increases by 0.2 year, or about $2\frac{1}{2}$ months.

    (d) Now 2005 corresponds to $t = 9$, so using the regression line we predict the life expectancy to be $0.2 \times 9 + 76.2 = 78$ years.

    (e) Now 1580 corresponds to $t = -416$, so from the regression line we expect the life expectancy to be $0.2 \times (-416) + 76.2 = -7$ years. This is nonsensical! The year 2300 corresponds to $t = 304$, so from the regression line we predict the life expectancy to be $0.2 \times 304 + 76.2 = 137$ years. This is unreasonable. In fact, both of these predictions are far out of the range of applicability of the regression line formula.

16. **XYZ Corporation stock prices**:

    (a) Let P be the stock price in dollars and t the time in months since January, 2005. Using regression gives the model $P = 0.547t + 43.608$.

(b) The figure below is the plot of the data with the regression line added.

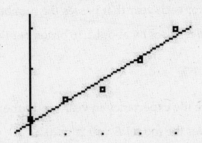

(c) The slope is 0.547 dollar per month, and this means that the stock price increases by about $0.55 each month.

(d) Now January, 2006 corresponds to $t = 12$, so using the regression line we predict the stock price to be $0.547 \times 12 + 43.608 = 50.172$, or about $50.17. Also January, 2007 corresponds to $t = 24$, so using the regression line we predict the stock price to be $0.547 \times 24 + 43.608 = 56.736$, or about $56.74.

17. **Crossing graphs**: We start with

$$4x - 2y = 9$$
$$x + y = 0$$

and solve each equation for y. For the first equation we have $4x - 2y = 9$, so $2y = 4x - 9$ and therefore $y = \dfrac{4x - 9}{2}$. For the second equation we have $x + y = 0$, so $y = -x$. The figure below shows the graph of both $y = \dfrac{4x - 9}{2}$ and $y = -x$ using a horizontal span of 0 to 2 and a vertical span of -5 to 0. The intersection point, which is the solution to the system, is at $x = 1.5$, $y = -1.5$.

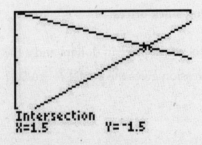

18. **Hand calculation**: To solve Exercise 17 by hand calculation, we take the second equation $x + y = 0$ and solve for y. As above for Exercise 17, this yields $y = -x$. Now we take the first equation $4x - 2y = 9$, substitute the expression above for y, and then solve for x:

$$
\begin{aligned}
4x - 2y &= 9 \\
4x - 2(-x) &= 9 \\
6x &= 9 \\
x &= \frac{9}{6} = 1.5.
\end{aligned}
$$

Now $y = -x = -1.5$.

19. **Bills**: Let T be the number of twenty-dollar bills and F the number of fifty-dollar bills. Since the total value is $400, we have $20T + 50F = 400$. Since there are 11 bills, we have $T + F = 11$. The second equation can easily be solved for F, giving $F = 11 - T$. Substituting into the first equation, we get

$$
\begin{aligned}
20T + 50F &= 400 \\
20T + 50(11 - T) &= 400 \\
20T + 50 \times 11 - 50T &= 400 \\
-30T &= 400 - 50 \times 11 \\
T &= \frac{400 - 50 \times 11}{-30} = 5.
\end{aligned}
$$

Thus there are 5 twenty-dollar bills and 6 fifty-dollar bills.

20. **Mixing paint**: Let B be the number of gallons of blue paint and Y the number of gallons of yellow paint. Since the total is 10 gallons, we have $B + Y = 10$. Since there is 3 times as much yellow paint as blue paint, we have $Y = 3B$. We insert the formula for Y from the second equation into the first:

$$
\begin{aligned}
B + Y &= 10 \\
B + 3B &= 10 \\
4B &= 10 \\
B &= \frac{10}{4} = 2.5.
\end{aligned}
$$

Thus you should use 2.5 gallons of blue paint and 7.5 gallons of yellow paint.

Solution Guide for Chapter 4: Exponential Functions

4.1 EXPONENTIAL GROWTH AND DECAY

E-1. Practice with exponents:

(a) Separating the terms with a and those with b, and using the quotient law, we have

$$\frac{a^3 b^2}{a^2 b^3} = \frac{a^3}{a^2} \times \frac{b^2}{b^3}$$
$$= a^{3-2} \times b^{2-3}$$
$$= a^1 \times b^{-1} = \frac{a}{b}.$$

This expression can also be written as ab^{-1}.

(b) Using the power law, we have

$$\left((a^2)^3\right)^4 = \left(a^{2\times3}\right)^4$$
$$= a^{(2\times3)\times4} = a^{24}.$$

(c) Separating the terms with a and those with b, and using the product law, we have

$$a^3 b^2 a^4 b^{-1} = (a^3 a^4) \times (b^2 b^{-1})$$
$$= a^{3+4} b^{2+(-1)} = a^7 b^1 = a^7 b.$$

E-3. Finding exponential functions:

(a) Since N is an exponential function with growth factor 6, $N = P \times 6^t$, and we want to find P. Now $N(2) = 7$, so $P \times 6^2 = 7$ and therefore $P = 7/(6^2) = 0.19$. Thus $N = 0.19 \times 6^t$.

(b) Since N is an exponential function, $N = Pa^t$. The two equations $N(3) = 4$ and $N(7) = 8$ tell us that

$$4 = Pa^3$$
$$8 = Pa^7.$$

Dividing the bottom equation by the top gives

$$\frac{8}{4} = \frac{Pa^7}{Pa^3} = \frac{a^7}{a^3} = a^4.$$

Thus $a^4 = 2$ and so $a = 2^{1/4} = 1.19$.

Returning to the equation $N(3) = 4$, we now have $P \times 1.19^3 = Pa^3 = 4$ and so $P = 4/(1.19^3) = 2.37$. Thus $N = 2.37 \times 1.19^t$.

(c) Since N is an exponential function with growth factor a (we are assuming that a is some known number) and $N(2) = 3$, we have $Pa^2 = 3$. Solving for P, we have $P = 3/(a^2) = 3a^{-2}$. Thus $N = 3a^{-2}a^t$, which can also be written as $N = 3a^{t-2}$.

(d) Since N is an exponential function, $N = Pa^t$. The two equations $N(2) = m$ and $N(4) = n$ tell us that

$$m = Pa^2$$

$$n = Pa^4.$$

Dividing the bottom equation by the top gives

$$\frac{n}{m} = \frac{Pa^4}{Pa^2} = \frac{a^4}{a^2} = a^2.$$

Thus $a^2 = \dfrac{n}{m}$ and so

$$a = \left(\frac{n}{m}\right)^{1/2} = \sqrt{\frac{n}{m}}.$$

Returning to the equation $N(2) = m$, we now have

$$P \times \left(\sqrt{\frac{n}{m}}\right)^2 = Pa^2 = m$$

Using the power law and dividing we get

$$P = \frac{m}{\left(\frac{n}{m}\right)} = \frac{m^2}{n}.$$

Thus

$$N = \frac{m^2}{n} \left(\sqrt{\frac{n}{m}}\right)^t.$$

E-5. **Finding initial value:** Since N is an exponential function with growth factor 1.94, $N = P \times 1.94^t$, and we want to find the initial value P. Now $N(5) = 6$, so $P \times 1.94^5 = 6$ and therefore $P = 6/(1.94^5) = 0.22$. Thus $N = 0.22 \times 1.94^t$.

S-1. **Function value from initial value and growth factor:** Since the growth factor is 2.4 and the initial value is $f(0) = 3$, $f(2) = 3 \times 2.4 \times 2.4 = 17.28$. Since f is an exponential function, we can get a formula for $f(x)$ in the form Pa^x, where a is the growth factor 2.4 and P is the initial value 3. Thus $f(x) = 3 \times 2.4^x$.

S-3. **Finding the growth factor:** The growth factor is the factor by which f changes for a change of 1 in the variable. If a is the growth factor, then

$$a = \frac{f(5)}{f(4)} = \frac{10}{8} = 1.25.$$

S-5. **Rate of change**: The rate of change of an exponential function is proportional to the function value.

S-7. **Percentage decay**:If a function has an initial value of 10 and decays by 4% per year, then it is an exponential function with $P = 10$ and $a = 1 - r = 1 - 0.04 = 0.96$. Thus an exponential function which describes this is 10×0.96^t, with t in years.

S-9. **Percentage change**: If a bank account grows by 9% each year, then the yearly growth factor is $a = 1 + r = 1 + 0.09 = 1.09$. The monthly growth factor, since a month is 1/12th of a year, is $1.09^{1/12} = 1.0072$, which is a growth of 0.72% per month.

S-11. **Percentage growth**: If the population grows by 3.5% each year, then the yearly growth factor is $a = 1 + r = 1 + 0.035 = 1.035$. The decade growth factor is $1.035^{10} = 1.4106$ since a decade is 10 years. This represents a growth of 41.06% per decade.

1. **Exponential growth with given initial value and growth factor**: The initial value is $P = 23$, and the growth factor is $a = 1.4$, so the formula for the function is

$$N(t) = 23 \times 1.4^t.$$

The table of values below leads us to choose a horizontal span of 0 to 5 and a vertical span of 0 to 130. The graph is shown below.

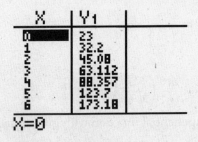
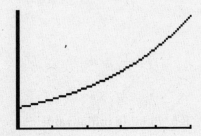

3. **A population with given per capita growth rate**:

   (a) Since the population grows at a rate of 2.3% per year, the yearly growth factor for the function is $a = 1 + .023 = 1.023$. Since initially the population is $P = 3$ million, the exponential function is

   $$N(t) = 3 \times 1.023^t.$$

   Here N is the population in millions and t is time in years.

   (b) The population after 4 years is expressed by $N(4)$ using functional notation. Its value, using the formula from Part (a), is $N(4) = 3 \times 1.023^4 = 3.29$ million.

5. **Unit conversion with exponential decay:**

   (a) Now $N(2) = 500 \times 0.68^2 = 231.2$, which means that after 2 years there are 231.2 grams of the radioactive substance present.

   (b) The yearly decay factor is $0.68 = 1 - 0.32$, so $r = 0.32$. Thus the yearly percentage decay rate is 32%.

   (c) Now $t = \dfrac{1}{12}$ corresponds to one month, so the monthly decay factor is

   $$0.68^{1/12} = 0.968.$$

   The monthly percentage decay rate is found using the same method as in Part (b): $0.968 = 1 - 0.032$, so $r = 0.032$. Thus the monthly percentage decay rate is 3.2%.

   (d) There are 31,536,000 seconds in a year, so one second corresponds to $\dfrac{1}{31,536,000}$ year. Thus the decay factor per second is

   $$0.68^{1/31,536,000} = 0.9999999878 = 1 - 0.0000000122.$$

   Thus $r = 0.0000000122$, and so the percentage decay rate per second is only 0.00000122%.

7. **Half-life of heavy hydrogen:** If we start with 100 grams of H_3 then, according to Example 4.1, the decay is modeled by

   $$A = 100 \times 0.783^t.$$

   As in Example 4.1, the half-life of H_3 is obtained by solving $100 \times 0.783^t = 50$ to get $t = 2.83$ years. So, after 2.83 years we will have 50 grams of H_3 left. This is the same length of time as found in Example 4.1, Part 4, for 50 grams of H_3 to decay to 25 grams.

   Since the amount we start with decays by half every 2.83 years, we see that after starting with 100 grams of H_3 we will have 50 grams after 2.83 years, 25 grams after $2 \times 2.83 = 5.66$ years, 12.5 grams after $3 \times 2.83 = 8.49$ years, and 6.25 grams after $4 \times 2.83 = 11.32$ years. This is four half-lives.

9. **Inflation**: The yearly growth factor for prices is given by $1 + 0.03 = 1.03$. We want to find the value of t for which $1.03^t = 2$, since that is when the prices will have doubled. We solve this using the crossing graphs method. The table of values below leads us to choose a horizontal span of 0 to 25 years and a vertical span of 0 to 3. In the graph below, the horizontal axis is years, and the vertical axis is the factor by which prices increase. We see that the two graphs cross at $t = 23.45$ years. Thus, prices will double in 23.45 years. Note that we did not use the amount of the fixed income. The answer, 23.45 years, is the same regardless of the income.

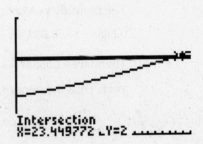

11. **The MacArthur-Wilson Theory of biogeography**: According to the theory, stabilization occurs when the rate of immigration equals the rate of extinction, that is, $I = E$. This is equivalent to solving the equation $4.2 \times 0.93^t = 1.5 \times 1.1^t$. This can be solved using a table or a graph. The graph below, which uses a horizontal span of 0 to 10 and a vertical span of 0 to 5, shows that the solution occurs when $t = 6.13$ years, at which time the immigration and extinction rates are each 2.69 species per year.

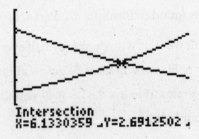

13. **The population of Mexico**:

    (a) The yearly growth factor for the population is $1 + 0.026 = 1.026$, and the initial population is 67.38 million. The population N (in millions) at the time t years after 1980 is

    $$N = 67.38 \times 1.026^t.$$

(b) The population of Mexico is 1983 is expressed as $N(3)$ in functional notation, since 1983 is $t = 3$ years after 1980. The value of $N(3)$ is $67.38 \times 1.026^3 = 72.77$ million.

(c) The population of Mexico is 90 million when $67.38 \times 1.026^t = 90$. We solve this equation using the crossing graphs method. The table below leads us to choose a horizontal span of 0 to 15 and a vertical span of 60 to 100. The curve is the graph of 67.38×1.026^t, and the thick line is the graph of 90. We see that the two cross at 11.28 years after 1980. That is just over a quarter of the way through 1991.

X	Y1	Y2
0	67.38	90
3	72.773	90
6	78.599	90
9	84.89	90
12	91.685	90
15	99.024	90
18	106.95	90

X=0

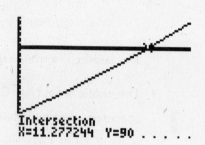

Intersection
X=11.277244   Y=90 . . . . .

15. **Tsunami waves in Crescent City:**

(a) The probability P is an exponential function of Y because for each change in Y of 1 year, the probability P decreases by 2%, so P exhibits constant percentage change.

(b) The decay factor for P is 1 minus the decay percentage, so $a = 1 - r = 1 - 0.02 = 0.98$ per year.

(c) The initial value of P, $P(0)$, represents the probability of no tsunami wave of height 15 feet or more striking over a period of 0 years. Since this is a certainty (as a probability), the initial value of P is 1.

(d) Since P is an exponential function of Y (by Part (a)) with a decay factor of 0.98 (by Part (b)) and an initial value of 1 (by Part (c)) the formula is $P = 1 \times 0.98^Y$, or simply $P = 0.98^Y$.

(e) The probability of no tsunami waves 15 feet or higher striking Crescent City over a 10-year period is $P(10) = 0.98^{10} = 0.82$. Thus the probability of no such tsunami wave striking over a ten-year period is 0.82 or 82%. Over a 100-year period, the probability is $P(100) = 0.98^{100} = 0.13$. Thus the probability of no such tsunami wave striking over a 100-year period is 0.13 or only 13%.

(f) To find the probability Q that at least one wave 15 feet or higher will strike Crescent City over a period of Y years, we first see that $P + Q = 1$ since $P + Q$ represents all possibilities. Thus $Q = 1 - P = 1 - 0.98^Y$.

17. **Grains of wheat on a chess board**: The growth factor here is 2, and the initial amount is 1. Thus the function we need is 1×2^t, or simply 2^t, where t is measured in days since the first day. Since we count the first square as being on the initial day ($t = 0$), the second day corresponds to $t = 1$, and so on; thus the 64th day corresponds to $t = 63$. Hence the king will have to put $2^{63} = 9.22 \times 10^{18}$ grains of wheat on the 64th square. If we divide 9.22×10^{18} by the number of grains in a bushel, namely 1,100,000, we get 8.38×10^{12} bushels. If each bushel is worth \$4.25, then the value of the wheat on the 64th square was

$$4.25 \times 8.38 \times 10^{12} = 3.56 \times 10^{13} = 35,600,000,000,000 \text{ dollars.}$$

This is over 35.6 trillion dollars, a sum that may trouble even a king.

19. **The photic zone**: Near Cape Cod, Massachusetts, the depth of the photic zone is about 16 meters. We know that the photic zone extends to a depth where the light intensity is about 1% of surface light, that is, $I = 0.01 I_0$. Since the Beer-Lambert-Bouguer law implies that I is an exponential function of depth, $I = I_0 \times a^d$. Since $I(16) = 0.01 I_0$, we have $I_0 \times a^{16} = 0.01 I_0$, so $a^{16} = 0.01$. Thus the decay factor is $a = 0.01^{1/16} = 0.750$, representing a 25.0% decrease in light intensity for each additional meter of depth.

21. **Headway on four-lane highways**:

   (a) A flow of 500 vehicles per hour is $\frac{500}{3600} = 0.14$ vehicle per second, so $q = 0.14$. Thus the probability that the headway exceeds 15 seconds is given by $P = e^{-0.14 \times 15} = 0.122$, or about 12%.

   (b) On a four-lane highway carrying an average of 500 vehicles per hour, $q = 0.14$, and so $P = e^{-0.14t}$. Using the law of exponents, this can be written as $P = e^{-0.14t} = (e^{-0.14})^t$. In this form it is clear that the decay factor is $a = e^{-0.14} = 0.87$.

23. **Continuous compounding**:

   (a) The limiting value of the function given by the formula

$$\left(1 + \frac{0.1}{n}\right)^n$$

   can be found using a table or a graph. Scanning down a table, for example, shows that the limiting value is 1.1052 to four decimal places.

   (b) Since the APR is 0.1, $e^{APR} = e^{0.1} = 1.1052$ to four decimal places..

   (c) Since the answers to Parts (a) and (b) are the same, this shows that e^{APR} is the yearly growth factor for continuous compounding in the case when the APR is 10%.

4.2 MODELING EXPONENTIAL DATA

E-1. **Testing for exponential data**: To test unevenly spaced data, we calculate the new y/old y ratio to see if it equals a raised to the difference in the x values. For those that are exponential, this allows us to find a, then we use a value from the data table to calculate the initial value P.

(a) Calculating the ratios of the y values with the appropriate powers of a, we get

$$a^3 = a^{5-2} = \frac{3072}{48} = 64$$

$$a^2 = a^{7-5} = \frac{49{,}152}{3072} = 16$$

$$a^4 = a^{11-7} = \frac{12{,}582{,}912}{49{,}152} = 256.$$

The three equations each give values for a, namely $64^{1/3} = 4$, $16^{1/2} = 4$, and $256^{1/4} = 4$. Since all three give the same value, the data are exponential with growth factor $a = 4$. The first data point allows us to find the initial value:

$$P \times a^2 = 48$$

$$P \times 4^2 = 48$$

$$P = 48/(4^2) = 3.$$

Thus the exponential model for the data is $y = 3 \times 4^x$.

(b) Calculating the ratios of the y values with the appropriate powers of a, we get

$$a^2 = a^{4-2} = \frac{1}{4} = 0.25$$

$$a^4 = a^{8-4} = \frac{0.0625}{1} = 0.0625$$

$$a^2 = a^{10-8} = \frac{0.015625}{0.0625} = 0.25.$$

The three equations each give values for a, namely $0.25^{1/2} = 0.5$, $0.0625^{1/4} = 0.5$, and $0.25^{1/2} = 0.5$. Since all three give the same value, the data are exponential with growth factor $a = 0.5$. The first data point allows us to find the initial value:

$$P \times a^2 = 4$$

$$P \times 0.5^2 = 4$$

$$P = 4/(0.5^2) = 16.$$

Thus the exponential model for the data is $y = 16 \times 0.5^x$.

(c) Calculating the ratios of the y values with the appropriate powers of a, we get

$$a^3 = a^{5-2} = \frac{972}{36} = 27$$

$$a^2 = a^{7-5} = \frac{8748}{972} = 9$$

$$a^2 = a^{9-7} = \frac{88{,}740}{8748} = 10.14.$$

The three equations each give values for a, namely $27^{1/3} = 3$, $9^{1/2} = 3$, and $10.14^{1/2} = 3.18$. Since the three do not give the same value, the data are not exponential.

E-3. **Testing exponential data**: Calculating the ratios of the y values with the appropriate powers of the growth factor (which we did not denote by a since that letter is already used in the data), we get

$$\text{Growth factor}^3 = \text{Growth factor}^{(a+3)-a} = \frac{b^3}{1} = b^3$$

$$\text{Growth factor}^2 = \text{Growth factor}^{(a+5)-(a+3)} = \frac{b^5}{b^3} = b^2$$

$$\text{Growth factor}^1 = \text{Growth factor}^{(a+6)-(a+5)} = \frac{b^6}{b^5} = b.$$

Since the three equations each give the same value for the growth factor, the data are exponential with growth factor b, and thus the exponential model is $y = P \times b^x$. The first data point allows us to find the initial value P:

$$P \times b^a = 1$$

$$P = 1/(b^a) = b^{-a}.$$

Thus the exponential model for the data is $y = b^{-a} \times b^x$, which can also be written as $y = b^{x-a}$.

S-1. **Finding the growth factor**: Since N is multiplied by 8 if t is increased by 1, the growth factor a is 8. Since the initial value P is 7, a formula for N is $N = P \times a^t = 7 \times 8^t$.

S-3. **Finding the growth factor**: Since N is multiplied by 62 if t is increased by 7, the growth factor a satisfies the equation $a^7 = 62$. Thus $a = 62^{1/7} = 1.803$. Since the initial value P is 12, a formula for N is $N = P \times a^t = 12 \times 1.803^t$.

S-5. **Testing exponential data**: Calculating the ratios of successive terms of y, we get $\frac{10}{5} = 2$, $\frac{20}{10} = 2$, and $\frac{40}{20} = 2$. Since the x values are evenly spaced and these ratios show a constant value of 2, the table does show exponential data.

S-7. **Modeling exponential data**: The data from Exercise S-4 show a constant ratio of 3 for x values increasing by 1's and so the growth factor is $a = 3$. The initial value of $P = 2.6$ comes from the first entry in the table. Thus an exponential model for the data is $y = 2.6 \times 3^x$.

S-9. **Testing exponential data**: Calculating the ratios of successive terms of y, we get $\dfrac{18}{6} = 3$, $\dfrac{54}{18} = 3$, and $\dfrac{162}{54} = 3$. Since the x values are evenly spaced and these ratios show a constant value of 3, the table does show exponential data.

S-11. **Testing exponential data**: Calculating the ratios of successive terms of y, we get $\dfrac{240}{1000} = 0.24$, $\dfrac{57.6}{240} = 0.24$, and $\dfrac{13.8}{57.6} = 0.24$. Since the x values are evenly spaced and these ratios show a constant value of 0.24, the table does show exponential data.

1. **Making an exponential model**: To show that this is an exponential function we must show that the successive ratios are the same.

t increment	0 to 1	1 to 2	2 to 3	3 to 4	4 to 5
$\dfrac{\text{new}}{\text{old}}$ ratio of $f(t)$	$\dfrac{3.95}{3.80} = 1.04$	$\dfrac{4.11}{3.95} = 1.04$	$\dfrac{4.27}{4.11} = 1.04$	$\dfrac{4.45}{4.27} = 1.04$	$\dfrac{4.62}{4.45} = 1.04$

Note that the ratios are not precisely equal, but they are equal when rounded to 2 decimal places. The table in the exercise tells us that the initial value is $P = 3.80$ (this corresponds to $t = 0$). Since t increases in units of 1, we don't need to adjust the units, and so the growth factor is $a = 1.04$. The formula for the function is

$$f(t) = 3.80 \times 1.04^t.$$

3. **Data that are not exponential**: To show that these data are not exponential, we show that some of the successive ratios are not equal. It suffices to compute the first two ratios:

$$\frac{26.6}{4.9} = 5.43 \text{ and } \frac{91.7}{26.6} = 3.45.$$

Since these ratios are not equal and yet the changes in t are the same, h is not an exponential function.

5. **Magazine sales**: The calculations of differences and ratios for the years in the data set are in the table below.

Interval	Differences	Ratios
1998–1999	$8.82 - 7.76 = 1.06$	$8.82/7.76 = 1.137$
1999–2000	$9.88 - 8.82 = 1.06$	$9.88/8.82 = 1.120$
2000–2001	$10.94 - 9.88 = 1.06$	$10.94/9.88 = 1.107$
2001–2002	$12.00 - 10.94 = 1.06$	$12.00/10.94 = 1.097$
2002–2003	$13.08 - 12.00 = 1.08$	$13.08/12.00 = 1.090$
2003–2004	$14.26 - 13.08 = 1.18$	$14.26/13.08 = 1.090$
2004–2005	$15.54 - 14.26 = 1.28$	$15.54/14.26 = 1.090$

From 1998 to 2002, the magazine sales exhibit a constant growth rate of 1.06 thousand dollars per year. From 2002 to 2005, sales grew at a constant proportional rate, that is, each year's sales are 1.09 times those of the previous year—this is a growth rate of 9% per year.

7. **An investment**:

   (a) The amount of money originally invested is the balance when $t = 0$. Hence the original investment was $1750.00.

   (b) Each successive ratio of new/old is 1.012, which shows the data is exponential. Furthermore, since t increases by single units, the common ratio 1.012 is the monthly growth factor. The formula for an exponential model is

   $$B = 1750.00 \times 1.012^t.$$

   Here t is time in months, and B is the savings balance in dollars.

   (c) The growth factor is $1.012 = 1 + 0.012$, so $r = 0.012$. The monthly interest rate is 1.2%.

   (d) To find the yearly interest rate we first have to find the yearly growth factor. There are 12 months in a year, so the yearly growth factor is $1.012^{12} = 1.154$. Now $1.154 = 1 + 0.154$, so $r = 0.154$, and thus the yearly interest rate is 15.4%.

   (e) The account will grow for 18 years or 216 months, so when she is 18 years old, her college fund account balance will be

   $$B(216) = 1750 \times 1.012^{216} = \$23,015.94.$$

   (f) The account will double the first time when it reaches $3500 and double the second time when it reaches $7000. That is, we need to solve the equations

   $$1750 \times 1.012^t = 3500 \text{ and } 1750 \times 1.012^t = 7000.$$

We do this using the crossing graphs method. The vertical span we want is 0 to 8000, and the table of values below for B leads us to choose 0 to 120 months for a horizontal span. In the right-hand figure below we have graphed B along with 3500 and 7000 (thick lines). The horizontal axis is months since the initial investment, and the vertical axis is account balance.

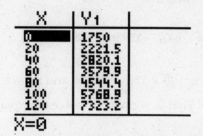
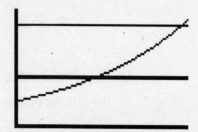

In the left-hand figure below, we have calculated the intersection of B with 3500, and we find that the balance doubles in 58.11, or about 58.1, months. In the right-hand figure below, we have calculated the intersection with 7000, and we see the account reaches \$7000 after 116.22 months. Thus the second doubling occurs 58.11, or about 58.1, months after the first one. (In fact, the account doubles every 58.11 months.)

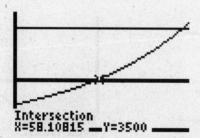
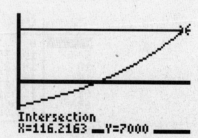

9. **A skydiver**:

(a) The successive ratios in D values are always 0.42, so the data can be modeled with an exponential function. Since the t values are measured every 5 seconds, 0.42 is the decay factor every 5 seconds. To find the decay factor every second we compute

$$0.42^{1/5} = 0.84.$$

The initial value is 176.00, and so the exponential model is

$$D = 176.00 \times 0.84^t.$$

(b) The decay factor per second is $0.84 = 1 - 0.16$, so $r = 0.16$, and the percentage decay rate is 16% per second. This means that the difference between the terminal velocity and the skydiver's velocity decreased by 16% each second.

(c) Now D represents the difference between terminal velocity of 176 feet per second and the skydiver's velocity, V. Thus $D = 176 - V$, and so $V = 176 - D$. Using the formula from Part (a) for D, we find that

$$V = 176.00 - 176.00 \times 0.84^t.$$

(d) Now 99% of terminal velocity is $0.99 \times 176 = 174.24$, and we want to know when the velocity reaches this value. That is, we want to solve the equation

$$176 - 176 \times 0.84^t = 174.24.$$

We do this using the crossing graphs method. We know that the velocity starts at 0 and increases toward 176, and so we use a vertical span of 160 to 190. The table below leads us to choose a horizontal span of 0 to 35. In the right-hand figure below, we have graphed velocity and the target velocity of 174.24 (thick line). We see that the intersection occurs at $t = 26.41$. The skydiver reaches 99% of terminal velocity after 26.41 seconds.

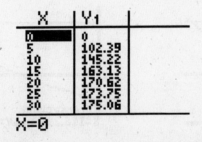

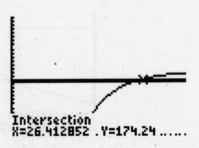

11. **An inappropriate linear model for radioactive decay:**

(a) As seen in the figure below, the data points do seem to fall on a straight line.

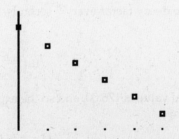

(b) In the left-hand figure below, we have calculated the regression line parameters, and the equation of the regression line is $U = -0.027t + 0.999$. In the right-hand figure, we have added the graph of the regression line to the data plot.

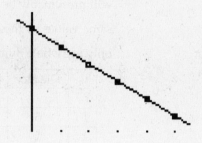

(c) Using the regression line equation for the decay, we see that the time required for decay to 0.5 gram is given by solving $-0.027t + 0.999 = 0.5$. This shows that the time required would be 18.48, or about 18.5, minutes.

(d) The regression line predicts that after 60 minutes there will be -0.621 gram of the uranium 239 left, which is impossible.

Two items are worth noting here. The first is that the data has a linear appearance because we have sampled over too short a time period. Sampling over a longer time period would have clearly shown that the data is not linear. It is crucial in any experiment to gather enough data to allow for sound conclusions. Second, it is always dangerous to propose mathematical models based solely on the appearance of data. In the case of radioactive decay, physicists have good scientific reasons for believing that radioactive decay is an exponential phenomenon. Collected data serves to verify this, and to allow for the calculation of a constant such as the half-life for specific substances.

13. **Rates vary**: To determine the interest rates, we calculate the ratios of new balances to old balances, as shown in the table below.

Time interval	Ratios of New/Old Balances
$t = 0$ to $t = 1$	$262.50/250.00 = 1.050$
$t = 1$ to $t = 2$	$275.63/262.50 = 1.050$
$t = 2$ to $t = 3$	$289.41/275.63 = 1.050$
$t = 3$ to $t = 4$	$302.43/289.41 = 1.045$
$t = 4$ to $t = 5$	$316.04/302.43 = 1.045$
$t = 5$ to $t = 6$	$330.26/316.04 = 1.045$

Thus the yearly interest rate was 5.0% for the first 3 years, then 4.5% after that.

15. **Stochastic population growth**:

(a) The table below shows the Monte Carlo method applied by rolling a die ten times. The faces showing for these ten rolls are: 5, 2, 6, 3, 6, 1, 2, 1, 5, and 4. Your answer will presumably be different since it is highly unlikely that your die will roll the same values as these(!). Since the numbers are the size of a population, we will only use whole numbers.

t	Die face	Population % Change	Population change	Population
0				500
1	5	Up 4%	+20	520
2	2	Down 1%	−5	515
3	6	Up 9%	+46	561
4	3	No change	0	561
5	6	Up 9%	+50	611
6	1	Down 2%	−12	599
7	2	Down 1%	−6	593
8	1	Down 2%	−12	581
9	5	Up 4%	+23	604
10	4	Up 2%	+12	616

(b) The exponential model is 500×1.02^t since the initial value is 500 and the growth rate is 2%. A graph of the data points from Part (a) together with this exponential model is shown below. Although the points are scattered about and do not lie on the exponential curve, they do follow, very roughly, the general upward trend of the curve.

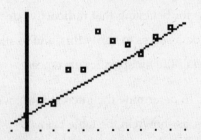

4.3 MODELING NEARLY EXPONENTIAL DATA

E-1. **Population growth**: Because the population has abundant resources and there are few predators, an exponential model is appropriate.

E-3. **Constant speed**: In this context the rate of change of the distance traveled is the speed, which is constant. Since the rate of change for the distance traveled is constant, a linear model is appropriate.

S-1. **Population growth**: Because the population has abundant resources, an exponential model should be appropriate.

S-3. **Exponential regression**: The exponential model is $y = 51.01 \times 1.04^x$.

S-5. **Exponential regression**: The exponential model is $y = 2.22 \times 1.96^x$. A plot of the exponential model, together with the data, is shown below.

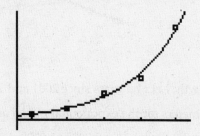

S-7. **Exponential regression**: The exponential model is $y = 6.35 \times 1.03^x$. A plot of the exponential model, together with the data, is shown below.

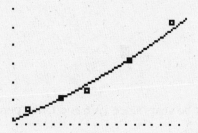

S-9. **Exponential regression**: The exponential model is $y = 2.39 \times 1.40^x$. A plot of the exponential model, together with the data, is shown below.

S-11. **Exponential regression**: The exponential model is $y = 3.17 \times 1.15^x$. A plot of the exponential model, together with the data, is shown below.

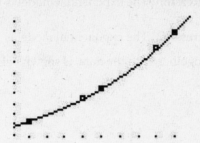

1. **Population growth**: Let t be years since 2001 and N the population, in thousands. The regression parameters for the exponential model are calculated in the figure below. We find the exponential model $N = 2.300 \times 1.090^t$.

```
ExpReg
 y=a*b^x
 a=2.300181189
 b=1.090147741
```

3. **Cable TV**: Let t be years since 1976 and C the percent with cable.

    (a) The plot of the data points is in the figure below. The horizontal axis is years since 1976, and the vertical axis is percent of homes with cable TV.

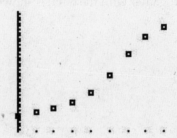

The plot suggests the shape of exponential growth, so it is reasonable to approximate the data with an exponential function.

(b) The regression parameters for the exponential model are calculated in the left-hand figure below. The exponential model is given by

$$C = 13.892 \times 1.155^t.$$

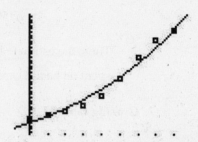

(c) We have added the graph of the exponential model in the right-hand figure above.

(d) Since the yearly growth factor is $1.155 = 1 + 0.155$, we have that $r = 0.155$. Thus the yearly percentage growth rate is 15.5%.

(e) Since 1987 is 11 years after 1976, we would predict that $13.892 \times 1.155^{11} = 67.79$ percent of the American households would have cable TV in 1987. So it appears that the executive's plan is reasonable.

5. **National health care spending**: Let t be years since 1950 and H the costs in billions of dollars.

(a) The plot of the data points is in the figure on the left below. The horizontal axis is years since 1950, and the vertical axis is cost. The plot suggests the shape of exponential growth, so it is reasonable to approximate the data with an exponential function.

(b) The regression parameters for the exponential model are calculated in the right-hand figure below. The exponential model is given by

$$H = 11.313 \times 1.104^t.$$

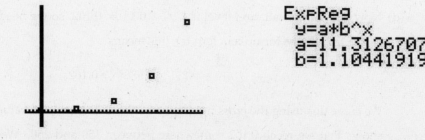

(c) The yearly growth factor is $1.104 = 1 + 0.104$, so $r = 0.104$. Thus the yearly percentage increase is 10.4%.

(d) The year 2000 is 50 years after 1950, so $t = 50$. Thus the amount spent in 2000 on health care is expressed in functional notation as $H(50)$. We would estimate that

$$H(50) = 11.313 \times 1.104^{50} = 1592.32 \text{ billion dollars.}$$

Thus, the estimate is that about 1592 billion dollars, or 1.592 trillion dollars, was spent on health care in the year 2000.

7. **Grazing rabbits:**

(a) The plot of the data points is shown below. The horizontal axis is the vegetation level V, and the vertical axis is the difference D. The plot suggests the shape of exponential decay, so it appears that D is approximately an exponential function of V.

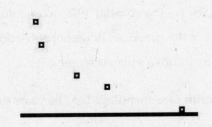

(b) We use regression to calculate the exponential model and find

$$D = 0.211 \times 0.988^{V}.$$

(c) The satiation level for the rabbit is 0.18 pound per day, and thus we have

$$D = \text{satiation level} - A = 0.18 - A.$$

Solving for A yields $A = 0.18 - D$, and, using the formula from Part (b), we find that

$$A = 0.18 - 0.211 \times 0.988^{V}.$$

(d) Now 90% of the satiation level is $0.90 \times 0.18 = 0.162$, so we need to find when $A = 0.162$. By the formula in Part (c), this means

$$0.18 - 0.211 \times 0.988^{V} = 0.162.$$

We solve this using the crossing graphs method. The table of values for A below shows that we reach 0.162 somewhere between 150 and 250. We use this for a horizontal span, and we use 0.1 to 0.2 for a vertical span. In the graph below, the horizontal axis is the vegetation level, and the vertical axis is the amount eaten. We have added the graph of 0.162 (thick line) and calculated the intersection point. Thus, the amount of food eaten by the rabbit will be 90% of satiation level when the vegetation level is $V = 203.89$ pounds per acre.

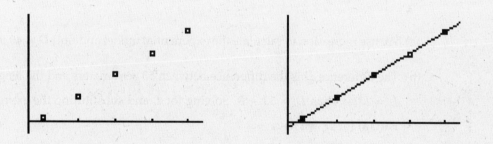

9. **Nearly linear or exponential data**: The left-hand figure below is the plot of the data from Table A. This appears to be linear. Computing the regression line gives $f = 19.842t - 16.264$, and adding it to the plot (right-hand figure) shows how very close to linear f is. We conclude that Table A is approximately linear, with a linear model of $f = 19.842t - 16.264$.

The plot of data from Table B in the figure below looks more like exponential data than linear data. We use regression to calculate the exponential model and find

$$g = 2.330 \times 1.556^t.$$

11. **Sound pressure:**

   (a) A plot of the data is shown below. The plot suggests the shape of exponential growth, so it appears that pressure is approximately an exponential function of loudness.

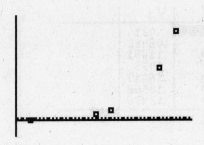

   (b) We use regression to calculate the exponential model and find $P = 0.0003 \times 1.116^D$.

   (c) When loudness D is increased by one decibel, the pressure P is multiplied by a factor of 1.116, that is, it increases by 11.6%.

13. **Growth in length of haddock:**

   (a) We use regression to calculate the exponential model and find $D = 40.909 \times 0.822^t$.

   (b) The difference D is the difference between 53 centimeters and the length at age t, $L = L(t)$. Thus $D = 53 - L$. Solving for L and substituting the expression from Part (a) for D, we have

$$
\begin{aligned}
D &= 53 - L \\
D + L &= 53 \\
L &= 53 - D \\
L &= 53 - 40.909 \times 0.822^t.
\end{aligned}
$$

   (c) The figure below shows a plot of the experimentally gathered data for the length L at ages 2, 5, 7, 13, and 19 years along with the graph of the model for L from Part (b). This graph shows that the 5-year-old haddock is a bit shorter than would be expected from the model for L.

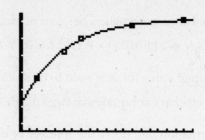

(d) If a fisherman has caught a haddock which measures 41 centimeters, then $L = 41$ and so its age t satisfies the equation $41 = 53 - 40.909 \times 0.822^t$. Solving for t, we find that $t = 6.26$. Thus the haddock is about 6.3 years old.

15. **Injury versus speed**:

(a) We use regression to calculate the exponential model and find $N = 16.726 \times 1.021^s$.

(b) Using the formula from Part (a), we find that $N(70) = 16.726 \times 1.021^{70} = 71.65$. In practical terms this means that if vehicles are traveling at 70 miles per hour, then we expect 71.65 persons injured (on average) per 100 vehicles involved in accidents.

(c) An increase in 1 mile per hour of speed s causes the number of people injured per 100 accident-involved vehicles N to be multiplied by 1.021, and so to increase by 2.1%.

17. **Walking in Seattle**:

(a) We use regression to calculate the exponential model and find $P = 84.726 \times 0.999^D$.

(b) The percentage of pedestrians who walk at least $D = 200$ feet from parking facilities is $P = 84.726 \times 0.999^{200} = 69.36\%$.

(c) For $D = 0$, the exponential model indicates that $P = 84.726$ and so only 84.726% of pedestrians walk at least 0 feet. The correct percentage is 100%. Thus this model is not appropriate to use for very small distances D.

19. **Frequency of earthquakes**:

(a) We use regression to calculate the exponential model and find $N = 24,670,000 \times 0.125^M$.

(b) To find the number of earthquakes per year of magnitude at least 5.5, we put $M = 5.5$ in the formula from Part (a) and find $N = 24,670,000 \times 0.125^{5.5} = 266.18$, or about 266 earthquakes per year.

(c) The number of earthquakes per year of magnitude 8.5 or greater predicted by the model is $N = 24{,}670{,}000 \times 0.125^{8.5} = 0.52$, or about 0.5 earthquake per year.

(d) The limiting value for N is seen to be 0 using a table or graph. In practical terms this means that earthquakes of large magnitude are very rare.

(e) In the situation described, the number of earthquakes predicted is multiplied by 0.125, which is a reduction of 87.5%. Thus there are 87.5% fewer earthquakes of magnitude $M + 1$ or greater than of magnitude M or greater.

21. **Laboratory experiment**: Answers will vary. Please go to the website for instructions and details.

23. **Research project**: Answers will vary. Please go to the website for instructions and details.

4.4 LOGARITHMIC FUNCTIONS

E-1. **Solving equations with logarithms**: In each case we apply the common logarithm to each side of the equation and use the properties of logarithms to solve for t.

(a)

$$
\begin{aligned}
5 \times 4^t &= 7 \\
\log(5 \times 4^t) &= \log 7 \\
\log 5 + \log(4^t) &= \log 7 \qquad \textbf{Product law} \\
\log 5 + t \log 4 &= \log 7 \qquad \textbf{Power law} \\
t &= \frac{\log 7 - \log 5}{\log 4} = 0.24.
\end{aligned}
$$

(b)

$$
\begin{aligned}
2^{t+3} &= 7^{t-1} \\
\log(2^{t+3}) &= \log(7^{t-1}) \\
(t+3)\log 2 &= (t-1)\log 7 \qquad \textbf{Power law} \\
t \log 2 + 3 \log 2 &= t \log 7 - \log 7 \\
t(\log 2 - \log 7) &= -\log 7 - 3 \log 2 \\
t &= \frac{-\log 7 - 3 \log 2}{\log 2 - \log 7} = 3.21.
\end{aligned}
$$

(c)

$$\frac{12^t}{9^t} = 7 \times 6^t$$

$$\log\left(\frac{12^t}{9^t}\right) = \log(7 \times 6^t)$$

$$\log(12^t) - \log(9^t) = \log 7 + \log(6^t) \quad \textbf{Quotient and Product laws}$$

$$t \log 12 - t \log 9 = \log 7 + t \log 6 \quad \textbf{Power law}$$

$$t(\log 12 - \log 9 - \log 6) = \log 7$$

$$t = \frac{\log 7}{\log 12 - \log 9 - \log 6} = -1.29.$$

(d)

$$a^{2t} = 3a^{4t}$$

$$\log(a^{2t}) = \log(3a^{4t})$$

$$2t \log a = \log 3 + 4t \log a \quad \textbf{Power and Product laws}$$

$$t(2 \log a - 4 \log a) = \log 3$$

$$t = \frac{\log 3}{2 \log a - 4 \log a} = -\frac{\log 3}{2 \log a}.$$

This assumes that $a \neq 1$, so that $\log a \neq 0$.

E-3. **Spectroscopic parallax again**: Exercise 13 from Exercise Set 4.4 below relates the spectroscopic parallax S to the distance D from Earth by the equation

$$S = 5 \log D - 5.$$

(a) If $D = 38.04$, then $S = 5 \log 38.04 - 5 = 2.90$. (This part does not use the laws of logarithms.)

(b) If $S = 1.27$, then $1.27 = 5 \log D - 5$. Now we solve for D:

$$5 \log D = 6.27 \quad \textbf{Adding 5}$$

$$\log D = \frac{6.27}{5} \quad \textbf{Dividing by 5}$$

$$D = 10^{6.27/5} = 17.95. \quad \textbf{By definition of log}$$

Thus the distance from Earth is 17.95 parsecs.

(c) If D is multiplied by 10, then $\log(10D) = \log 10 + \log D = 1 + \log D$ and so the spectroscopic parallax S is increased by $5 \times 1 = 5$ units.

(d) If D is multiplied by 3.78, then $\log(3.78D) = \log 3.78 + \log D = 0.577 + \log D$ and so the spectroscopic parallax S is increased by $5 \times 0.577 = 2.89$ units.

E-5. Substitutions:

(a) Substituting $y = 2^x$ into $4^x - 5 \times 2^x + 6 = 0$ yields $y^2 - 5y + 6 = 0$ since $(2^x)^2 = 2^{2x} = (2^2)^x = 4^x$. Now $y^2 - 5y + 6$ factors as $(y-2)(y-3)$ and so $y = 2$ or $y = 3$. Substituting for y, we see that $2^x = 2$ or $2^x = 3$. The first gives $x = 1$, and the second, using logarithms, gives $x = \dfrac{\log 3}{\log 2} = 1.58$.

(b) Substituting $y = 5^x$ into $25^x - 7 \times 5^x + 12 = 0$ yields $y^2 - 7y + 12 = 0$ since $(5^x)^2 = 5^{2x} = (5^2)^x = 25^x$. Now $y^2 - 7y + 12$ factors as $(y-3)(y-4)$ and so $y = 3$ or $y = 4$. Substituting for y, we see that $5^x = 3$ or $5^x = 4$. Using logarithms, the first gives $x = \dfrac{\log 3}{\log 5} = 0.68$ and the second gives $x = \dfrac{\log 4}{\log 5} = 0.86$.

(c) Substituting $y = 3^x$ into $9^x - 3^{x+1} + 2 = 0$ yields $y^2 - 3y + 2 = 0$ since $(3^x)^2 = 3^{2x} = (3^2)^x = 9^x$ and $3^{x+1} = 3^x 3^1 = 3 \times 3^x$. Now $y^2 - 3y + 2$ factors as $(y-1)(y-2)$ and so $y = 1$ or $y = 2$. Substituting for y, we see that $3^x = 1$ or $3^x = 2$. Using logarithms, the first gives $x = \dfrac{\log 1}{\log 3} = 0$ and the second gives $x = \dfrac{\log 2}{\log 3} = 0.63$.

E-7. Solving logarithmic equations: To solve these logarithmic equations, we exponentiate each of the equation and utilize the identity $e^{\ln x} = x$.

(a)

$$
\begin{aligned}
\ln(x+2) &= 5 \\
e^{\ln(x+2)} &= e^5 \\
x+2 &= e^5 \\
x &= e^5 - 2 = 146.41
\end{aligned}
$$

(b)

$$
\begin{aligned}
\ln(x+5) &= 2\ln(x-1) = \ln((x-1)^2) \\
e^{\ln(x+5)} &= e^{\ln(x-1)^2} \\
x+5 &= (x-1)^2 = x^2 - 2x + 1 \\
0 &= x^2 - 3x - 4 = (x-4)(x+1)
\end{aligned}
$$

Thus $x = 4$ or $x = -1$. Unfortunately $x = -1$ is not actually a solution since $2\ln(x-1)$ has no value when $x = -1$. The only solution is $x = 4$.

(c)

$$\ln(3x - 10) = \ln(x + 2) + \ln(x - 5)$$

$$e^{\ln(3x-10)} = e^{\ln(x+2)+\ln(x-5)}$$

$$3x - 10 = e^{\ln(x+2)}e^{\ln(x-5)}$$

$$3x - 10 = (x + 2)(x - 5) = x^2 - 3x - 10$$

$$0 = x^2 - 6x = x(x - 6)$$

Thus $x = 0$ or $x = 6$. Unfortunately $x = 0$ is not actually a solution since $\ln(3x - 10)$ has no value when $x = 0$. The only solution is $x = 6$.

S-1. The Richter scale: If one earthquake reads 4.2 on the Richter scale and another reads 7.2, then the 7.2 earthquake is $10^{7.2-4.2} = 10^3 = 1000$ times as powerful as the 4.2 earthquake.

S-3. The decibel scale: If one sound has a relative intensity of 1000 times that of another, then the common logarithm of the more intense sound is 3 more than that of the other, since $1000 = 10^3$. Since the decibel level is 10 times the logarithm, the decibel level is increased by $10 \times 3 = 30$ decibels. Thus the more intense sound has a decibel level 30 decibels more than the other.

S-5. Calculating logarithms:

(a) Since $1000 = 10^3$, $\log 1000 = 3$.

(b) Since $1 = 10^0$, $\log 1 = 0$.

(c) Since $\dfrac{1}{10} = 10^{-1}$, $\log \dfrac{1}{10} = -1$.

S-7. How the logarithm increases:

(a) Now $\log(10x) = \log 10 + \log x = 1 + 6.6 = 7.6$.

(b) Now $\log(1000x) = \log 1000 + \log x = 3 + 6.6 = 9.6$.

(c) Now $\log \dfrac{x}{10} = \log x - \log 10 = 6.6 - 1 = 5.6$.

S-9. Solving exponential equations:

(a) If $10^t = 5$, then $\log 10^t = \log 5$. Now $\log 10^t = t$, so $t = \log 5$, and so $t = 0.70$.

(b) If $10^t = a$, then $\log 10^t = \log a$. Now $\log 10^t = t$, so $t = \log a$.

(c) If $a^t = b$, then $\log a^t = \log b$. Now $\log a^t = t \log a$. Thus $t = \dfrac{\log b}{\log a}$.

S-11. **Solving logarithmic equations**: If $t = \log(2x)$, then $10^t = 2x$ and so $x = 0.5 \times 10^t$.

1. **Earthquakes in Alaska and Chile**:

   (a) The New Madrid earthquake of Example 4.8 had a Richter magnitude of 8.8, whereas the 1964 Alaska earthquake had a Richter magnitude of 8.4. The New Madrid earthquake is therefore $10^{8.8-8.4} = 10^{0.4} = 2.51$ times as strong as the Alaska earthquake.

   (b) The Chilean earthquake was 1.4 times as powerful as the Alaska quake. We want to write 1.4 as a power of 10. Solving the equation $1.4 = 10^t$, we find that $t = \log 1.4 = 0.15$, and so the Chilean earthquake was $10^{0.15}$ times as powerful as the Alaska quake. Thus the Richter scale reading for the Chilean earthquake was 0.15 more than that of the Alaska earthquake, so it was $8.4 + 0.15 = 8.55$.

3. **Moore's law**:

   (a) If a chip were introduced in the year 2008, then that chip would be $10^{0.2 \times 8} = 40$ times as fast as a chip introduced in 2000, such as the Pentium 4.

   (b) The speed of a chip will be 10,000 times the speed of the Pentium 4 when $10,000 = 10^{0.2t}$, where t is years since 2000, according to Moore's law. Now $10,000 = 10^4$, so $4 = 0.2t$, and so $t = \dfrac{4}{0.2} = 20$. Thus that limit will be reached in the year 2020.

   (c) If the fastest speed possible is about 10^{40} times that of the Pentium 4, then, according to Moore's law, this limit will be reached when $10^{40} = 10^{0.2t}$, where t is years since 2000. Thus $40 = 0.2t$ and so $t = \dfrac{40}{0.2} = 200$, and so that limit will be reached in the year 2200.

5. **The pH scale**:

   (a) Rain in the eastern United States has a pH level of 3.8, which is 1.8 less than the 5.6 pH level of normal rain. Thus the eastern United States rain is $10^{1.8} = 63.10$ times as acidic as normal rain.

   (b) A pH of 5 is 0.6 less than a pH of 5.6, so such water is $10^{0.6} = 3.98$ times as acidic as normal water.

7. **Weight gain**:

   (a) The figure below shows a graph of G against M using a horizontal span of 0 to 0.4 and a vertical span of 0 to 0.05.

   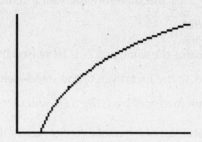

   (b) If $M = 0.3$, then $G = 0.067 + 0.052 \log 0.3 = 0.04$ unit. This could also have been solved using a table or graph.

   (c) A zookeeper wants $G = 0.03$, so $0.03 = 0.067 + 0.052 \log M$. This can be solved directly: $0.052 \log M = 0.03 - 0.067 = -0.037$, so $\log M = \dfrac{-0.037}{0.052}$ and therefore $M = 10^{-0.037/0.052} = 0.19$ unit should be the daily milk-energy intake.

   (d) The graph is increasing and concave down; thus higher values of M produce smaller and smaller effects on the value of G, supporting the quotation from the study.

9. **Age of haddock**:

   (a) The figure below shows a graph of age T versus length L using a horizontal span of 25 to 50 and a vertical span of 0 to 15.

   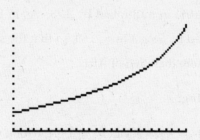

   (b) In functional notation the age of a haddock that is 35 centimeters long is expressed by $T(35)$. The value of $T(35)$ is given by the formula, for example, as $T(35) = 19 - 5 \ln(53 - 35) = 4.55$ years old.

   (c) If a haddock is 10 years old, then $T = 10$ and so $10 = 19 - 5 \ln(53 - L)$. Solving for L directly, we see that $-5 \ln(53 - L) = 10 - 19 = -9$, so $\ln(53 - L) = \dfrac{-9}{-5} = 1.8$. Thus $53 - L = e^{1.8}$, so $53 = e^{1.8} + L$ and finally $L = 53 - e^{1.8} = 46.95$ centimeters.

11. **Stand density:**

   (a) If the stand has $N = 500$ trees per acre, and the diameter of a tree of average size is $D = 7$ inches, then, according to the formula, $\log SDI = \log 500 + 1.605 \log 7 - 1.605 = 2.45$ and therefore the stand-density index is $SDI = 10^{2.45} = 281.84$, or about 282.

   (b) If D remains the same and N is increased by a factor of 10, then $\log N$ will increase by 1 and so the logarithm of the stand-density index will increase by 1. Since the logarithm increases by 1, the SDI will increase by a factor of $10^1 = 10$.

   (c) If the diameter of a tree of average size is 10 inches, then

   $$\log SDI = \log N + 1.605 \log 10 - 1.605 = \log N + 1.605 \times 1 - 1.605 = \log N,$$

   and therefore $SDI = N$.

13. **Spectroscopic parallax:**

   (a) If the distance to the star Kaus Astralis is $D = 38.04$ parsecs, then its spectroscopic parallax is $S = 5 \log 38.04 - 5 = 2.90$.

   (b) If the spectroscopic parallax for the star Rasalhague is $S = 1.27$, then $1.27 = 5 \log D - 5$, so $5 \log D = 1.27 + 5 = 6.27$, and therefore $\log D = \dfrac{6.27}{5}$. Thus the distance to Rasalhague is $D = 10^{6.27/5} = 17.95$ parsecs.

   (c) If distance D is multiplied by 10, then $\log D$ is increased by 1, and so the spectroscopic parallax S is increased by $5 \times 1 = 5$ units.

   (d) Because the star Shaula is 3.78 times as far away as the star Atria, the distance D for Atria in multiplied by 3.78. As in Part (c), the spectroscopic parallax S is increased by $5 \log 3.78 = 2.89$, so that the spectroscopic parallax for Shaula is 2.89 units more than that of Atria.

15. **Rocket staging:**

   (a) Since $R_1 = R_2 = 3.4$ and $c = 3.7$, the total velocity attained by this two-stage craft is $v = 3.7 \times \ln 3.4 + 3.7 \times \ln 3.4 = 9.06$ kilometers per second.

   (b) Since the velocity v from Part (a) is greater than 7.8 kilometers per second, this craft can achieve a stable orbit.

17. **Relative abundance of species:**

(a) Since $S = 197$ and $N = 6814$, $\dfrac{S}{N} = \dfrac{197}{6814} = 0.0289$ for this collection.

(b) The figure below shows a graph of the function $\dfrac{x-1}{x}\ln(1-x)$ using a horizontal span of 0 to 1 and a vertical span of 0 to 1.

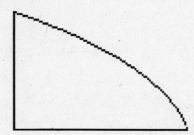

(c) Using crossing graphs, we can determine the value of x for which $\dfrac{x-1}{x}\ln(1-x) = 0.0289$. We find that $x = 0.9945$.

(d) For this collection $\alpha = \dfrac{N(1-x)}{x} = \dfrac{6814(1-0.9945)}{0.9945} = 37.68$.

(e) The number of species of moth in the collection represented by 5 individuals is
$$\alpha\left(\frac{x^5}{5}\right) = 37.68\left(\frac{0.9945^5}{5}\right) = 7.33, \text{ or about 7 species of moths.}$$

(f) The figure below shows a graph of the number of species represented by n individuals, that is $\alpha\left(\dfrac{x^n}{n}\right) = 37.68\left(\dfrac{0.9945^n}{n}\right)$ against n, using a horizontal span of 0 to 20 and a vertical span of 0 to 40.

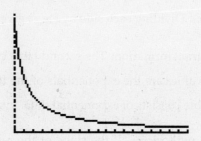

4.5 CONNECTING EXPONENTIAL AND LINEAR DATA

E-1. **Verifying the quotient law:** To verify the quotient law for logarithms $\ln\dfrac{x}{y} = \ln x - \ln y$, suppose that $\ln x = p$ and $\ln y = q$. Then $x = e^p$ and $y = e^q$. Hence

$$\frac{x}{y} = \frac{e^p}{e^q} = e^{p-q}.$$

Thus $p - q$ is the power of e that gives $\dfrac{x}{y}$, and therefore

$$\ln\frac{x}{y} = p - q = \ln x - \ln y.$$

E-3. **An important fact about logarithms**: Since $\ln x$ is the power of e which gives x, $\ln e^x$ is the power of e which gives e^x and that is x (!).

E-5. **More calculations using laws of logarithms**:

(a) $\ln e^3 = 3 \ln e = 3 \times 1 = 3$

(b)

$$\ln\left(\frac{\sqrt{e}}{e^4}\right) = \ln(\sqrt{e}) - \ln(e^4) = \ln(e^{1/2} - 4\ln e = \frac{1}{2}\ln e - 4\ln e = \frac{1}{2} - 4 = -\frac{7}{2} \text{ or} - 3.5$$

(c) $\ln\dfrac{1}{e} = \ln 1 - \ln e = 0 - 1 = -1$

(d)

$$\ln\ln\left(e^{(e^4)}\right) = \ln\left(\ln\left(e^{(e^4)}\right)\right)$$
$$= \ln\left(e^4(\ln e)\right)$$
$$= \ln(e^4) = 4\ln e = 4 \times 1 = 4$$

E-7. **Change of base**: Consider $p = \log_a b$ and $q = \log_b x$. This means that $a^p = b$ and $b^q = x$. Thus, substituting a^p for b, we have $(a^p)^q = x$, and so $a^{pq} = x$. By the definitions of p and q, this means that

$$a^{(\log_a b)(\log_b x)} = x,$$

as we needed to show. By definition of $\log_a$, the equation above means that $\log_a x = (\log_a b)(\log_b x)$. Dividing each side by $\log_a b$ gives the change-of-base formula.

S-1. **Exponential transformation**: The second table consists of the same input values, while the function values are the exponentials of the function values of linear data; therefore the second table consists of exponential data.

S-3. **Slope and growth factor**: If the slope of the regression line for the natural logarithm of exponential data is $m = -0.77$, then the decay factor for the exponential function is $a = e^m = e^{-0.77} = 0.463$.

S-5. **Exponential regression**: Calculating the natural logarithm of the data, we obtain the table

x	1	2	3	4	5
$\ln y$	1.411	2.163	2.955	3.353	4.170

The regression line is calculated as $\ln y = 0.671x + 0.798$.

S-7. **Exponential regression**: Calculating the natural logarithm of the data, we obtain the table

x	4.2	7.9	10.8	15.5	20.2
$\ln y$	2.015	2.092	2.140	2.322	2.510

The regression line is calculated as $\ln y = 0.031x + 1.849$. The exponential regression model is therefore $y = P \times a^x$, where $P = e^{1.849} = 6.35$ and $a = e^{0.031} = 1.03$. Thus the exponential model is $y = 6.35 \times 1.03^x$.

S-9. **Exponential regression**: Calculating the natural logarithm of the data, we obtain the table

x	1	2	3	4	5
$\ln y$	1.308	1.459	1.808	2.208	2.610

The regression line is calculated as $\ln y = 0.335x + 0.873$. The exponential regression model is therefore $y = P \times a^x$, where $P = e^{0.873} = 2.39$ and $a = e^{0.335} = 1.40$. Thus the exponential model is $y = 2.39 \times 1.40^x$.

S-11. **Exponential regression**: Calculating the natural logarithm of the data, we obtain the table

x	2	5	6	9	10
$\ln y$	1.435	1.856	2.001	2.425	2.557

The regression line is calculated as $\ln y = 0.141x + 1.155$. The exponential regression model is therefore $y = P \times a^x$, where $P = e^{1.155} = 3.17$ and $a = e^{0.141} = 1.15$. Thus the exponential model is $y = 3.17 \times 1.15^x$.

1. **Population**: Because the logarithm of exponential data is linear, a plot of the logarithm of the population values should be linear.

3. **Population growth**: Let t be years since 2001 and N the population (in thousands).

   (a) The plot of the natural logarithm of the data points is in the figure below.

Since these points are close to being on a line, it is reasonable to approximate the original data with an exponential function.

(b) The regression line parameters for the natural logarithm of the data are calculated in the left-hand figure below. We find that $\ln N = 0.042t + 0.445$. We have added this graph to the data plot in the right-hand figure below.

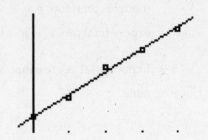

5. **Population:** Since $\ln N = 0.039t - 0.693$, an exponential model for N is $P \times a^t$, where $P = e^{-0.693} = 0.50$ and $a = e^{0.039} = 1.04$. Thus an exponential model is $N = 0.50 \times 1.04^t$.

7. **Sales growth:**

(a) Since $\ln S = 0.049t + 2.230$, an exponential model for S is $P \times a^t$, where $P = e^{2.230} = 9.30$ and $a = e^{0.049} = 1.05$. Thus an exponential model is $S = 9.30 \times 1.05^t$.

(b) By Part (a), the growth factor is 1.05, so sales grow by 5% each year.

(c) Using the formula from Part (a), $S(6) = 9.30 \times 1.05^6 = 12.46$. In practical terms, this means 6 years after 2005, that is, at the start of 2011, sales will be 12.46 thousand dollars.

(d) We expect sales to reach a level of 12 thousand dollars when $12 = 9.30 \times 1.05^t$. Solving, for example by using a table or graph, we find that $t = 5.22$. Thus we expect sales to reach $12,000 when $t = 5.22$, or about mid-March of 2010.

9. **Cable TV:** Let t be years since 1976 and C the percent with cable.

(a) The plot of the natural logarithm of the data points is in the figure below.

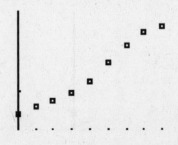

Since these points are close to being on a line, it is reasonable to approximate the original data with an exponential function.

(b) The regression line parameters for the natural logarithm of the data are calculated in the left-hand figure below. We find that $\ln C = 0.144t + 2.631$. We have added this graph to the data plot in the right-hand figure below.

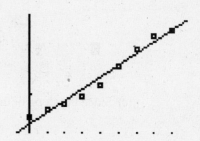

(c) The slope of the regression line is 0.144, and so the yearly growth factor is

$$a = e^{0.144} = 1.155.$$

The exponential initial value is

$$P = e^{2.631} = 13.888.$$

The exponential model is given by

$$C = 13.888 \times 1.155^t.$$

11. **National health care spending**: Let t be years since 1950 and H the costs in billions of dollars.

   (a) The plot of the natural logarithm of the data is shown below. Here we used the number of years since 1950 for the horizontal axis. The data points look like they lie on a straight line, so it is reasonable to model the original data with an exponential function.

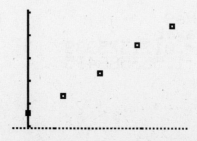

(b) The left-hand figure below shows that the equation of the regression line is $\ln H = 0.099t + 2.426$. The graph is added to the data plot in the right-hand figure below.

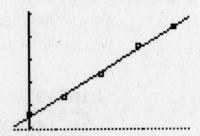

(c) Since the slope of the line is 0.099, the yearly growth factor is

$$a = e^{0.099} = 1.104.$$

Therefore $r = 0.104$, so the yearly percentage increase is 10.4%.

13. **Grazing rabbits:**

(a) Below is a plot of $\ln D$ against V.

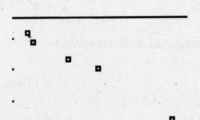

Since the points roughly fall on a straight line, it appears that D is approximately an exponential function of V.

(b) From the left-hand figure below, the regression line is $\ln D = -0.012V - 1.554$. We have added it to the data plot in the right-hand figure.

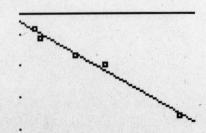

(c) The exponential function which approximates D is found from the regression line. The decay factor is $a = e^{-0.012} = 0.988$, and the initial value is $P = e^{-1.554} = 0.211$, so the function is

$$D = 0.211 \times 0.988^V.$$

(d) The satiation level for the rabbit is 0.18 pound per day, and thus we have

$$D = \text{satiation level} - A = 0.18 - A.$$

Solving for A yields $A = 0.18 - D$, and, using the formula from Part (c), we find that

$$A = 0.18 - 0.211 \times 0.988^V.$$

15. **Nearly linear or exponential data**: The left-hand figure below is the plot of the data from Table A. This appears to be linear. Computing the regression line $f = 19.84t - 16.26$ and adding it to the plot (right-hand figure) shows how very close to linear f is.

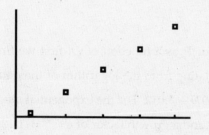 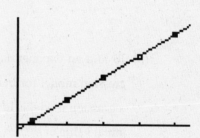

We conclude that Table A is approximately linear, with a linear model of $f = 19.84t - 16.26$.

The plot of data from Table B in the left-hand figure below looks more like exponential data than linear data. We plot the logarithm of the data in the right-hand figure, and this falls on a straight line, which indicates that the data is indeed exponential.

The equation of the regression line is $\ln g = 0.442t + 0.846$. The exponential model therefore has a growth factor of $a = e^{0.442} = 1.56$ and an initial value of $P = e^{0.846} = 2.33$, which gives the formula

$$g = 2.33 \times 1.56^t.$$

17. **Sound pressure:**

(a) A plot the natural logarithm of the data is shown below. Since a plot of the natural logarithm of P against D appears linear, it is reasonable to model pressure as an exponential function of loudness.

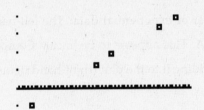

(b) To find an exponential model of P as a function of D, first we find a linear regression model for $\ln P$. Calculating from the logarithm of the data, we find the regression model is $\ln P = 0.109D - 8.072$. For the exponential model, we use an initial value of $e^{-8.072} = 0.0003$ and a growth factor of $e^{0.109} = 1.12$; therefore the exponential model is $P = 0.0003 \times 1.12^D$.

(c) When loudness D is increased by one decibel, the pressure P is multiplied by a factor of 1.12, that is, it increases by 12%.

19. **Catalogue sales:** Let t be years since 2004 and C catalogue sales in billions of dollars.

(a) The plot of the natural logarithm of the data is shown below. We find the regression model to be $\ln C = 0.065t + 4.963$.

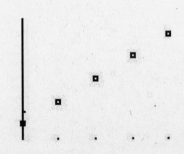

(b) For the exponential model we use an initial value of $e^{4.963} = 143.022$ and a growth factor of $e^{0.065} = 1.067$. Therefore the exponential model is $C = 143.022 \times 1.067^t$.

Chapter 4 Review Exercises

1. **Percentage growth**: The population grows by 4.5% each year, so the yearly growth factor is $a = 1 + r = 1 + 0.045 = 1.045$. Since a month is 1/12th of a year, the monthly growth factor is $1.045^{1/12} = 1.0037$. That is a growth of 0.37% per month.

2. **Percentage decline**: The population declines by 0.5% each year, so the yearly decay factor is $a = 1 - r = 1 - 0.005 = 0.995$. Since a decade is 10 years, the decade decay factor is $0.995^{10} = 0.9511$. Since $1 - 0.9511 = 0.0489$, that is a decline of 4.89% per decade.

3. **Credit cards**:

   (a) The decay factor is $a = (1 + 0.02)(1 - 0.05) = 0.969$.

   (b) Because the decay factor is $a = 0.969$, we have $r = 1 - 0.969 = 0.031$. Hence the balance decreases by 3.1% each month.

   (c) Because 3 years is 36 months, the balance after 3 years of payments will be $1000 \times 0.969^{36} = 321.85$ dollars.

4. **More on credit cards**:

   (a) We have $r = 0.015$ and $m = 0.04$, so the base is $a = (1 + 0.015)(1 - 0.04) = 0.9744$.

   (b) Because the decay factor is $a = 0.9744$, we have $r = 1 - 0.9744 = 0.0256$. Hence the balance decreases by 2.56% each month.

   (c) We have $r = 0.015$ and $m = 0.01$, so the base is $a = (1 + 0.015)(1 - 0.01) = 1.00485$. This number is greater than 1, so the balance will increase each month. This is not a realistic situation.

5. **Testing exponential data**: Calculating the ratios of successive terms of y, we get $\dfrac{30.0}{25.0} = 1.2$, $\dfrac{36.0}{30.0} = 1.2$, and $\dfrac{43.2}{36.0} = 1.2$. Since the x values are evenly spaced and these ratios show a constant value of 1.2, the table does show exponential data.

6. **Modeling exponential data**: The data from Exercise 5 show a constant ratio of 1.2 for x values increasing by 1's, and so the growth factor is $a = 1.2$. The initial value of $P = 25.0$ comes from the first entry in the table. Thus an exponential model for the data is $y = 25 \times 1.2^x$.

7. **Credit card balance**:

   (a) Each successive ratio of new/old is 0.97, so this is exponential data. Because n increases by single units, the decay factor is $a = 0.97$. The initial value of 500.00 comes from the first entry in the table. Thus an exponential model for the data is
   $B = 500 \times 0.97^n$.

   (b) From the first entry in the table the initial charge is $500.00.

   (c) Because 2 years is 24 months, the balance after 2 years of payments will be $500 \times 0.97^{24} = 240.71$ dollars.

8. **Inflation**:

   (a) Each successive ratio of new/old is 1.03, so this is exponential data.

   (b) Let t be the time in years since the start of 2002 and P the price in dollars. Because t increases by single units, by Part (a) the growth factor is $a = 1.03$. The initial value of 265.50 comes from the first entry in the table. Thus an exponential model for the data is $P = 265.5 \times 1.03^t$.

   (c) We want to find the first whole number t for which P is greater than 325. From a table of values or a graph we find the number to be $t = 7$. Thus at the beginning of 2009 the price will surpass $325.

9. **Exponential regression**: The exponential model is $y = 20.97 \times 1.34^x$.

10. **Exponential regression**: The exponential model is $y = 10.96 \times 0.80^x$. A plot of the exponential model, together with the original data, is shown below.

11. **Sales:**

   (a) The exponential model for A is $A = 8.90 \times 1.02^t$, and the exponential model for B is $B = 2.30 \times 1.27^t$.

   (b) By Part (a) the growth factor for A is 1.02, so the sales for *Alpha* are growing by 2% per year. By Part (a) the growth factor for B is 1.27, so the sales for *Beta* are growing by 27% per year.

   (c) We want to find the first whole number t for which B is greater than A. From a table of values or a graph we find the number to be $t = 7$. Thus at the beginning of 2009 sales for *Beta* will overtake sales for *Alpha*.

12. **Credit card payments:**

   (a) The exponential model for B is $B = 520 \times 0.9595^n$.

   (b) From Part (a) the initial charge was $520.00.

   (c) From Part (a) the decay factor is 0.9595. Because $m = 0.05$ we know that

   $$(1 + r)(1 - 0.05) = 0.9595,$$

   and thus

   $$1 + r = \frac{0.9595}{1 - 0.05}.$$

   Hence

   $$r = 1 - \frac{0.9595}{1 - 0.05} = 0.01.$$

   Thus the monthly finance charge is 1%.

13. **Solving logarithmic equations:** If $\log x = -1$ then $x = 10^{-1} = 0.1$.

14. **Comparing logarithms:** If $\log x = 3.5$ and $\log y = 2.5$ then $\log y$ is 1 less than $\log x$. Hence y is 10^{-1} times x, so $y = 0.1x$.

15. **Comparing earthquakes:** The Double Spring Flat quake was $6.0 - 5.5 = 0.5$ higher on the scale than the Little Skull Mountain quake. Hence the Double Spring Flat quake was $10^{0.5} = 3.16$ times as powerful as the Little Skull Mountain quake.

16. **Population growth:**

   (a) We put $N = 200$ in the formula and get $T = 25 \log 200 - 50 = 7.53$. Thus the population reaches a size of 200 after 7.53 years.

(b) We want to find the value of N for which $T = 0$. Thus we need to solve the equation $25 \log N - 50 = 0$ for N. We find $25 \log N = 50$ or $\log N = \dfrac{50}{25} = 2$. Hence $N = 10^2 = 100$. Thus the initial population size was 100.

17. **Exponential regression**: Calculating the natural logarithm of the data, we obtain the table

x	1	5	7	10	11
$\ln y$	3.336	4.508	5.093	5.971	6.264

The regression line is calculated as $\ln y = 0.293x + 3.043$. The exponential regression model is therefore $y = P \times a^x$, where $P = e^{3.043} = 20.97$ and $a = e^{0.293} = 1.34$. Thus the exponential model is $y = 20.97 \times 1.34^x$.

18. **Exponential regression**: Calculating the natural logarithm of the data, we obtain the table

x	1	2	3	4	5
$\ln y$	2.175	1.946	1.723	1.504	1.281

The regression line is calculated as $\ln y = -0.223x + 2.395$. The exponential regression model is therefore $y = P \times a^x$, where $P = e^{2.395} = 10.97$ and $a = e^{-0.223} = 0.80$. Thus the exponential model is $y = 10.97 \times 0.80^x$.

19. **Economic output**: If a linear model fits the logarithm of the data for economic output, then the economist should use an exponential function to fit the original data.

20. **Account balance**: Let t be the time in months since the start of the year and B the balance in dollars. For example, the beginning of February corresponds to $t = 1$.

   (a) The regression line for the natural logarithm of the data is calculated as $\ln B = 0.0025t + 7.1309$.

   (b) The slope of the regression line is 0.0025, and so the yearly growth factor is $a = e^{0.0025} = 1.0025$. The exponential initial value is $P = e^{7.1309} = 1250.00$. The exponential model is given by $B = 1250 \times 1.0025^t$.

   (c) The monthly growth factor is 1.0025, so the monthly interest rate is 0.25%.

   (d) To estimate the balance at the beginning of March we put $t = 2$ in the formula and obtain

$$B = 1250 \times 1.0025^2 = 1256.26 \text{ dollars.}$$

Solution Guide for Chapter 5: A Survey of Other Common Functions

5.1 POWER FUNCTIONS

E-1. More skid marks: Parts (a) and (b) of Exercise 5 do not use homogeneity, so we need to consider only Part (c). We want to slow to a speed that will cut the emergency stopping distance in half. Since we are originally traveling at 60 miles per hour, the stopping distance is $L = \dfrac{1}{30h}60^2$, since the speed is 60. The new stopping distance will be $\dfrac{1}{2}L$, so if S is the new reduced speed, then

$$\frac{1}{30h}S^2 = \frac{1}{2}L = \frac{1}{2}\left(\frac{1}{30h}60^2\right).$$

Multiplying both sides by $30h$ and cancelling yields $S^2 = \dfrac{1}{2}60^2$. Thus $S = \sqrt{\dfrac{1}{2}60^2}$, which equals 42.43 miles per hour. This is between 42 and 43 miles per hour.

E-3. Calculating a limit: To calculate the $\lim_{x\to\infty} \dfrac{2 \times 3^x + x^{1000}}{3^x}$, we use the hint to rewrite the fraction as a sum of two fractions:

$$\frac{2 \times 3^x + x^{1000}}{3^x} = \frac{2 \times 3^x}{3^x} + \frac{x^{1000}}{3^x}.$$

In the first fraction, the 3^x terms cancel, leaving only the 2. For the second fraction, we know from the discussion in the text that

$$\lim_{x\to\infty} \frac{x^p}{a^x} = 0$$

when a is greater than 1. In our case, $p = 1000$ and $a = 3$, so that limit is zero. Thus $\lim_{x\to\infty} \dfrac{2 \times 3^x + x^{1000}}{3^x} = 2$.

E-5. Uniqueness of homogeneity: Suppose that a function f has the property that $f(tx) = t^k f(x)$ for all positive x and t. This means that in particular this is true when $x = 1$. When $x = 1$, the property is $f(t) = t^k f(1)$ for all t. Writing $c = f(1)$, we have $f(t) = ct^k$, so f is a power function with power k.

S-1. **Graph of a power function**: Graphing a power function with negative k, such as x^{-2}, shows that the graph is decreasing. See Key Idea 5.1.

S-3. **Homogeneity**: In this case f is a power function with power $k = 1.47$. By the homogeneity property, if x is increased by a factor of t, then f is increased by a factor of $t^{1.47}$. In this case, x is tripled, so $t = 3$. Therefore f is increased by a factor of $3^{1.47}$, that is, by a factor of 5.03.

S-5. **Homogeneity**: In this case f is a power function with power $k = 3.11$. By the homogeneity property, if x is increased by a factor of t, then f is increased by a factor of $t^{3.11}$. Since x increases from 3.6 to 5.5, x is increased by a factor of $\dfrac{5.5}{3.6}$ and so f is increased by a factor of $\left(\dfrac{5.5}{3.6}\right)^{3.11} = 3.74$, that is, $f(5.5)$ is 3.74 times as large as $f(3.6)$.

S-7. **Homogeneity**: In this case f is a power function with power k. By the homogeneity property, if x is increased by a factor of t, then f is increased by a factor of t^k. Since x increases from 1.76 to 6.6, 1.76 is increased by a factor of $t = \dfrac{6.6}{1.76}$. Thus f is increased by a factor of $t^k = \left(\dfrac{6.6}{1.76}\right)^k$. Since $f(6.6)$ is 6.2 times as large as $f(1.76)$,

$$6.2 = t^k = \left(\frac{6.6}{1.76}\right)^k.$$

Solving for k yields $k = 1.38$.

S-9. **Constant term**: Since $f(x) = cx^{-1.32}$ and $f(5) = 11$, $11 = c \times 5^{-1.32}$. Thus $c = \dfrac{11}{5^{-1.32}}$, which equals 92.05, so $c = 92.05$.

S-11. **Power**: From 1 to 2, 1 is increased by a factor of $t = \dfrac{2}{1} = 2$. Thus f is increased by a factor of $t^k = 2^k$. Since $f(2)$ is 10 times the size of $f(1)$,

$$10 = t^k = 2^k.$$

Solving for k yields $k = 3.32$.

1. **The role of the constant term in power functions with positive power**: We use a horizontal span of 0 to 5 as suggested, and the table of values below leads us to choose a vertical span of 0 to 100. In the right-hand figure, x^2 is the bottom graph, $2x^2$ is the next one up, and $4x^2$ is the top graph.

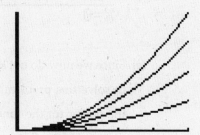

Larger values of c make the function increase faster.

3. **Speed and stride length**: We use the homogeneity property of power functions to solve this. If one animal has a stride length 3 times that of another, then it will run faster by a factor of

$$3^{\text{Function power}} = 3^{1.7} = 6.47,$$

or about 6.5 times as fast.

5. **Length of skid marks versus speed**:

    (a) We are on dry concrete, so the value of the friction coefficient is $h = 0.85$. Thus the function we are dealing with is

    $$L = \frac{1}{30 \times 0.85}S^2 = \frac{S^2}{25.5}.$$

    If the speed is $S = 55$ miles per hour, then the skid marks will have length

    $$L = \frac{55^2}{25.5} = 118.63 \text{ feet}.$$

    (b) Again we are on dry concrete pavement, so we use the same function as in Part (a). We need to know what speed will result in skid marks which are 230 feet long. That is, we need to solve the equation

    $$\frac{S^2}{25.5} = 230.$$

    We will solve this using the crossing graphs method. We use a horizontal span of 0 to 80 miles per hour, and the table of values for L below leads us to choose a vertical span of 0 to 300. We have graphed L and the target length of 230 (thick line). We see that the intersection occurs at a speed of $S = 76.58$ miles per hour. The driver would appear to be in danger of receiving a citation.

X	Y₁	Y₂
0	0	230
20	15.686	230
40	62.745	230
60	141.18	230
80	250.98	230
100	392.16	230
120	564.71	230

X=0

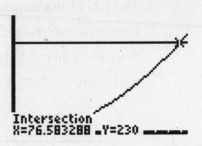

Intersection
X=76.583288 ⌐Y=230 ▬▬▬

(c) Since we now do not know the value of h, we must use the homogeneity property to solve this problem. We want to change the distance by a factor of 0.5. Since the power for the function is 2, changing the speed by a factor of f changes the distance L by a factor of f^2. We want to cut the speed in half, so we want to choose f so that

$$f^2 = 0.5.$$

This may be solved by the crossing graphs method or simply by noting that the solution is

$$f = \sqrt{0.5} = 0.7071.$$

Hence we need to change our speed from 60 miles per hour to

$$60 \times 0.7071 = 42.43 \text{ miles per hour.}$$

The new speed should be between 42 and 43 miles per hour.

7. **Life expectancy of stars:**

(a) Since E is a power function with negative exponent, as M increases, E decreases, so a more massive star has a shorter life expectancy than a less massive star.

(b) Since Spica has a mass of 7.3 solar masses, $M = 7.3$, and therefore $E = M^{-2.5} = 7.3^{-2.5}$, which equals 0.0069, or about 0.007, of a solar lifetime. Since Earth's life expectancy is about 10 billion years, Spica's life expectancy is about $10 \times 0.0069 = 0.069$ billion years or about 70 million years.

(c) In functional notation, the life expectancy of a main-sequence star with mass $M = 0.5$ is $E(0.5)$. Its value is $E = 0.5^{-2.5}$ or about 5.66 solar lifetimes. This is about 56.6 billion years.

(d) Because Vega has a life expectancy of 6.36 billion years and 1 solar lifetime is 10 billion years, Vega has a life expectancy of 0.636 of a solar lifetime. Since $E = 0.636$, we have the equation $M^{-2.5} = 0.636$, which we can solve for M. This yields $M = 1.20$, so the mass of Vega is 1.20 solar masses.

(e) If one main-sequence star is twice as massive as another, then M is changed by a factor of 2, and therefore E is changed by a factor of $2^{-2.5} = 0.18$. Thus the life expectancy of the larger star is 0.18 times that of the smaller.

9. **Tsunami waves and breakwaters**: The wave height h beyond the breakwater is a power function of the width ratio R with power 0.5.

   (a) We measure H, W, w, and h in feet. Since the wave has height 8 feet in a channel of width 5000 before the breakwater, $H = 8$ and $W = 5000$. Since the breakwater has width 3000 feet, $w = 3000$, and so $R = \dfrac{w}{W} = \dfrac{3000}{5000} = 0.6$. The height of the wave beyond the breakwater is $h = HR^{0.5} = 8 \times 0.6^{0.5} = 6.20$ feet.

   (b) If the channel width is cut in half, then R is multiplied by $\dfrac{1}{2}$. Thus the new height is $\left(\dfrac{1}{2}\right)^{0.5} = 0.71$ times the original height.

11. **Terminal velocity**: Terminal velocity T is a power function with power 0.5, so if the length L changes by a factor of t, then T changes by a factor of $t^{0.5}$.

   (a) The man is 36 times as long as the mouse, so $t = 36$. The terminal velocity of the man is therefore $36^{0.5} = 6$ times that of the mouse.

   (b) If the terminal velocity of the man is 120 miles per hour, then, from Part (a), that of the mouse is $\dfrac{120}{6} = 20$ miles per hour.

   (c) Since a squirrel is 7 inches long, it is only $\dfrac{7}{72}$ times the length of the 6-foot man, so the terminal velocities differ by a factor of $\left(\dfrac{7}{72}\right)^{0.5}$. Thus the terminal velocity of a squirrel is $120 \times \left(\dfrac{7}{72}\right)^{0.5} = 37.42$ miles per hour.

13. **Newton's law of gravitation**:

   (a) If the distance d is changed by a factor of 0.5, then the gravitational force will be changed by a factor of

   $$0.5^{\text{Function power}} = 0.5^{-2} = 4.$$

   Hence if the distance between the asteroids is halved, the force of gravity will be 4 times as strong.

   If the distance is reduced to one quarter of its original value, it is changed by a factor of 0.25. That gives a change in force by a factor of

   $$0.25^{\text{Function power}} = 0.25^{-2} = 16.$$

   Thus if the distance is reduced to one quarter of its original value, the force of gravity will be 16 times stronger.

(b) We want to know the value of c that will make

$$2,000,000 = c \times 300^{-2}.$$

This is a linear equation in c which we can solve by dividing each side by 300^{-2}:

$$c = \frac{2,000,000}{300^{-2}} = 1.8 \times 10^{11}.$$

Hence we know that, for these asteroids, the gravitational force is given by

$$F = 1.8 \times 10^{11} \times d^{-2}.$$

When $d = 800$, we get that the force is

$$F = 1.8 \times 10^{11} \times 800^{-2} = 281,250 \text{ newtons.}$$

This could also be solved used the homogeneity property of power functions. The first step is to note that increasing the distance from 300 to 800 kilometers is an increase by a factor of $800/300 = 2.67$.

(c) We use a horizontal span of 0 to 1000 as suggested. The table of values below for $F = 1.8 \times 10^{11} \times d^{-2}$ leads us to choose a vertical span of 0 to $20,000,000 = 2 \times 10^7$. In the graph below, the horizontal axis is distance between the centers, and the vertical axis is force of gravity.

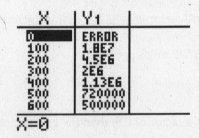
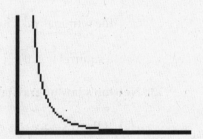

The graph shows that when the asteroids are relatively close, gravitational force is very strong. (It is in fact so strong that when bodies of planetary size get too close, the force of gravity will tear one or both of them to pieces.) As they move farther apart, the gravitational force decreases toward zero.

15. **Giant ants and spiders**:

(a) Weight is a power function of length, with power 3. Thus, if the length of the ant is increased by a factor of 500, then its weight is increased by a factor of $500^3 = 125,000,000$.

(b) Cross-sectional area of limbs is a power function of length, with power 2. Thus, if the length of the ant is increased by a factor of 500, then the cross-sectional area of its legs is increased by a factor of $500^2 = 250,000$.

(c) Using the hint and the results of Parts (a) and (b), we see that the pressure on a leg increases by a factor of $\dfrac{125,000,000}{250,000} = 500$. The poor ant will be flattened.

17. **Time to failure**: If the building collapses in 30 seconds at one intensity, put $t_1 = 30$ and denote that intensity as I_1. Tripling that intensity means putting $I_2 = 3I_1$, so

$$\frac{t_1}{t_2} = \left(\frac{I_2}{I_1}\right)^2 = \left(\frac{3I_1}{I_1}\right)^2 = 9.$$

Since $t_1 = 30$, we have $\dfrac{30}{t_2} = 9$. Solving for t_2 yields 3.33 seconds for the building to collapse.

5.2 MODELING DATA WITH POWER FUNCTIONS

E-1. **Finding power functions**: We wish to find power functions passing through certain points, so the functions will be of the form $y = cx^k$ and we are to determine c and k.

(a) For the power function to pass through both $(2, 7)$ and $(6, 56)$, two equations must hold:

$$7 = c \times 2^k$$
$$56 = c \times 6^k.$$

Dividing the second equation by the first gives

$$\frac{56}{7} = \frac{c \times 6^k}{c \times 2^k} = \left(\frac{6}{2}\right)^k$$

or

$$8 = 3^k.$$

We can solve this using logarithms:

$$\ln 8 = \ln 3^k$$
$$\ln 8 = k \ln 3$$
$$\frac{\ln 8}{\ln 3} = k$$
$$1.89 = k.$$

We put this value back into the first equation to solve for c:

$$7 = c \times 2^{1.89}$$

$$\frac{7}{2^{1.89}} = c$$

$$1.89 = c.$$

The power function is therefore $y = 1.89x^{1.89}$. It is purely coincidental that the c and k values are the same.

(b) For the power function to pass through both $(a, 12)$ and $(10a, 1200)$, two equations must hold:

$$12 = c \times a^k$$

$$1200 = c \times (10a)^k.$$

Dividing the second equation by the first gives

$$\frac{1200}{12} = \frac{c \times (10a)^k}{c \times a^k} = \left(\frac{10a}{a}\right)^k$$

or

$$100 = 10^k.$$

We could solve using logarithms, but since we know that $10^2 = 100$, clearly $k = 2$. To do this using common logarithms, for example, we compute as follows:

$$\log 100 = \log 10^k$$

$$\log 100 = k \log 10$$

$$\frac{\log 100}{\log 10} = k$$

$$2 = k.$$

We put this value back into the first equation to solve for c. (Note that c must depend on a, since different values of a will require use of different power functions.) We get:

$$12 = c \times a^2$$

$$\frac{12}{a^2} = c.$$

The power function is therefore $y = \left(\frac{12}{a^2}\right) x^2$.

E-3. **Linear to power functions**: We wish to show that if $\ln y = \ln c + k \ln x$, then $y = cx^k$. To do this, we exponentiate both sides of the equation and use the properties of exponents:

$$\begin{aligned} \ln y &= \ln c + k \ln x \\ e^{\ln y} &= e^{\ln c + k \ln x} \\ y &= e^{\ln c} \times e^{k \ln x} \\ y &= c(e^{\ln x})^k \\ y &= cx^k. \end{aligned}$$

S-1. **Logarithmic conversion of both x and y**: If we start from power data and logarithmically convert both rows, then the resulting data is linear.

S-3. **Getting the power**: The slope of the regression line for the logarithm of the data is the same as the power, so that power is 3.

S-5. **Modeling power data**: To find the power model, we convert the data into linear data using the logarithm. To do this, we put the original data in the 3rd and 4th columns, the logarithms in the 1st and 2nd columns (see the figure on the left below), and then calculate the linear relation using linear regression. This is shown below on the right.

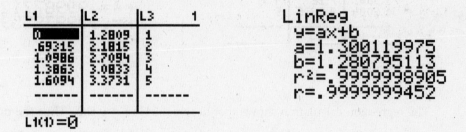

The regression calculation shows that $\ln f = 1.30 \ln x + 1.28$. Thus the power model uses $k = 1.30$ and $c = e^{1.28} = 3.60$, and so $f = 3.60x^{1.30}$. This fits the original data, as shown in the figure below, where we have graphed the original data from the 3rd and 4th columns together with the power model. This can also be seen from a table of values.

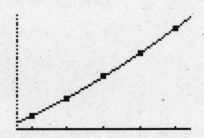

S-7. **Modeling almost power data**: To find the power model, we convert the data into linear data using the logarithm. To do this, we put the original data in the 3rd and 4th columns, the logarithms in the 1st and 2nd columns (see the figure on the left below), and then calculate the linear relation using linear regression. This is shown below on the right.

L1	L2	L3	1
0	.78846	1	
.69315	3.9455	2	
1.0986	5.8257	3	
1.3863	7.12	4	
1.6094	8.0639	5	
------	------	------	

L1(1)=0

```
LinReg
y=ax+b
a=4.540030781
b=.8016327937
r²=.9998771219
r=.999938559
```

The regression calculation shows that $\ln f = 4.54 \ln x + 0.802$. Thus the power model uses $k = 4.54$ and $c = e^{0.802} = 2.23$, and so $f = 2.23x^{4.54}$. This closely fits the original data, as shown in the figure below, where we have graphed the original data from the 3rd and 4th columns together with the power model.

S-9. **Modeling almost power data**: To find the power model, we convert the data into linear data using the logarithm. To do this, we put the original data in the 3rd and 4th columns, the logarithms in the 1st and 2nd columns (see the figure on the left below), and then calculate the linear relation using linear regression. This is shown below on the right.

```
L1        L2       L3        1        LinReg
█▉▊█▊     1.7228   .3                 y=ax+b
.26236    .69315   1.3                a=-.8969818772
.78846    -.0034   2.2                b=.7194011162
1.1939    -.2614   3.3                r²=.9768570041
1.411     -.6733   4.1                r=-.9883607662
------    ------   ------

L1(1)=-1.20397280...
```

The regression calculation shows that $\ln f = -0.90\ln x + 0.719$. Thus the power model uses $k = -0.90$ and $c = e^{0.719} = 2.05$, and so $f = 2.05x^{-0.90}$. This closely fits the original data, as shown in the figure below, where we have graphed the original data from the 3rd and 4th columns together with the power model.

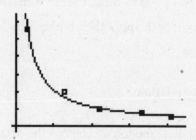

S-11. **Modeling almost power data**: To find the power model, we convert the data into linear data using the logarithm. To do this, we put the original data in the 3rd and 4th columns, the logarithms in the 1st and 2nd columns (see the figure on the left below), and then calculate the linear relation using linear regression. This is shown below on the right. The regression calculation shows that $\ln f = -1.59\ln x + 0.83$. Thus the power model uses $k = -1.59$ and $c = e^{0.83} = 2.29$, and so $f = 2.29x^{-1.59}$.

```
L1        L2       L3        2        LinReg
0         █▊▊█▊    1                  y=ax+b
.69315    -.2744   2                  a=-1.588772971
1.0986    -.9163   3                  b=.8294652476
1.3863    -1.386   4
1.6094    -1.715   5

------    ------   ------

L2(1)=.8329091229...
```

1. Hydroplaning:

(a) From the figure below, the equation of the regression line is $\ln V = 0.50 \ln p + 2.34$.

```
LinReg
y=ax+b
a=.4984425235
b=2.342347626
```

■

(b) From Part (a), the power is 0.5, and the coefficient c is $e^{2.34} = 10.4$. Thus the formula is $V = 10.4p^{0.5}$.

(c) We want to compare $V(35)$ with $V(60)$. From the formula in Part (b) we get $V(35) = 10.4 \times 35^{0.5} = 61.5$, while $V(60) = 10.4 \times 60^{0.5} = 80.6$. Thus the critical speed for hydroplaning is 61.5 mph for the car and 80.6 mph for the bus. Since both are traveling at 65 mph, the car is in danger of hydroplaning, while the bus is not. Thus, if both apply their brakes, the car might hydroplane and hit the bus from the rear.

3. Mass-luminosity relation:

(a) To find the power model, we convert the data into linear data using the logarithm. To do this, we put the original data in the 3rd and 4th columns, the logarithms in the 1st and 2nd columns (see the figure on the left below), and then calculate the linear relation using linear regression. This is shown below on the right.

```
L1      L2      L3      1
1.9874  6.9565  7.3
1.1314  4.0073  3.1
0       .09531  1
0       0       1
-1.772  -6.215  .17
------  ------  ------

L1(1)=1.987874348...
```

```
LinReg
y=ax+b
a=3.50493207
b=.0244633946
r²=.9999189337
r=.999959466
```

■

The regression calculation shows that $\ln L = 3.50 \ln M + 0.02$. Thus the power model uses $k = 3.5$ and $c = e^{0.02} = 1$ to one decimal place, and so $L = M^{3.5}$ is the power model.

(b) Since Kruger 60 has a mass of 0.11 solar mass, its relative luminosity is $L(0.11)$ in functional notation. The value of $L(0.11)$ is $0.11^{3.5} = 4.41 \times 10^{-4}$ or 0.000441, or about 0.0004.

(c) Since Wolf 359 has a relative luminosity of about 0.0001, for that star $L = 0.0001$ and so $M^{3.5} = 0.0001$. Solving for M, for example by $M = 0.0001^{1/3.5}$, shows that $M = 0.072$, or about 0.07, of a solar mass.

(d) If one star is 3 times as massive an another, then M changes by a factor of 3 and so, by the homogeneity property with power 3.5, L changes by a factor of $3^{3.5} = 46.77$. Thus the more massive star has a relative luminosity 46.77, or about 47, times that of the less massive star.

5. **Speed in flight versus length:**

(a) Table 5.3 indicates that larger animals fly faster, since as L increases so does F.

(b) As the figure below at the left indicates, the regression line for $\ln F$ against $\ln L$ is $\ln F = 0.34 \ln L + 3.026$. This gives $k = 0.34$ and $c = e^{3.026} = 20.61$, so the desired formula is $F = 20.61 L^{0.34}$.

(c) The graph of $F = 20.61 L^{0.34}$ is shown below at the right. The horizontal span is 0 to 300, and the vertical span is 0 to 150.

```
LinReg
 y=ax+b
 a=.3383005338
 b=3.025729482
```

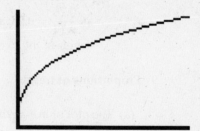

(d) The graph is concave down and increasing. This means that the flying speed F increases as the length L increases, but at a decreasing rate.

(e) If L is increased by a factor of 10, then F is increased by a factor of $10^{0.34} = 2.19$. Thus the longer bird should fly 2.19 times as fast.

7. **Metabolism:**

   (a) As the figure below indicates, the formula for the regression line of $\ln B$ against $\ln W$ is $\ln B = 0.75 \ln W + 3.69$. Since the slope of the regression line is 0.75, then $k = 0.75$. Since the vertical intercept of the regression line is 3.69, then $c = e^{3.69} = 40.04$. So the power function is $B = 40.04 W^{0.75}$.

   ```
   LinReg
   y=ax+b
   a=.7450302162
   b=3.687311995
   ```

   ■

   (b) i. By Part (a), the basic energy needed for survival (B in our notation) is proportional not to the weight W but to $W^{0.75}$. Thus to see how an animal is eating relative to its energy requirements, we should divide its intake by $W^{0.75}$, not by W.

   ii. The metabolic weight of a 2.76-pound animal is $2.76^{0.75} = 2.14$, and the metabolic weight of a 126.8-pound animal is $126.8^{0.75} = 37.79$.

   iii. We use the metabolic weights found in Part ii above. For the rabbit the ratio is $\dfrac{0.18}{2.14} = 0.08$, and for the sheep it is $\dfrac{2.8}{37.79} = 0.07$. So the daily consumption levels of the two animals are about the same on this basis.

9. **Proportions of trees:**

   (a) The plot of $\ln h$ against $\ln d$ is in the left-hand figure below.

   (b) As shown in the right-hand figure below, the regression line is $\ln h = 0.66 \ln d + 3.58$. The graph of this line is shown in the left-hand figure below.

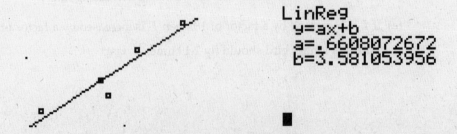

   ```
   LinReg
   y=ax+b
   a=.6608072672
   b=3.581053956
   ```

   (c) The plains cottonwood is taller for its diameter since the point in the plot corresponding to the cottonwood lies above the regression line, while that for the willow lies below the line.

(d) From the regression line, $k = 0.66$ and $c = e^{3.58} = 35.87$. Thus $h = 35.87d^{0.66}$.

(e) i. The powers of d for the trees and the critical height function are about the same, since $\frac{2}{3} = 0.67$ to two decimal places. However, the coefficient is 35.87 for the trees, while the critical height function has a coefficient of 140, which is about four times as large.

  ii. No tree from the table is taller than its critical buckling height. This can be determined by looking at a table of values for $140d^{2/3}$.

11. **Self-thinning**:

(a) As time went by, the density decreased, which means that there were fewer plants in the plot.

(b) As the figure below indicates, the regression line for $\ln w$ against $\ln p$ is given by $\ln w = -1.48 \ln p + 8.47$. From the regression line we get that $k = -1.48$ and $c = e^{8.47} = 4769.52$, so $w = 4769.52 p^{-1.48}$.

```
LinReg
 y=ax+b
 a=-1.481804242
 b=8.46927465
```

■

(c) We use the homogeneity property of power functions. The power here is -1.48, so if the density p changes by a factor of $\frac{1}{2}$ then the weight w changes by a factor of $\left(\frac{1}{2}\right)^{-1.48} = 2.79$.

(d) The total yield is $y = w \times p = 4769.52 p^{-1.48} \times p = 4769.52 p^{-0.48}$. The power is negative, so y decreases as p increases. On the other hand, as time increases, p decreases, as noted in Part (a). As p decreases, y increases; so, as time increases, the yield also increases.

13. **Cost of transport:**

(a) In general, as the weight increases, the cost of transport decreases. The only exception is the white rat.

(b) The plot of $\ln C$ against $\ln W$ is in the left-hand figure below.

(c) As shown in the right-hand figure below, the regression line is $\ln C = -0.40 \ln W + 2.15$.

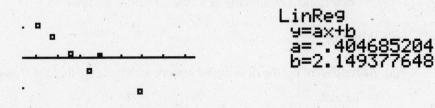

The graph of this line is shown below.

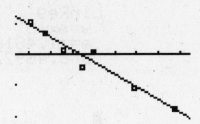

(d) We evaluate $\ln C$ when $\ln W = \ln (20{,}790) = 9.94$ using the plot in Part (c), as shown in the figure below. The result is $\ln C = -1.83$. Thus the formula predicts that the cost of transport will be $C = e^{-1.83} = 0.16$. However, the actual cost of transport is $C = 0.43$. Thus, the penguin has a higher cost of transport than would be predicted given its weight. This does confirm the stereotype of penguins as awkward waddlers.

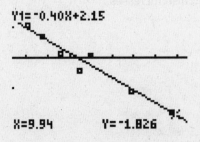

(e) From the regression line, $k = -0.40$ and $c = e^{2.15} = 8.58$, so the formula is $C = 8.58W^{-0.40}$.

15. **Exponential growth rate and generation time:**

(a) The plot of $\ln r$ against $\ln T$ is shown below. The plot is reasonably close to linear. Hence it is reasonable to model r as a power function of T.

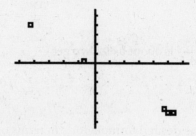

(b) As the figure below indicates, the regression line for the data plotted in Part (a) is $\ln r = -1.0 \ln T - 0.17$ if we round the slope to one decimal place.

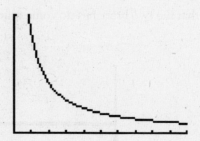

```
LinReg
 y=ax+b
 a=-.9560993251
 b=-.165381232
```

(c) From Part (b) the power is $k = -1.0$, and $c = e^{-0.17} = 0.8$, rounding to one decimal place. Thus the formula is $r = 0.8T^{-1}$. The graph is shown below using a horizontal span from 0 to 10 and a vertical span from 0 to 1.

(d) From Example 5.6, the generation time in years of an organism is modeled by $1.1L^{0.8}$, if L is the length in feet. In this exercise we measure generation time T in days, so we multiply the model by 365:

$$T = 365 \times 1.1L^{0.8} = 401.5L^{0.8}.$$

This relationship holds over a wide range of organisms. From Part (c) we know that, for our group of lower organisms, $r = 0.8T^{-1}$. Since $T^{-1} = \dfrac{1}{T}$, this can be written as $r = \dfrac{0.8}{T}$. We substitute into this formula for r the model for T in terms of L to get

$$r = \frac{0.8}{T} = \frac{0.8}{401.5L^{0.8}}.$$

Since $\dfrac{0.8}{401.5}$ is about 0.002, we get

$$r = \frac{0.002}{L^{0.8}}.$$

This is the desired formula for r in terms of L.

17. **Laboratory experiment**: Answers will vary. Please see the website for more information.

5.3 COMBINING AND DECOMPOSING FUNCTIONS

E-1. **Further investigation of the case** $c = -0.75$: In the plot we used a vertical span of -1 to 1. It appears that the two branches do come together.

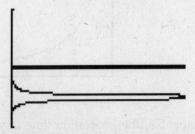

E-3. **Graphical analysis of another sequence**: Choosing $c = 0.5$ gives a sequence that decreases to 0, as the graph on the left below illustrates. (We included the first 5 iterates and used a vertical span of 0 to 0.1.) For the choice $c = 3.1$ the sequence seems to split into two sequences, as the graph on the right below illustrates. (We included the first 50 iterates and used a vertical span of 0 to 1.)

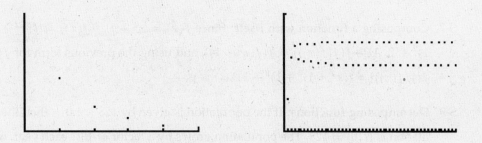

For the choice $c = 3.6$ the sequence appears to be chaotic, as the graph below illustrates. (We included the first 500 iterates and used a vertical span of 0 to 1.)

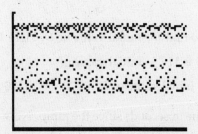

S-1. **Formulas for composed functions**: In each case we have a formula for w as a function of s and a formula for s as a function of t. To find a formula for w as a function of t, we simply replace s by its expression in terms of t.

(a) Since $s = t - 3$ and $w = s^2 + 1$, w can also be written as $w = s^2 + 1 = (t - 3)^2 + 1$. This expresses w as a function of t.

(b) Since $s = t^2 + 2$ and $w = \dfrac{s}{s + 1}$, w can also be written as $w = \dfrac{s}{s + 1} = \dfrac{t^2 + 2}{(t^2 + 2) + 1}$, which expresses w as a function of t. This expression can be simplified to $w = \dfrac{t^2 + 2}{t^2 + 3}$.

(c) Since $s = e^t - 1$ and $w = \sqrt{2s + 3}$, w can also be written as $w = \sqrt{2s + 3} = \sqrt{2(e^t - 1) + 3}$, which expresses w as a function of t. This expression can be simplified to $w = \sqrt{2e^t + 1}$ since $2(e^t - 1) + 3 = 2e^t - 2 + 3 = 2e^t + 1$.

S-3. **Limiting values**: To find the limiting value of $7 + a \times 0.6^t$, we note that $a \times 0.6^t$ is a function which represents exponential decay, so the limiting value of $a \times 0.6^t$ is 0. Thus the limiting value of $7 + a \times 0.6^t$ is 7.

S-5. **Adding functions**: Since the function f is the sum of two temperatures, its formula may be written as the sum of the formulas for the two temperatures. Thus $f = (t^2 + 3) + \dfrac{t}{t^2 + 1}$.

S-7. **Composing a function with itself**: Since $f(x) = x^2 + 1$, $f(f(x)) = f(x^2 + 1) = (x^2 + 1)^2 + 1$. Also $f(f(f(x))) = f(f(x^2 + 1))$, and using the previous form for $f(f(x))$ gives $f(f(f(x))) = ((x^2 + 1)^2 + 1)^2 + 1$.

S-9. **Decomposing functions**: If the population is given by 128×1.07^t, then the initial population is $N(0) = 128$. The population grows by a factor of 1.07 each year, which represents a growth of 7% each year.

S-11. **Combining functions**: Because $r(t) = 1 + 2t$ and $A(r) = \pi r^2$, we have $A(r(t)) = \pi(1 + 2t)^2$. Thus the formula is $A = \pi(1 + 2t)^2$.

1. **A skydiver**:

   (a) We want to find the initial value of D. Here $D = T - v$ and $T = 176$, so the initial value of D is $176 -$ Initial value of v. Now v starts at 0, so the initial value of v is 0; thus the initial value of D is $176 - 0 = 176$ feet per second.

   (b) Let t be time in seconds since the jump. Now D is an exponential function of time, so it can be written as $D = Pa^t$. Here P is the initial value of D, so from Part (a), $P = 176$. Thus $D = 176a^t$. On the other hand, $v(2) = 54.75$, so $D(2) = 176 - 54.75 = 121.25$, and thus $176a^2 = 121.25$. Solving for a yields $a = \sqrt{\dfrac{121.25}{176}} = 0.83$, so an exponential formula for D is $D = 176 \times 0.83^t$.

   (c) To find a formula for v, note that we have $D = T - v$. Thus $D + v = T$ and so $v = T - D$. On the other hand, $T = 176$ and $D = 176 \times 0.83^t$, so a formula for v is $v = 176 - 176 \times 0.83^t$.

   (d) The velocity 4 seconds into the fall is expressed as $v(4)$ in functional notation. Its value is $v(4) = 176 - 176 \times 0.83^4 = 92.47$ feet per second.

3. **Immigration**: We have a population N given by a formula

$$N = \frac{v}{r}(a^t - 1),$$

where v is a positive number, r is a number which is negative, and a is a positive number less than 1. Now a^t represents exponential decay (since a is less than 1), so a^t has a limiting value of 0. Thus the limiting value of the population N is $\frac{v}{r}(0-1) = -\frac{v}{r}$.

5. **Biomass of haddock**:

   (a) Now D is the difference between the maximum length, which is 21 inches, and the length L, so $D(t) = 21 - L(t)$ for all t. The initial value of L is 4 inches, so $L(0) = 4$, and therefore the initial value of D is $D(0) = 21 - L(0) = 21 - 4 = 17$ inches.

   (b) To find a formula for L, we start with $D = 21 - L$. This says that $D + L = 21$ and therefore $L = 21 - D$. To obtain a formula for L in terms of t we need a formula for D in terms of t.

   We know from Part (a) that D is an exponential function with initial value 17, so $D = 17a^t$ for some a. Since a 6-year-old haddock is about 15.8 inches long, when $t = 6$, then $D = 21 - L = 21 - 15.8 = 5.2$. Using the formula for D, we have $5.2 = 17a^6$. Solving for a gives $a = 0.82$. (For example, since $\frac{5.2}{17} = a^6$, $a = \left(\frac{5.2}{17}\right)^{1/6} = 0.82$). Thus $D = 17 \times 0.82^t$. Since $L = 21 - D$, then using the formula for D above, we find that $L = 21 - 17 \times 0.82^t$.

   (c) Weight W is written as function of L by the formula $W = 0.000293L^3$, and L is written as a function of t by the formula in Part (b) above, so W can be written as a function of t by function composition:

$$
\begin{aligned}
W &= 0.000293L^3 \\
W &= 0.000293(21 - 17 \times 0.82^t)^3.
\end{aligned}
$$

   (d) The total biomass B is the product of the number of fish N and the weight per fish W. Each of N and W can be expressed as functions of the age t, so B can be also:

$$
\begin{aligned}
B &= N \times W \\
B &= \left(1000e^{-0.2t}\right) \times \left(0.000293(21 - 17 \times 0.82^t)^3\right).
\end{aligned}
$$

(e) The maximum value of the biomass B can be determined by graphing B as a function t using the formula from Part (d). We are to consider ages up to 10 years, so the horizontal span should be from 0 to 10. Looking at a table of values (see the figure on the left below), we see that a vertical span of 0 to 400 should nicely display the graph. The figure on the right below shows the graph with the maximum calculated.

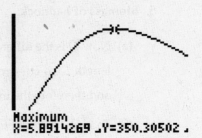

The maximum occurs at $t = 5.89$ with $B = 350.31$. So the biomass is at a maximum when the cohort is 5.89 years old, or about 5.9 years old.

7. **Waiting at a stop sign**: In this exercise $T = 5$, so the formula for the average delay D, in seconds, at a stop sign to enter a highway is given by

$$D = \frac{e^{5q} - 1 - 5q}{q},$$

where q is the flow rate, in cars per second, of traffic on the highway.

(a) If the flow rate is 500 cars per hour, then $q = \dfrac{500}{60 \times 60} = 0.14$ car per second, and so the delay is

$$D = \frac{e^{5 \times 0.14} - 1 - 5 \times 0.14}{0.14} = 2.24 \text{ seconds.}$$

(b) The service rate s, in cars per second, is a function of D given by the formula $s = D^{-1}$. To represent s as a function of the flow rate q, we substitute the formula for D in terms of q into the formula for s in terms of D:

$$
\begin{aligned}
s &= D^{-1} \\
s &= \left(\frac{e^{5q} - 1 - 5q}{q} \right)^{-1} \\
s &= \frac{q}{e^{5q} - 1 - 5q}.
\end{aligned}
$$

(c) If the service rate is 5 cars per minute, then $s = \dfrac{5}{60} = 0.083$ car per second. To solve the equation

$$
\begin{aligned}
s &= \frac{q}{e^{5q} - 1 - 5q} \\
0.083 &= \frac{q}{e^{5q} - 1 - 5q}
\end{aligned}
$$

we can use the crossing-graphs method as shown below with a horizontal span of q from 0 to 1 cars per second and a vertical span of s from 0 to 0.15 car per second.

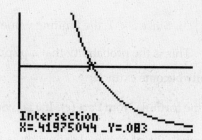

```
Intersection
X=.41975044 _Y=.083
```

This shows that when $s = 0.083$, $q = 0.42$ car per second, or 25.2 cars per minute.

9. **Probability of extinction:**

(a) We are given that the probability of extinction by time t is

$$P(t) = \frac{d(e^{rt} - 1)}{be^{rt} - d}$$

and that $d = 0.24$, $b = 0.72$, and so $r = b - d = 0.48$. Substituting these values, we have

$$P(t) = \frac{0.24(e^{0.48t} - 1)}{0.72e^{0.48t} - 0.24}.$$

Scanning a table of values, for example the table below, we see quickly that the limiting value of P is 0.33, so there is a 0.33 or 33% probability that the population will eventually become extinct.

X	Y₁
0	0
10	.3315
20	.33332
30	.33333
40	.33333
50	.33333
60	.33333

X=0

(b) We have a formula for Q expressing Q as a function of P, namely $Q = P^k$, and we already have a formula expressing P in terms of t, b, d, and r from Part (a). To find a formula for Q in terms of these latter variables and k, we substitute the expression for P from Part (a) into the formula for Q in terms of P:

$$Q = P^k$$
$$Q = \left(\frac{d(e^{rt} - 1)}{be^{rt} - d} \right)^k.$$

(c) We are told that if $b > d$, then Q can be expressed as

$$Q = \left(\frac{d(1 - a^t)}{b - da^t}\right)^k,$$

where $a < 1$. Since $a < 1$, the limiting value of a^t is zero, so the limiting value of Q is $\left(\frac{d}{b}\right)^k$. This is the probability that a population starting with k individuals will eventually become extinct.

(d) We use the formula from Part (c). If b is twice as large as d, then $\frac{d}{b} = \frac{1}{2} = 0.5$, and so the probability of eventual extinction is $\left(\frac{d}{b}\right)^k = 0.5^k$.

(e) As a function of k, the limiting value of 0.5^k is zero since $0.5 < 1$. In practical terms, this means that for a large initial population, the probability that the population will eventually become extinct is very small.

11. **Applications of the litter formula:**

(a) When $t = \frac{3}{k}$, then, replacing t by $3/k$ in the formula for A, we find that

$$A = R - Re^{-kt} = R - Re^{-k \times 3/k} = R - Re^{-3} = R - 0.05R = 0.95R,$$

so the amount of litter present A is 95% of R, the limiting value.

(b) By Part (a), we know that 95% of the limiting value is reached when $t = \frac{3}{k}$. In this case $k = 0.003$ per year, so 95% of the limiting value is reached in $t = \frac{3}{k} = \frac{3}{0.003} = 1000$ years.

(c) To reach 95% of the limiting value for forest litter in the Congo takes $t = \frac{3}{4} = 0.75$ of a year; in the Sierra Nevada Mountains it takes $t = \frac{3}{0.063} = 47.62$ years.

13. **Gray wolves in Michigan:**

(a) Since the gray wolves recolonized in the Upper Peninsula of Michigan it is reasonable to assume that this area is both suitable habitat and has far fewer wolves than it had historically. Under these conditions, we would expect exponential growth.

(b) We use regression to find an exponential model since the data is not exactly exponential. We find the model $W = 10.94 \times 1.39^t$, where t is the number of years since 1990 and W is the number of wolves.

(c) The figure below is a graph of the data, with W as a function of t, together with the exponential model from Part (b).

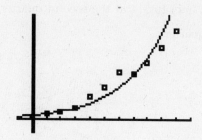

The data looks as if it follows two different exponential functions, neither of which is the exponential model we graphed. It would be better to use a piecewise-defined function.

(d) We find an exponential model for 1990 through 1996 by omitting later data. Using regression we get the exponential model $W = 7.95 \times 1.59^t$, where t is time in years since 1990. This formula holds for t from 0 through 6. We use the remaining data, from 1997 through 2000, to calculate the exponential model for W as a function of T, where T is time in years since 1997. Using regression we obtain $W = 112.22 \times 1.24^T$. This formula holds for T from 0 through 3, which is the same as t from 7 through 10. In fact, $T = t - 7$, and so the second exponential model can also be written as $W = 112.22 \times 1.24^{t-7}$.

(e) A single formula for W as a piecewise-defined function, using the formulas from Part (d), is

$$W = \begin{cases} 7.95 \times 1.59^t & \text{for} \quad 0 \le t \le 6 \\ 112.22 \times 1.24^{t-7} & \text{for} \quad 7 \le t \le 10 \end{cases}$$

Graphing these two functions as a single piecewise-defined function shows an excellent fit with the data, as in the figure below.

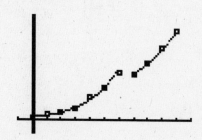

15. **Quarterly pine pulpwood prices:**

(a) The graph should start in the first quarter at $18.50, decrease steadily (so in a straight line) to $14 at the second quarter, and then rise steadily to $18 by the fourth quarter:

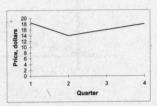

(b) Since the price P is decreasing steadily, it has a constant rate of change and so is a linear function of the quarter.

(c) From the first to second quarter, P decreases by $4.50 over one quarter from a first quarter value of $18. Thus the slope of P is -4.50. Now $P = 18.50$ when $t = 1$, so the value when $t = 0$ would be $18.50 - (-4.5) = 23$, and thus the formula is $P = -4.50t + 23$. This formula is only valid for t from 1 to 2.

(d) Since the price P is increasing steadily, it has a constant rate of change and so is a linear function of t.

(e) From the second to fourth quarter, P increases by $4 over two quarters from $14. Thus the slope of P is $\dfrac{4}{2} = 2$. Now $P = 14$ when $t = 2$, so the value when $t = 0$ would be $14 - 2 \times 2 = 10$, and so the formula is $P = 2t + 10$. This formula is only valid for t from 2 to 4.

(f) Using the formulas from Parts (c) and (e), we can write a single formula for P as a piecewise-defined function of t as follows:

$$P = \begin{cases} -4.5t + 23 & \text{for} \quad 1 \le t \le 2 \\ 2t + 10 & \text{for} \quad 2 \le t \le 4 \end{cases}$$

5.4 QUADRATIC FUNCTIONS AND PARABOLAS

E-1. **Solving equations by completing the square:**

(a) To solve $x^2 + 6x = 8$ by completing the square, we add $\dfrac{b^2}{4}$ to each side, where $b = 6$; thus we need to add $\dfrac{6^2}{4} = 9$ to each side: $x^2 + 6x + 9 = 8 + 9$. The left side is $(x+3)^2$, so we now have $(x+3)^2 = 17$. Thus $x + 3 = \pm\sqrt{17}$, and so $x = -3 \pm \sqrt{17}$ are the solutions.

(b) To solve $x^2 - 2x - 5 = 0$ by completing the square, we add 5 to each side and then add $\dfrac{b^2}{4}$ where $b = -2$, that is, we add $\dfrac{(-2)^2}{4} = 1$. Thus we add $5 + 1 = 6$ to each side, which gives $x^2 - 2x + 1 = 6$. The left side is $(x-1)^2$, so $(x-1)^2 = 6$. Thus $x - 1 = \pm\sqrt{6}$, and so $x = 1 \pm \sqrt{6}$ are the solutions.

(c) To solve $2x^2 = x + 4$ by completing the square, we subtract x from each side of the equation, in order to collect the x-terms, and then we divide both sides by 2, since that is the coefficient of x^2:

$$
\begin{aligned}
2x^2 &= x + 4 \\
2x^2 - x &= 4 \\
x^2 - \frac{1}{2}x &= 2.
\end{aligned}
$$

Now we proceed, as in Parts (a) and (b), to add to each side $\dfrac{b^2}{4}$ where $b = -\dfrac{1}{2}$, that is, we add $\dfrac{(-1/2)^2}{4} = \dfrac{1}{16}$. This yields $x^2 - \dfrac{1}{2}x + \dfrac{1}{16} = 2 + \dfrac{1}{16}$. Now the left side is $\left(x - \dfrac{1}{4}\right)^2$ and the right is $\dfrac{33}{16}$, so $\left(x - \dfrac{1}{4}\right)^2 = \dfrac{33}{16}$. Thus $x - \dfrac{1}{4} = \pm\sqrt{\dfrac{33}{16}}$, and so $x = \dfrac{1}{4} \pm \sqrt{\dfrac{33}{16}}$ are the solutions. Thus can be simplified to $x = \dfrac{1}{4} \pm \dfrac{\sqrt{33}}{4}$ since $\sqrt{16} = 4$.

(d) To solve this equation, we need to first simplify it by expanding the terms and collecting the x^2 and x terms together:

$$
\begin{aligned}
(x+1)^2 + (x+2)^2 &= (x+4)^2 \\
(x^2 + 2x + 1) + (x^2 + 4x + 4) &= x^2 + 8x + 16 \\
2x^2 + 6x + 5 &= x^2 + 8x + 16 \\
x^2 - 2x &= 11.
\end{aligned}
$$

As in Part (b), we add 1 to each side to complete the square: $x^2 - 2x + 1 = 12$. Since $x^2 - 2x + 1 = (x-1)^2$, we have $(x-1)^2 = 12$. Thus $x - 1 = \pm\sqrt{12}$, and therefore $x = 1 \pm \sqrt{12}$ are the solutions.

E-3. Complex solutions: The quadratic formula says that the solutions to $ax^2 + bx + c = 0$ are

$$x = \frac{-b \pm \sqrt{b^2 - 4ac}}{2a}.$$

(a) To solve $x^2 + 2x + 3 = 0$ we use $a = 1$, $b = 2$, and $c = 3$, so the solutions are

$$
\begin{aligned}
x &= \frac{-2 \pm \sqrt{2^2 - 4 \times 1 \times 3}}{2 \times 1} \\
&= \frac{-2 \pm \sqrt{-8}}{2} = \frac{-2 \pm 2\sqrt{2}i}{2} = -1 \pm \sqrt{2}i
\end{aligned}
$$

since $\sqrt{-8} = \sqrt{2^2 \times 2 \times (-1)} = 2\sqrt{2}\sqrt{-1} = \pm 2\sqrt{2}i$.

(b) To solve $x^2 = x - 8$, we subtract $x - 8$ in order to rewrite the equation in the form $ax^2 + bx + c = 0$. Subtracting yields $x^2 - x + 8 = 0$, so we use $a = 1$, $b = -1$, and $c = 8$. Thus the solutions are

$$
\begin{aligned}
x &= \frac{-(-1) \pm \sqrt{(-1)^2 - 4 \times 1 \times 8}}{2 \times 1} \\
&= \frac{1 \pm \sqrt{-31}}{2} = \frac{1 \pm \sqrt{31}i}{2}
\end{aligned}
$$

since $\sqrt{-31} = \pm\sqrt{31}i$.

(c) To solve $2x^2 + 3x + 4 = 0$ we use $a = 2$, $b = 3$, and $c = 4$, so the solutions are

$$
\begin{aligned}
x &= \frac{-3 \pm \sqrt{3^2 - 4 \times 2 \times 4}}{2 \times 2} \\
&= \frac{-3 \pm \sqrt{-23}}{4} = \frac{-3 \pm \sqrt{23}i}{4}
\end{aligned}
$$

since $\sqrt{-23} = \pm\sqrt{23}i$.

E-5. High powers of i: To calculate i^{233}, we start with low powers and look for a pattern. Now $i^2 = -1$, $i^3 = i^2 i = -i$, and $i^4 = (i^2)^2 = (-1)^2 = 1$, so $i^5 = i^4 i = i$ and so on, that is, the powers repeat in cycles of 4 as $i, -1, -1, 1, \ldots$. Now $233 = 58 \times 4 + 1$, so $i^{233} = i^{58 \times 4 + 1} = i^{58 \times 4} i^1 = (i^4)^{58} i = i$, or observe that $58 \times 4 + 1$ takes you to the beginning of the cycle, so that power of i is i.

E-7. The square roots of i: To show that $\pm\left(\dfrac{1}{\sqrt{2}} + \dfrac{1}{\sqrt{2}}i\right)$ are the square roots of i, it suffices to show that the square of that expression equals i:

$$
\begin{aligned}
\left(\pm\left(\frac{1}{\sqrt{2}} + \frac{1}{\sqrt{2}}i\right)\right)^2 &= \left(\frac{1}{\sqrt{2}}\right)^2 + 2 \times \frac{1}{\sqrt{2}} \times \left(\frac{1}{\sqrt{2}}i\right) + \left(\frac{1}{\sqrt{2}}i\right)^2 \\
&= \frac{1}{2} + 2\frac{1}{2}i + \frac{1}{2}i^2 = \frac{1}{2} + i + \frac{1}{2}(-1) = i.
\end{aligned}
$$

S-1. **The rate of change**: The rate of change of a quadratic function is a linear function.

S-3. **Testing for quadratic data**: To see if the data in the table are quadratic, we calculate the second-order differences:

x	1	2	3	4	5
y	1	5	9	24	37
First-order difference	4	4	15	13	
Second-order difference	0	11	−2		

Since the second-order differences are not equal, the data are not quadratic.

S-5. **Quadratic formula**: The quadratic formula gives the solutions of $ax^2 + bx + c = 0$ as

$$x = \frac{-b \pm \sqrt{b^2 - 4ac}}{2a}.$$

To solve $-2x^2 + 2x + 5 = 0$, we use $a = -2$, $b = 2$, and $c = 5$, so

$$x = \frac{-2 \pm \sqrt{2^2 - 4 \times -2 \times 5}}{2 \times (-2)} = \frac{-2 \pm \sqrt{44}}{-4}$$

are the solutions. This expression can be simplified to $x = \dfrac{1 \pm \sqrt{11}}{2} = \dfrac{1}{2} \pm \dfrac{\sqrt{11}}{2}$ and gives the values $x = -1.16$ and $x = 2.16$.

S-7. **Single-graph method**: To solve the quadratic equation $3x^2 - 5x + 1 = 0$ using the single-graph method, we graph the function with a suitable window. Using a window with a horizontal span of -2 to 2 and a vertical span of -5 to 5, we can find the roots. One root is shown in the figure on the left below, while the other is shown on the right:

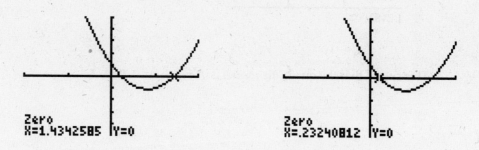

These graphs show that the solutions are $x = 1.43$ and $x = 0.23$.

S-9. **Quadratic regression**: To find a quadratic model for this data set using regression, we enter the data in the first two columns, as shown in the figure below on the left, and use the calculator to find directly the quadratic regression coefficients, as shown in the figure on the right below.

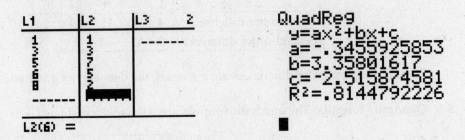

This shows that the quadratic model is $f(x) = -0.35x^2 + 3.36x - 2.52$.

S-11. **Quadratic regression**: To find a quadratic model for this data set using regression, we enter the data in the first two columns, as shown in the figure below on the left, and use the calculator to find directly the quadratic regression coefficients, as shown in the figure on the right below.

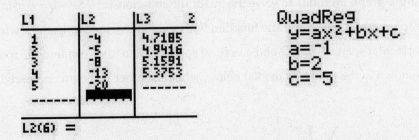

This shows that the quadratic model is $f(x) = -x^2 + 2x - 5$.

1. **Sales growth**:

   (a) To make a graph of G versus s we use a horizontal span of 0 to 4 since the model is valid up to a sales level of 4 thousand dollars. Using a table of values, we determine a vertical span of 0 to 1.3, and show the graph below.

   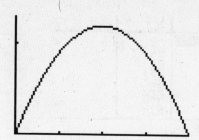

   (b) When the sales level is \$2260, $s = 2.26$ since s is the sales level in thousands. The rate of growth in represented in functional notation by $G(2.26)$. Using the formula to evaluate this, we find that the growth is $G(2.26) = 1.2 \times 2.26 - 0.3(2.26)^2 = 1.18$ thousand dollars per year. This can also be found from a table of values or the graph.

   (c) The rate of growth G is maximized at the maximum point of the graph. This is shown on the graph below:

   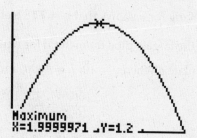

   The maximum growth rate is achieved when $s = 2.00$, that is, at a sales level of \$2000.

3. **Cox's formula:**

(a) To graph the quadratic function $4v^2 + 5v - 2$ using a horizontal span from 0 to 10, a table of values suggests a vertical span from 0 to 400. The table of values is shown below on the left while the graph is shown on the right.

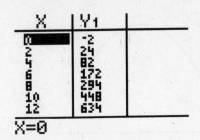
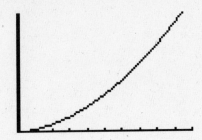

(b) If the velocity v is 0.25 foot per second, then the value of $4v^2 + 5v - 2$ is -0.5. This would mean that the right side of Cox's formula is negative, whereas the left side of Cox's formula is $\dfrac{Hd}{L}$ and H, d, and L are all positive numbers. Thus it is not possible for v to be 0.25 foot per second.

(c) If $d = 4$, $L = 1000$, and $H = 50$, then Cox's formula implies that

$$\frac{50 \times 4}{1000} = \frac{4v^2 + 5v - 2}{1200}.$$

Evaluating the left-hand side and multiplying by 1200, we see that $4v^2 + 5v - 2 = 240$. Solving for v shows that $v = 7.18$ feet per second.

(d) To find the largest pipe diameter d for which v does not exceed 10 feet per second, we calculate d when $v = 10$, $L = 1000$, and $H = 50$:

$$\frac{50d}{1000} = \frac{4 \times 10^2 + 5 \times 10 - 2}{1200}.$$

Evaluating, we see that $d = 7.47$ inches is the largest such pipe diameter.

5. **Surveying vertical curves:**

(a) Since the rate of change of grade is constant, a quadratic model is appropriate.

(b) In this case, $g_1 = 1.35$, $g_2 = -1.75$, $L = 5$, and $E = 1040.63$, so the formula for the elevations of the vertical curve is

$$
\begin{aligned}
y &= \frac{-1.75 - 1.35}{2 \times 5} x^2 + 1.35x + 1040.63 - \frac{1.35 \times 5}{2} \\
&\doteq -0.31x^2 + 1.35x + 1037.255.
\end{aligned}
$$

(c) Using the formula from Part (b), the stations correspond to $x = 0$ through $x = 5$. The six stations are given in the table below:

Station number	0	1	2	3	4	5
Elevation	1037.26	1038.30	1038.72	1038.52	1037.70	1036.26

Here the elevations are in feet.

(d) The highest point of the road on the vertical curve is at the maximum point on the graph of y. From the graph (see the figure below – we used a window with a horizontal span from 0 to 5 and a vertical span from 1035 to 1040) we find the maximum at station $x = 2.1774$, that is 217.74 feet along the vertical curve with an elevation of $E = 1038.72$ feet.

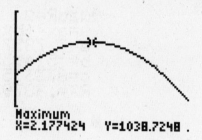

Maximum
X=2.177424   Y=1038.7248 .

7. **Data that is not quadratic**: We compute the first-order and second-order differences:

x	0	1	2	3	4
P	5	8	17	38	77
First-order difference	3	9	21	39	
Second-order difference	6	12	18		

Since the second-order differences are not constant, the data cannot be modeled by a quadratic function.

9. **Linear and quadratic data**: We compute the first-order and second-order differences in Table A:

x	0	1	2	3	4
f	10	17	26	37	50
First-order difference	7	9	11	13	
Second-order difference	2	2	2		

Since the second-order differences are constant, the data can be modeled by a quadratic function. To find the formula for the quadratic function, we use the standard form $f = ax^2 + bx + c$ and the relations

$$a = \frac{1}{2} \times \text{second-order difference}$$

$$b = \text{initial first-order difference} - a$$

$$c = \text{initial value.}$$

Thus $a = \frac{1}{2} \times 2 = 1$, $b = 7 - a = 7 - 1 = 6$, and $c = 10$, so $f = x^2 + 6x + 10$ is the formula.

Successive differences in Table B are always 7, so Table B represents a linear function with slope 7 (since x increases by single units). The initial value is 10, so $g = 7x + 10$.

11. **Traffic accidents:**

(a) Using quadratic regression as shown in the figure below, we see that the quadratic model is $R = 7.79s^2 - 514.36s + 8733.57$.

```
QuadReg
y=ax²+bx+c
a=7.785714286
b=-514.3571429
c=8733.571429
R²=.9879330701
∎
```

(b) Using the model from Part (a), we find that $R(50) = 7.79 \times 50^2 - 514.36 \times 50 + 8733.57 = 2490.57$, which means that commercial vehicles driving at night on urban streets at 50 miles per hour have traffic accidents at a rate of about 2491 per 100,000,000 vehicle-miles.

(c) The speed at which traffic accidents are minimized corresponds to the minimum point on the graph. This is calculated in the figure below:

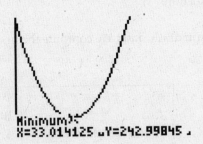

```
Minimum
X=33.014125  Y=242.99845
```

This shows that the minimum is when $s = 33.01$ or about 33 miles per hour.

13. **Women employed outside the home:**

(a) Entering the data, and using t as the number of years since 1942, we see the data shown in the figure below on the left and the calculation of the quadratic regression coefficients shown on the right.

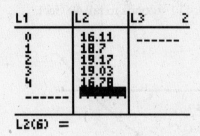

```
QuadReg
y=ax²+bx+c
a=-.735
b=3.107
c=16.154
R²=.9837621119
```

Using N as the number of women employed outside the home, in millions, the quadratic model is $N = -0.735t^2 + 3.107t + 16.154$.

(b) In functional notation, the number of women working outside the home in 1947 is $N(5)$. The quadratic model from Part (a) gives that value as $N(5) = 13.31$ million.

(c) The table below summarizes the quadratic prediction versus the actual numbers for 1947 and 1948:

Year	t	Quadratic prediction	Actual value
1947	5	13.31	16.90
1948	6	8.34	17.58

Here the figures are in millions. Clearly the predictions of the quadratic model are far from accurate, so a quadratic model is probably not appropriate for this time period. Another way to see this is to observe that the original data table shows a maximum in 1944, and the additional information in Part (c) shows that after 1946 the number started to increase. A quadratic model can't have both a maximum and a minimum.

15. **A falling rock:**

(a) We want to find t so that $16t^2 + 3t = 400$. To solve this using the quadratic formula, we first subtract 400 from each side to get $16t^2 + 3t - 400 = 0$. This is of the form $at^2 + bt + c = 0$ with $a = 16$, $b = 3$, and $c = -400$. Plugging these values into the quadratic formula, we get:

$$t = \frac{-b + \sqrt{b^2 - 4ac}}{2a} = \frac{-3 + \sqrt{3^2 - 4 \times 16 \times (-400)}}{2 \times 16} = 4.91$$

$$t = \frac{-b - \sqrt{b^2 - 4ac}}{2a} = \frac{-3 - \sqrt{3^2 - 4 \times 16 \times (-400)}}{2 \times 16} = -5.09.$$

The second of these times has no physical meaning, but from the first we see that it takes the rock 4.91 seconds to fall 400 feet.

(b) We use the crossing graphs method to find t so that $D(t) = 400$. In the figure below we have graphed $D = 16t^2 + 3t$ and the constant 400 using a horizontal span of 0 to 6 and a vertical span of 0 to 600. We see that the intersection occurs at $t = 4.91$. Thus, it takes the rock 4.91 seconds to fall 400 feet.

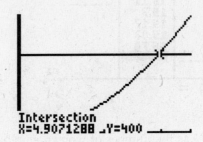

```
Intersection
X=4.9071288  Y=400
```

17. **Profit:**

(a) Total cost C is simply the fixed cost of $2000 plus the variable cost of $30 per widget times the number of widgets. Because N is the number of widgets, $C = 30N + 2000$.

(b) The table of price p as a function of number of widgets N shows a constant decrease in price of 0.50 per 50 additional widgets sold, so p should be a linear function of N. The slope is $\dfrac{-0.50}{50} = -0.01$ dollar per widget, so $p = -0.01N + b$ for some initial value b. Now when $N = 200$, then $p = 41$, according to the table, so $41 = -0.01 \times 200 + b$, and thus $b = 43$. Thus the formula is $p = -0.01N + 43$.

(c) The total revenue R is Price $\times$ Quantity sold, so $R = p \times N = (-0.01N + 43)N$ using the formula for p from Part (b).

(d) The profit P is Revenue $-$ Total cost, so $P = R - C = (-0.01N + 43)N - (30N + 2000)$. This is a quadratic function since, if multiplied out, there would be a term $-0.01N^2$.

(e) The manufacturer breaks even when $P = 0$. Solving the equation $(-0.01N + 43)N - (30N + 2000) = 0$, we find two solutions: N is 178.30 or 1121.70 widgets.

5.5 HIGHER-DEGREE POLYNOMIALS AND RATIONAL FUNCTIONS

E-1. **Finding the degree of a polynomial**: Since the polynomial has no complex zeros and no repeated zeros, the degree is exactly the number of real zeros, that is, the number of times the graph crosses the horizontal axis. Thus the degree of this polynomial is 5.

E-3. **Finding the polynomial**: The polynomial has zeros 2, 3, and 4, so by the factor theorem, the polynomial is divisible by $(x-2)(x-3)(x-4)$, which also has degree 3 and leading coefficient 1. Thus the polynomial must be precisely $(x-2)(x-3)(x-4)$. This can be written out as $x^3 - 9x^2 + 26x - 24$.

E-5. **Getting a polynomial from points**: In the three cases below we are looking for a quadratic $ax^2 + bx + c$ to pass through the three given points. Substituting the values for x into the quadratic gives three linear equations in terms of a, b and c which we solve by elimination (see pages 244-245 of the text for details about solving by elimination in general).

(a) For $y = ax^2 + bx + c$ to pass through $(1,5)$, $(-1,3)$, and $(2,9)$, the following equations need to hold:

$$a + b + c = 5$$
$$a - b + c = 3$$
$$4a + 2b + c = 9$$

Subtracting the first equation from the second, and subtracting 4 times the first equation from the third, yield:

$$a + b + c = 5$$
$$-2b = -2$$
$$-2b - 3c = -11$$

The second equation shows that $b = 1$. Substituting into the third equation, we find $-2 - 3c = -11$ and so $c = 3$. Substituting into the first equation, we get $a + 1 + 3 = 5$ and so $a = 1$. Thus the desired quadratic is $x^2 + x + 3$.

(b) For $y = ax^2 + bx + c$ to pass through $(1,5)$, $(2,4)$, and $(3,19)$, the following equations need to hold:

$$a + b + c = 5$$
$$4a + 2b + c = 4$$
$$9a + 3b + c = 19$$

Subtracting 4 times the first equation from the second, and subtracting 9 times the first equation from the third, yield:

$$a + b + c = 5$$
$$-2b - 3c = -16$$
$$-6b - 8c = -26$$

Subtracting 3 times the second equation from the third yields:

$$
\begin{aligned}
a + b + c &= 5 \\
-2b - 3c &= -16 \\
c &= 22
\end{aligned}
$$

Thus $c = 22$. Substituting into the second equation, we find $-2b - 3 \times 22 = -16$ and so $b = -25$. Substituting into the first equation, we get $a - 25 + 22 = 5$ and so $a = 8$. Thus the desired quadratic is $8x^2 - 25x + 22$.

(c) For $y = ax^2 + bx + c$ to pass through $(1, -2)$, $(0, -5)$, and $(-1, -2)$, the following equations need to hold:

$$
\begin{aligned}
a + b + c &= -2 \\
c &= -5 \\
a - b + c &= -2
\end{aligned}
$$

Thus $c = -5$. Substituting into the first and third equations yields:

$$
\begin{aligned}
a + b &= 3 \\
a - b &= 3
\end{aligned}
$$

Adding the equations, we find $2a = 6$, so $a = 3$. Substituting in either equation shows that $b = 0$. Thus the desired quadratic is $3x^2 - 5$.

S-1. **Recognizing polynomials**: A polynomial is a function which can be written as a sum of power functions where each power is a non-negative whole number.

(a) $x^8 - 17x + 1$ is a polynomial of degree 8.

(b) $\sqrt{x} + 8$ is not a polynomial since $\sqrt{x} = x^{0.5}$ has a fractional power.

(c) $9.7x - 53.1x^4$ is a polynomial of degree 4.

(d) $x^{3.2} - x^{2.3}$ is not a polynomial since the powers are not integers.

S-3. **Rational functions**: A rational function is a function which can be written as a ratio of polynomial functions.

S-5. **Cubic regression**: To use cubic regression, we enter the data in the first two columns (see the figure on the left below) and then use the calculator to determine the cubic regression coefficients (see the figure on the right below).

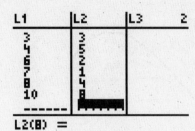

The cubic regression model is $y = 0.07x^3 - 1.11x^2 + 4.92x - 2.95$.

S-7. **Quartic regression**: Entering the data as in Exercise S-5, we use the calculator to determine the quartic regression coefficients (see the figure below)

```
QuarticReg
 y=ax4+bx3+…+e
 a=-.0056965351
 b=.1971500914
 c=-2.000770522
 d=7.259095035
↓e=-4.577797582
■
```

The quartic regression model is $y = -0.01x^4 + 0.20x^3 - 2.00x^2 + 7.26x - 4.58$.

S-9. **Finding poles**: The poles of $\dfrac{x}{x^2 - 3x + 2}$ are the values of x for which the denominator $x^2 - 3x + 2$ is zero but the numerator x is not zero. In this case, $x^2 - 3x + 2 = 0$ when $x = 2$ and $x = 1$, so these are the poles.

S-11. **Finding poles**: The poles of $\dfrac{3x - 1}{x^2 + 2x}$ are the values of x for which the denominator $x^2 + 2x$ is zero but the numerator $3x - 1$ is not zero. In this case, $x^2 + 2x = 0$ when $x = 0$ and $x = -2$, so these are the poles.

1. **Production:**

   (a) To include values of n up to 1.5 thousand units, we use a horizontal span from 0 to 1.5. A corresponding vertical span from 0 to 2.2 yields the following graph

   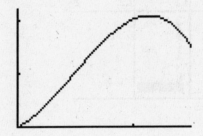

   (b) In functional notation, the amount produced if the input is 1.45 thousand units is $T(1.45)$. Using the formula, graph, or a table of values, we see that $T(1.45) = 1.66$ thousand units of product.

   (c) Tracing on the graph, we estimate the point of inflection at $n = 0.5$ and $T = 1$. This is shown in the figure below on the left.

   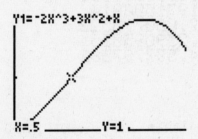   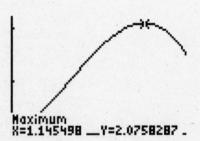

   (d) The maximum point on the point is when $n = 1.15$ and $T = 2.08$, as shown in the figure above on the right. Thus the maximum amount produced is $T = 2.08$ thousand units.

3. **Traffic accidents:**

   (a) Calculating the cubic regression coefficients, we find a cubic model for the data: $C = 0.000422s^3 - 0.0294s^2 + 0.562s - 1.65$. Here we rounded to two decimal places after the first non-zero digit, as specified in the problem.

   (b) Using the cubic model from Part (a), we can calculate $C(42)$ as 1.36. This means that if commercial vehicles travel at 42 miles per hour at night on urban streets, the cost due to traffic accidents will be 1.36 cents per vehicle-mile.

(c) Graphing the cubic model, we find the minimum from the graph. Using a horizontal span of s from 20 to 50 miles per hour and a corresponding vertical span of C from -2 to 4, the graph is:

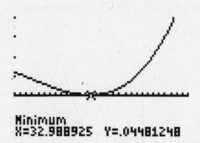

Minimum
X=32.988925 Y=.04481248

This shows the minimum to be when $s = 33$ miles per hour.

5. **Population genetics**:

(a) The quadratic equation $\lambda^2 = \frac{1}{2}\lambda + \frac{1}{4}$ can be solved by any of a number of techniques. For example, using crossing graphs with a horizontal span of λ from -1 to 1 and a vertical span from -1 to 1, we find the two solutions (see the figures below).

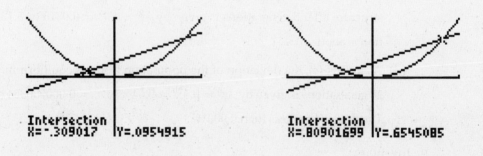

Intersection
X=-.309017 Y=.0954915

Intersection
X=.80901699 Y=.6545085

Here we see that the solutions are $\lambda = -0.31$ and $\lambda = 0.81$. Since λ_1 is the larger solution, $\lambda_1 = 0.81$.

(b) According to the problem, the proportion of the population which will be heterozygous (so not homozygous) after 5 generations is given by $\lambda_1^5 = 0.81^5 = 0.3487$ or 34.87%. Thus the homozygous proportion will be $100\% - 34.87\% = 65.13\%$, or about 65.1%.

(c) As in Part (b), the proportion of the population which will be heterozygous (so not homozygous) after 20 generations is given by $\lambda_1^{20} = 0.81^{20} = 0.0148$ or 1.48%. Thus the homozygous proportion will be $100\% - 1.48\% = 98.52\%$, or about 98.5%.

7. **Population genetics — second cousins:**

   (a) The quartic equation $\lambda^4 = \frac{1}{8}\lambda^2 + \frac{1}{32}\lambda + \frac{1}{64}$ can be solved by any of a number of techniques. For example, using crossing graphs with a horizontal span of λ from -0.75 to 0.75 and a vertical span from $-0 - .10$ to 0.20, we find two solutions (see the figures below).

   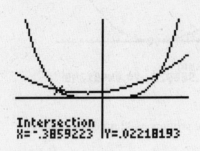   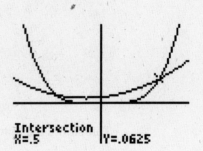

   Here we see that the solutions are $\lambda = -0.39$ and $\lambda = 0.50$. Since λ_1 is the larger solution, $\lambda_1 = 0.50$.

   (b) In general we expect four solutions to a quartic equation, although we know that there are at most four solutions.

   (c) According to the problem, the deviation of the population from its equilibrium structure after 5 generations is given by $\lambda_1^5 = 0.5^5 = 0.0313$ or 3.13%. The deviation is about 3.1%.

   (d) As in Part (c), the deviation of the population from its equilibrium structure after 20 generations is given by $\lambda_1^{20} = 0.5^{20} = 9.54 \times 10^{-7} = 0.000000954$ or 0.0000954%. The deviation is less than 0.0001%.

9. **Inventory:**

   (a) We use a horizontal span of 0 to 10 and a vertical span of 0 to 8500. The graph is shown below.

   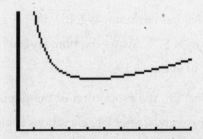

   (b) From the graph we see that, near the pole, E gets larger and larger as Q gets smaller and smaller. This says that the yearly inventory expense gets larger and larger as the order size gets smaller and smaller. This makes sense: if the order sizes are very small then the dealer will have to place many orders, and so the total cost will be very high.

11. **Traffic signals**:

(a) The graph of

$$n = 1 + \frac{v}{30} + \frac{90}{v}$$

is shown below on the left, using a horizontal span of v from 30 to 80 and a vertical span of n from 4.4 to 5.1.

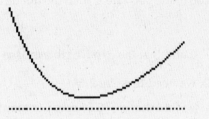   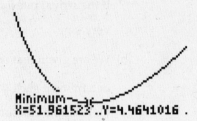

(b) In functional notation, the length of the yellow light when the approach speed is 45 feet per second is $n(45)$. The value of $n(45)$ is obtained using the formula given, a table of values, or the graph; we find that $n(45) = 4.5$ seconds.

(c) At v tends towards 0, n increases sharply. In practical terms, as the approach speed becomes very slow, the time needed to cross the intersection increases greatly. Therefore the yellow light needs to be much longer.

(d) Finding the minimum point on the graph, as shown in the figure above on the right, we see that the minimum occurs when $v = 51.96$ feet per second and $n = 4.46$ seconds. The minimum length of time for a yellow light is 4.46 seconds.

13. **Gliding pigeon**:

(a) The graph of $s = \dfrac{u^3}{2500} + \dfrac{25}{u}$ is shown below on the left using a horizontal span of u from 0 to 15 and a vertical span of s from 0 to 9.

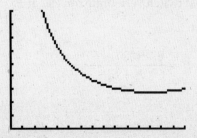   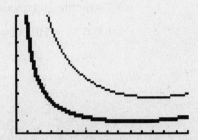

(b) Using the formula, a table of values, or a graph, we find that $s(10) = 2.9$ meters per second. In practical terms, a pigeon gliding with an airspeed of 10 meters per second will sink at a rate of 2.9 meters per second.

(c) The graph above on the right shows the performance diagrams for the falcon and the pigeon (the darker graph is that of the falcon). Since the falcon's graph lies below that of the pigeon, for the same airspeed, the falcon sinks much more slowly, so the falcon is the more efficient glider.

15. **Queues**:

(a) The rational function $L = \dfrac{a^2}{s(s-a)}$ has poles at $s = 0$ and $s - a$, that is $s = a$. Thus if the arrival rate a is very close to the gap rate s, then the queue length L becomes very long.

(b) We are given that $w = \dfrac{a}{s(s-a)}$ and that the gap length s is 3, so $w = \dfrac{a}{3(3-a)}$. To determine a such that $w = 2$, we need to solve $2 = \dfrac{a}{3(3-a)}$. Using any method, for example crossing graphs, we find that $a = 2.57$ cars per minute.

17. **Change in London travel time**: Using the values $v_0 = 28$, $a = 0.008$, and $d = 2$ for central London, the formula for the additional travel time if one additional vehicle is added to flow is

$$t' = \frac{60 \times 0.008 \times 2}{(28 - 0.008q)^2}.$$

(a) If $q = 1000$ then $t' = 0.0024$, so the addition of one vehicle increases travel time (for all vehicles) by 0.0024 minute, or about 0.14 second.

(b) If $q = 3400$ then $t' = 1.5$, so the addition of one vehicle increases travel time (for all vehicles) by 1.5 minutes.

(c) When traffic is light, additional cars have little effect on travel time, but when traffic flow is closer to 3500 vehicles per hour, additional traffic flow greatly increases travel time.

19. **Hill's law**:

(a) Using the given values for a fast twitch vertebrate muscle of $F_\ell = 300$, $a = 81$, and $b = 6.75$ in Hill's law, we obtain

$$S = \frac{6.75(300 - F)}{F + 81}.$$

(b) Graphing S using a horizontal span of F from 0 to 300 and a vertical of S from 0 to 25, we obtain the figure below.

(c) As the force F increases to 300, the muscle's contraction speed S decreases towards 0.

(d) Here $S = 0$ when $F = 300$, so the muscle does not contract at all when the maximum force is applied to the muscle.

(e) The horizontal asymptote of S is the limiting value. Using a table or graph, or formally calculating a limit, we see that the limiting value of S is -6.75. Thus S has a horizontal asymptote at $S = -6.75$. This makes no sense in terms of the muscle: as F gets large, it becomes greater than the maximum force; moreover, a negative value of S would not represent a muscle contraction.

(f) Here S has a vertical asymptote at $F = -81$. A negative force is not the situation for which the equation applies.

Chapter 5 Review Exercises

1. **Homogeneity**: In this case f is a power function with power $k = 3.2$. By the homogeneity property, if x is increased by a factor of t, then f is increased by a factor of $t^{3.2}$. In this case, x is doubled, so $t = 2$. Therefore f is increased by a factor of $2^{3.2}$, that is, by a factor of 9.19.

2. **Power**: From 1 to 10, 1 is increased by a factor of $t = 10$. Thus f is increased by a factor of $t^k = 10^k$. Since $f(10)$ is twice the size of $f(1)$,

$$2 = t^k = 10^k.$$

Solving for k yields $k = 0.30$.

3. **Flow rate**: The radius R is a power function with power 0.25, so if the flow rate F changes by a factor of t, then R changes by a factor of $t^{0.25}$.

   (a) If the flow rate is doubled, then $t = 2$, so the radius changes by a factor of $2^{0.25} = 1.19$.

   (b) If the radius is tripled, then $3 = t^{0.25}$. Solving for t yields $t = 81$, so the flow rate is multiplied by 81.

4. **Falling object**:

   (a) For Earth the formula is $T = 0.25s^{0.5}$. We put in $s = 16$ and find $T = 0.25 \times 16^{0.5} = 1$. Thus it takes 1 second for an object to fall 16 feet on Earth.

   (b) We know that $T = 5$ when $s = 20$, so $5 = c \times 20^{0.5}$. Thus $c = \dfrac{5}{20^{0.5}} = 1.12$.

   (c) The time T is a power function with power 0.5. By the homogeneity property, if s is increased by a factor of t, then T is increased by a factor of $t^{0.5}$. In this case, s is multiplied by 4 in going from 10 feet to 40 feet, so $t = 4$. Therefore T is increased by a factor of $4^{0.5}$, that is, by a factor of 2. Hence it takes an object $2 \times 1 = 2$ seconds to fall 40 feet. This can also be found by first solving for c using the method of Part (b) and then plugging $s = 40$ into the resulting formula.

5. **Power formula**: The slope of $\ln f$ as a linear function of $\ln x$ is the same as the power, so the value of k is 5.

6. **Modeling almost power data**: To find the power model, we convert the data into linear data using the logarithm. To do this, we put the original data in the 3rd and 4th columns, the logarithms in the 1st and 2nd columns, and then calculate the linear relation using linear regression. The regression calculation shows that $\ln f = -1.20 \ln x + 1.252$. Thus the power model uses $k = -1.2$ and $c = e^{1.252} = 3.50$, and so $f = 3.5x^{-1.2}$.

7. **Gas cost**:

   (a) To find the power model, we convert the data into linear data using the logarithm. To do this, we put the original data in the 3rd and 4th columns, the logarithms in the 1st and 2nd columns, and then calculate the linear relation using linear regression. The regression calculation shows that $\ln D = -1.00 \ln g + 3.092$. Thus the power model uses $k = -1$ and $c = e^{3.092} = 22.02$, and so $D = 22.02g^{-1}$.

   (b) The distance D is a power function with power -1. Now an increase of 50% in the price of gas corresponds to multiplying g by 1.5 or $3/2$. By the homogeneity property, if g is multiplied by 1.5, then D is multiplied by $1.5^{-1} = (3/2)^{-1} = 2/3$. Thus the distance is multiplied by 2/3, or reduced by 33.33%.

(c) If a gallon of gas costs \$1 then the distance can you drive on \$1 worth of gas is the distance you can drive on 1 gallon of gas, and that is your gas mileage. Thus we want to find the value of D when $g = 1$. Now if we put $g = 1$ into the formula from Part (a) we find $D = 22.02 \times 1^{-1} = 22.02$. Thus your gas mileage is 22.02, or about 22, miles per gallon. Another way to do this is to note that the gas mileage is the product of the cost g with the distance D (check the units!), and by the formula $D = 22.02g^{-1}$ from Part (a) we have $g \times D = g \times 22.02g^{-1} = 22.02$.

8. **Falling rocks:**

(a) To find the power model, we convert the data into linear data using the logarithm. To do this, we put the original data in the 3rd and 4th columns, the logarithms in the 1st and 2nd columns, and then calculate the linear relation using linear regression. The regression calculation shows that $\ln T = 0.50 \ln s - 1.616$. Thus the power model uses $k = 0.5$ and $c = e^{-1.616} = 0.20$, and so $T = 0.2s^{0.5}$.

(b) We put $s = 70$ into the formula from Part (a) and find $T = 0.2 \times 70^{0.5} = 1.67$. Thus it takes 1.67 seconds for a rock to fall 70 feet.

(c) We want to find the value of s for which $T = 2$, so (by the formula from Part (a)) we want to solve the equation $0.2s^{0.5} = 2$ for s. Using a table of values or a graph (or solving by hand), we find $s = 100$. Thus a rock falls 100 feet in 2 seconds.

(d) The time T is a power function with power 0.5. By the homogeneity property, if s is multiplied by 2, then T is multiplied by $2^{0.5} = 1.41$. Thus the time is multiplied by 1.41.

9. **Composing functions:** Because $x(t) = t - 1$ and $y(x) = 3x^2 + 5x$, we have $y(x(t)) = 3(t-1)^2 + 5(t-1)$. Thus the formula is $y = 3(t-1)^2 + 5(t-1)$.

10. **Decomposing functions:** The balance is $B = 120 \times 1.02^t$, and this is an exponential function. The initial value is 120, so the balance at the start of 2005 was \$120. The growth factor is 1.02, so the yearly percentage growth rate is 2%, and this is the annual interest rate.

11. **Population growth**:

   (a) We show the graph below with a horizontal span of 0 to 200 and a vertical span of 0 to 0.015.

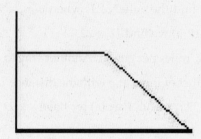

   (b) Because the population level N is between 100 and 200, we use the formula

$$R = 0.02 \left(1 - \frac{N}{200}\right).$$

   Putting in $N = 150$ gives

$$R = 0.02 \left(1 - \frac{150}{200}\right) = 0.005.$$

   Thus the per capita growth rate is 0.005 per year.

12. **Volume**:

   (a) We put $1.5 + 0.1t$ in place of V in the formula for R and find

$$R = \left(\frac{3(1.5 + 0.1t)}{4\pi}\right)^{1/3}.$$

   (b) We put $t = 2$ into the formula from Part (a) and find

$$R = \left(\frac{3(1.5 + 0.1 \times 2)}{4\pi}\right)^{1/3} = 0.74.$$

   Thus the radius is 0.74 inch.

13. **Quadratic formula**: The quadratic formula gives the solutions of $ax^2 + bx + c = 0$ as

$$x = \frac{-b \pm \sqrt{b^2 - 4ac}}{2a}.$$

   To solve $2x^2 - x - 1 = 0$, we use $a = 2$, $b = -1$, and $c = -1$. The solutions are

$$x = \frac{-(-1) \pm \sqrt{(-1)^2 - 4 \times 2 \times (-1)}}{2 \times 2} = \frac{1 \pm \sqrt{9}}{4} = \frac{1 \pm 3}{4}.$$

   This expression gives the values $x = 1$ and $x = -1/2$.

14. **Quadratic regression**: To find a quadratic model for this data set using regression, we enter the data in the first two columns and use the calculator to find directly the quadratic regression coefficients. This gives the quadratic model $f(x) = -3x^2 + 5x - 2$.

15. **Data that are not quadratic**: To show that the data in the table are not exactly quadratic, we calculate the second-order differences:

x	0	1	2	3	4
$f(x)$	0	0	-6	-24	-60
First-order difference	0	-6	-18	-36	
Second-order difference	-6	-12	-18		

Since the second-order differences are not equal, the data are not quadratic.

16. **Chemical reaction**:

   (a) Quadratic regression gives the model $R = 0.005x^2 - 0.75x + 25$.

   (b) Using the model from Part (a), we find that $R(24) = 0.005 \times 24^2 - 0.75 \times 24 + 25 = 9.88$ moles per cubic meter per second. This means that the reaction rate at a concentration of 24 moles per cubic meter is 9.88 moles per cubic meter per second.

   (c) We want to find the value of x for which $R = 6$. We can do this using the crossing-graphs method or the quadratic formula. The result is $s = 32.28$ moles per cubic meter. Thus the reaction rate is 6 moles per cubic meter per second at a concentration of 32.28 moles per cubic meter.

17. **Cubic regression**: To use cubic regression, we enter the data in the first two columns and then use the calculator to determine the cubic regression coefficients. This gives the cubic model $y = 0.1x^3 - 0.2x^2 + 3$.

18. **Finding poles**: The poles of $\dfrac{2x - 5}{x^2 + 4x + 3}$ are the values of x for which the denominator $x^2 + 4x + 3$ is zero but the numerator $2x - 5$ is not zero. In this case, $x^2 + 4x + 3 = 0$ when $x = -1$ and $x = -3$, so these are the poles.

19. **Expanding balloon**:

   (a) Calculating the cubic regression coefficients, we find a cubic model for the data:
   $V = 0.03t^3 + 0.50t^2 + 2.51t + 4.19$.

   (b) The volume after 5 seconds is expressed in functional notation as $V(5)$. Plugging $t = 5$ into the model from Part (a) gives the value 32.99 cubic inches.

20. **Travel time**:

   (a) The graph is shown below. We used a horizontal span of 0 to 70 and a vertical span of 0 to 50.

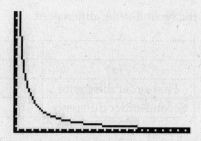

   (b) The value of $T(25)$ can be found by putting $s = 25$ into the formula given or by using a table of values (or the graph). The result is $T(25) = 4$ hours. This means that 4 hours is the time required to drive 100 miles if the average speed is 25 miles per hour.

   (c) At s tends towards 0, T increases sharply. In practical terms, as the average speed becomes very slow, the time needed to drive 100 miles increases greatly.

Solution Guide for Chapter 6: Rates of Change

6.1 VELOCITY

S-1. Velocity: The rate of change in directed distance is velocity.

S-3. Sign of velocity: When the graph of directed distance is decreasing, velocity is negative, so the graph of velocity is below the horizontal axis.

S-5. Constant velocity: When the graph of directed distance is a straight line, then the rate of change is constant, so the velocity is constant. Thus the graph of velocity is the graph of a constant function, that is, it is a horizontal line.

S-7. A car: We know that directed distance is a linear function. The slope of that linear function is its rate of change, which is its velocity. Thus the slope is 60 miles per hour.

S-9. A rock:

(a) The rock is going upwards 1 second after the toss since it has not yet reached its peak. Directed distance is increasing, so the velocity is positive.

(b) The rock has reached its peak 2 seconds after the toss. Directed distance is momentarily not changing, so the velocity is 0.

(c) The rock is going downwards 3 seconds after the toss, since it has reached its peak and is on its way down. Directed distance is decreasing, so the velocity is negative.

S-11. Change direction: When the graph of directed distance is increasing velocity is positive, and when the graph of directed distance is decreasing velocity is negative. Thus if the graph of directed distance switches from increasing to decreasing then velocity switches from positive to negative (passing through 0).

1. **From New York to Miami again**: In the left-hand picture below, we have marked the extreme locations and the places where the distance north is zero. Note that while the plane is on the ground at Miami its distance north from Richmond is negative. Hence it is marked below the horizontal axis. In the right-hand picture we have completed the graph.

Next we make the graph of the velocity of the plane relative to Richmond. The velocity is zero when the plane is at rest at New York and Miami. The velocity is negative when the plane is headed south toward Miami and positive when the plant is headed north toward New York. We have marked these regions in the left-hand picture below. In the right-hand figure, we have completed the graph of velocity. Note that the graph is below the horizontal axis where the graph of distance is decreasing and above the horizontal axis where the graph of distance is increasing. Note also that the graph of velocity is the same as the one shown in Figure 6.14 in the text.

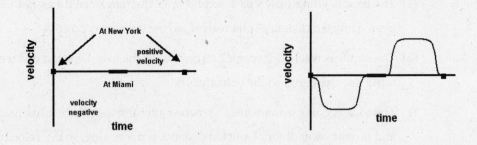

3. **The rock with a formula:**

(a) To select a graphing window, we look at the table of values below. We see that the rock is back on the ground before 2 seconds and that it reaches its peak somewhere around 14 feet. Allowing a little extra room, we set our window using a horizontal span of 0 to 2 and a vertical span of 0 to 20. The graph below shows time on the horizontal axis and height on the vertical axis.

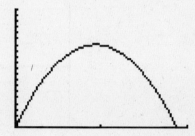

(b) The rock reaches it maximum height at the peak of the graph. From the left-hand figure below, we see that the rock reaches a maximum height of 14.06 feet at the time 0.94 seconds after the toss.

(c) The rock strikes the ground where the graph crosses the horizontal axis. That is, we need to find the root (or zero) of the function. From the right-hand figure below, we see that the rock strikes the ground at 1.875 seconds, or about 1.88 seconds, after the toss.

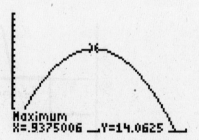

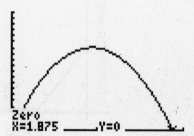

(d) When the rock is moving up from the ground, the velocity will be positive. The velocity will be zero at the point of greatest height, and it will be negative when the rock is dropping back toward the ground. In the left-hand figure below, we have marked the corresponding regions. The graph is completed in the right-hand figure below. (As in Exercise 2, the graph is a straight line. Students may have no way of knowing this, and other graphs could be acceptable.)

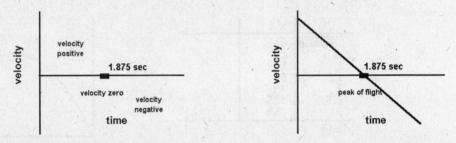

5. **Walking and running**: We measure location as west from your home. Since you live east of campus, your initial location is a positive number, so the graph starts above the horizontal axis. You are walking home at a constant rate, so the graph of location decreases in a straight line until it reaches the horizontal axis (home), where it stays for 5 minutes. After that, you are going back west, running now, so the graph rises quickly in a straight line until it reaches the level corresponding to your destination, where it remains, as a horizontal line, for 10 minutes. The graph of location is below on the left.

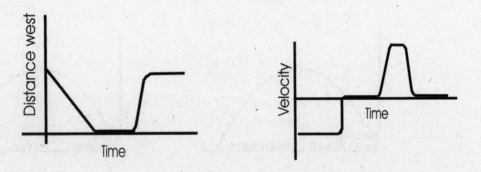

Velocity is the rate of change of location. When the graph of location is a straight line, then velocity is constant, so it forms a horizontal line. As you walk home, the velocity is negative, since the graph of location is decreasing, so initially the graph of velocity is a horizontal line below the axis. When you reach home, velocity is 0, so the graph lies on the horizontal axis. When you run back west, velocity is constant, but positive, so the graph of velocity is a horizontal line above the axis. It should be higher above the axis than the line below the axis was below it, since you run faster than you walk.

Finally, when you rest for 10 minutes, location is not changing, so velocity is 0 and the graph is on the horizontal axis. The graph of velocity is above on the right.

7. **Gravity on Earth and on Mars**: Since Mars is much smaller than Earth, the acceleration due to gravity will be greater on Earth than on Mars. So the rock on Earth will hit the ground much faster than the rock on Mars. Since we are measuring the distance from the top of the cliff, the beginning position is zero, and distance increases until the rock hits the ground. Our graph is shown in the left-hand figure below.

The velocity of both rocks will start at zero and then be positive since the rocks are moving away from the top of the cliff. Since the acceleration on Earth is greater than that on Mars, the velocity on Earth will be greater. Our graph is shown in the right-hand figure below.

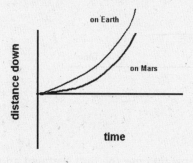
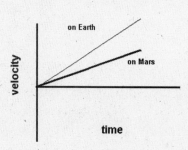

9. **Making up a story about a car trip:**

   (a) A graph of velocity that matches the information in the exercise is shown in the left-hand figure below.

   (b) The distance from home increases slightly at first and then levels off when the velocity is zero. Then the distance increases at a greater rate as the velocity increases. When the velocity is constant, the distance is still increasing, but now as a linear function. When the velocity is negative, the distance begins to decrease. Our graph of distance is shown in the right-hand figure below.

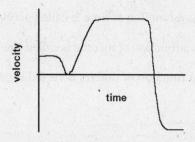
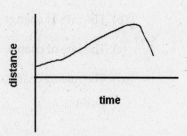

(c) Answers to this part will vary greatly. Here is one example of an answer: I started out of my driveway very slowly and then came to a stop at a traffic light at the end of the street. I pulled out on to the freeway and rapidly increased my speed to match the other cars, and the speed became constant when I set my cruise control at the posted speed limit. Just then I remembered that I had forgotten my briefcase, so I slowed down, looking for an exit. When I found one I turned around and drove back home fast.

11. **A car moving in an unusual way**:

(a) The graph of the velocity should be a horizontal line above the horizontal axis where the graph of position is going up. It jumps immediately to a horizontal line below the horizontal axis when the graph starts down.

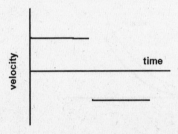

(b) The car's velocity is a positive constant over the first segment and then instantly negative after the peak in the distance graph. Thus it appears that the car has changed instantly from traveling forward to traveling backward without slowing down for the change. It is not possible to drive a car in that manner.

6.2 RATES OF CHANGE FOR OTHER FUNCTIONS

S-1. **Meaning of rate of change**:

(a) The rate of change of directed distance as a function of time is called velocity.

(b) The rate of change of velocity as a function of time is called acceleration.

(c) The rate of change of tax due as a function of income is called the marginal tax rate.

(d) The rate of change of profit as a function of dollars invested is called the marginal profit.

S-3. **Sign of the derivative**:

(a) If the function f is increasing, then its rate of change is positive, so $\dfrac{df}{dx}$ is positive.

(b) If the graph of f has reached a peak, then its rate of change is momentarily 0, so $\dfrac{df}{dx}$ is 0.

(c) If the function f is decreasing, then its rate of change is negative, so $\dfrac{df}{dx}$ is negative.

(d) If the graph of f is a horizontal line, then the value of f is not changing, so $\dfrac{df}{dx}$ is zero.

S-5. **A value for the rate of change**: The slope of f is precisely its rate of change, so since $\dfrac{df}{dx}$ is 10, the slope is also 10.

S-7. **Marginal tax rate**: If you earn some extra income working at $20 per hour, you will get to keep $20 less the taxes on $20. If your marginal tax rate is 34%, then you get to keep $100\% - 34\% = 66\%$ of $20, that is, $0.66 \times 20 = 13.20$ dollars of the hourly wage.

S-9. **Graph of f**: If the graph of $\dfrac{df}{dx}$ is below the horizontal axis, then $\dfrac{df}{dx}$ is negative, so the graph of f is decreasing.

S-11. **Gasoline prices**: When the price reaches a minimum the rate of change is momentarily 0, so $\frac{dP}{dt}$ is 0 at that time.

1. **Estimating rates of change**: The graph of $f(x) = x^3 - 5x$ is shown at the left below with a horizontal span of -3 to 3 and a vertical span of -10 to 10.

(a) The figure at the left below shows that the function is increasing at $x = 2$. Thus $\dfrac{df}{dx}$ is positive there.

(b) Now $\dfrac{df}{dx}$ is negative wherever the function is decreasing. The figure at the right below shows that the function begins to decrease at $x = -1.29$, and it decreases until $x = 1.29$. Thus any point on the graph for which x lies between -1.29 and 1.29 is correct.

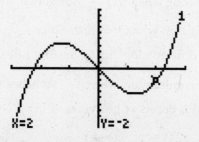

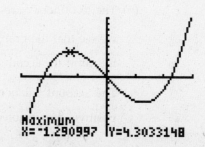

3. **Mileage for an old car**:

   (a) The expression $\frac{dM}{dt}$ gives us the change in the car's gas mileage we expect over a year's time.

   (b) Since we expect that the gas mileage of the car will decrease as the car gets older, then we expect $\frac{dM}{dt}$ to be negative.

5. **Hiking**:

   (a) The expression $\frac{dE}{dt}$ is the change in elevation we expect over one unit of time.

   (b) When $\frac{dE}{dt}$ is a large positive number, then elevation is increasing quickly, and so we might be climbing up a hill.

   (c) When $\frac{dE}{dt}$ is briefly zero we might have reached the top of a hill, or we might have stopped to catch our breath.

   (d) When $\frac{dE}{dt}$ is a large negative number, then elevation is decreasing quickly, and so we might be descending from the top of a hill.

7. **Health plan**:

   (a) We want to make the graph of $10e^{0.1t} - 12$. Since we are interested in the first 5 years of the plan, we use a horizontal span of 0 to 5. The table below leads us to choose a vertical span of -3 to 5. The horizontal axis of the graph is years since the beginning of the plan, and the vertical axis is the rate of change in account balance.

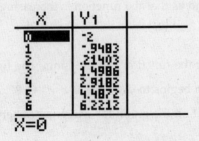   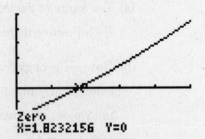

   (b) The account balance B is decreasing whenever $\frac{dB}{dt}$ is negative. In the graph above, we see that the graph of $\frac{dB}{dt}$ is zero when $t = 1.82$. Since $\frac{dB}{dt}$ is negative before this time, the account balance is decreasing during the first 1.82 years.

   (c) The account balance B is at its minimum when $\frac{dB}{dt}$ switches from negative to positive. As we noted above, this occurs at 1.82 years.

9. **The acceleration due to gravity**: Acceleration is the rate of change in velocity with respect to time. In symbols this means that $\dfrac{dV}{dt}$ is acceleration. Since acceleration $\dfrac{dV}{dt}$ has a constant value of 32 feet per second per second, velocity V is a linear function of t with slope 32.

11. **A population of bighorn sheep**:

    (a) The expression $\dfrac{dN}{dt}$ is the change in the sheep population we expect over one unit of time. In more familiar terms, it is the rate of population growth.

    (b) We expect $\dfrac{dN}{dt}$ to be positive when the conditions are favorable for the sheep because we would expect the population to grow.

    (c) The population would grow less rapidly, making $\dfrac{dN}{dt}$ smaller than before. If the problem is acute, the population might stop increasing, making $\dfrac{dN}{dt}$ zero, or it might decline, making $\dfrac{dN}{dt}$ negative.

    (d) The population will change slowly if at all. Thus $\dfrac{dN}{dt}$ will be near zero.

13. **Visiting a friend**: Assume that the car travels at the speed limit whenever this is possible. The distance graph should be a straight line with slope 35 until the stop sign is reached. The graph then changes to a straight line with slope 65. The trip back should be a reflection of the trip to the friend's home. This is shown in the left-hand graph below.

    The velocity is constant at 35 up to the stop sign, constant at 65 from the stop sign to the friend's home, constant at −65 from the friend's home back to the stop sign, and finally constant at −35 on the final part of the trip. The graph of velocity is shown in the right-hand figure below.

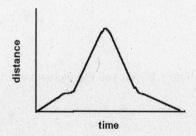

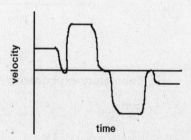

Acceleration is zero where the velocity is constant; that is for most of the graph. Filling in the spaces between is simply a matter of remembering that acceleration is positive when velocity is increasing and negative where velocity is decreasing. Our graph is shown below.

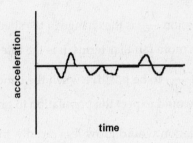

15. **An airplane**: The distance from Atlanta increases until the plane nears Dallas. While the airplane is circling, its distance from Atlanta increases and then decreases with each circle. For much of the trip, we expect velocity to be constant, and so distance should be linear. This is reflected in the left-hand figure below.

The velocity is constant over most of the trip. Near the airport, we need to remember that velocity is positive when distance is increasing and negative when distance is decreasing. Velocity is shown in the right-hand figure below.

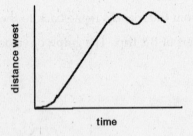 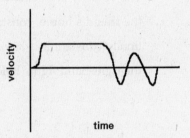

17. **Laboratory experiment**: Answers will vary. Please see the website for further information.

6.3 ESTIMATING RATES OF CHANGE

S-1. **Rate of change for a linear function**: If f is the linear function $f = 7x - 3$, then its slope is 7. The rate of change of f is therefore 7, so the value of $\dfrac{df}{dx}$ is 7.

S-3. **Rate of change from data**: If $f(3) = 8$ and $f(3.005) = 7.972$, then we can estimate the value of $\dfrac{df}{dx}$ at $x = 3$ using the average rate of change. The average rate of change is

$$\frac{f(3.005) - f(3)}{3.005 - 3} = \frac{7.972 - 8}{0.005} = -5.6,$$

so we estimate the value of $\dfrac{df}{dx}$ at $x = 3$ as -5.6.

S-5. **Estimating rates of change**: To estimate the value of $\dfrac{df}{dx}$ for $f(x) = \dfrac{1}{x^2}$ at $x = 4$, we calculate the average rate of change in f from $x = 4$ to $x = 4.0001$ using the formula for f:

$$\frac{f(4.0001) - f(4)}{4.0001 - 4} = \frac{(1/4.0001^2) - (1/4^2)}{0.0001} = \frac{0.0624968751 - 0.0625}{0.0001} = -0.0312488282.$$

Based on this, we estimate the value of $\dfrac{df}{dx}$ at $x = 3$ to be about -0.03.

S-7. **Estimating rates of change with the calculator**: Using a horizontal span for x from 1 to 4 and a vertical span for $y = x + \dfrac{1}{x}$ from 0 to 5, we can see from the figure below that the calculated rate of change is 0.89.

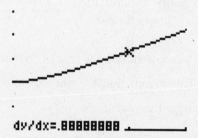

S-9. **Estimating rates of change with the calculator**: Using a horizontal span for x from 0 to 4 and a vertical span for $y = 3^{-x}$ from -1 to 1, we can see from the figure below that the calculated rate of change is -0.04.

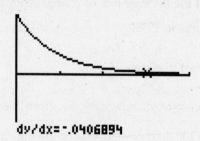

S-11. **Limiting value of a rate of change**: The graph from Exercise S-7 looks like a straight line for large values of x, so we expect the rate of change to be nearly constant there. We enlarge the horizontal span to go from 0 to 100 and the vertical span to go from 0 to 110, and then we calculate $\frac{df}{dx}$ at $x = 100$. As shown in the figure below, the value we find for the rate of change is 0.9999, which is about 1. Thus $\frac{df}{dx}$ is near 1 for large values of x.

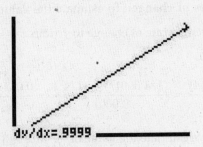

1. **Population growth**:

   (a) The average rate of change from 1955 to 1960 is

   $$\frac{\text{Change in population}}{\text{Elapsed time}} = \frac{2793 - 678}{5} = 423 \text{ reindeer per year.}$$

   Using this to approximate the rate of change at 1955, we get $\frac{dN}{dt} = 423$ reindeer per year. This means that we can expect the population to grow by about 423 reindeer during the year 1955.

   (b) For each year past 1955 we expect the population to increase by 423 reindeer, so in 1957 we expect the population to be around $678 + 2 \times 423 = 1524$ reindeer.

   (c) If the reindeer population is more closely exponential, then our answer is too large. The reason is that the rate of change after 1955 is larger than it is in 1955 if the data is exponential, since then the function is increasing at an increasing rate. This means that the average rate of change from 1955 to 1960 is larger than the actual rate of change in 1955.

3. **Deaths from heart disease**:

   (a) The average rate of change for H_m from 1980 to 1985 is $\frac{651.9 - 746.8}{1985 - 1980} = -18.98$ deaths per 100,000 per year, and this is the approximate value of $\frac{dH_m}{dt}$ in 1980.

   (b) The number in Part (a) means that the number of deaths per 100,000 males caused by heart disease is decreasing in 1980 by about 18.98 per year.

(c) Since the number of deaths is decreasing by about 18.98 per year then in 1983 we would expect there to be about $746.8 - 3 \times 18.98 = 689.86$ deaths per 100,000 in 1983.

(d) The average rate of change in H_f from 1980 to 1985 is $\dfrac{250.3 - 272.1}{1985 - 1980} = -4.36$ deaths per 100,000 per year, and this is the approximate value of $\dfrac{dH_f}{dt}$ in 1980.

(e) In 1980, the number of males dying from heart disease was much greater than the number of females dying from heart disease; however, for males that number was decreasing much more rapidly than the number for females.

5. **Falling with a parachute**:

(a) Based on the table of values at the left below, we use a horizontal span of 0 to 10 and a vertical span of 0 to 200. In the figure at the right below we have made the graph.

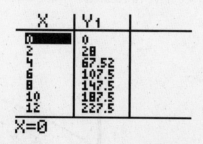
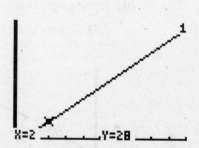

(b) On the graph above we have evaluated S at $t = 2$, and we see that the parachutist falls 28 feet in 2 seconds.

(c) In the graph below we have calculated $\dfrac{dS}{dt}$ at $t = 2$, and the result is 19.20. This tells us that, after he falls 2 seconds, we expect the parachutist to fall 19.20 feet over the next second. In other words, after falling 2 seconds his velocity is 19.20 feet per second.

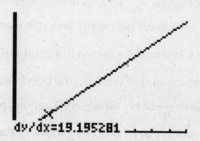

7. **A yam baking in the oven:**

   (a) We use a horizontal span of 0 to 45 as indicated. Based on the table of values at the left below, we use a vertical span of 0 to 300. In the figure at the right below we have made the graph.

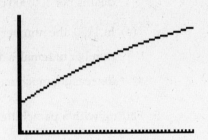

   (b) In the graph below at the left we have calculated $\frac{dY}{dt}$ at $t = 10$, and the result is 5.32 degrees per minute.

   (c) In the graph below at the right we have calculated $\frac{dY}{dt}$ at $t = 30$, and the result is 3.57 degrees per minute.

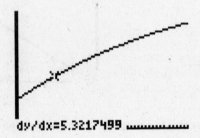 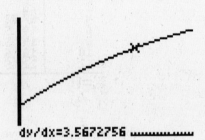

   (d) A comparison of the answers in Parts (b) and (c) shows that the temperature of the yam increases more slowly the longer it has been in the oven. This is also indicated by the fact that the graph is concave down.

9. **A pond:**

   (a) The expression $\frac{dG}{dt}$ tells us the change (measured in gallons) in the amount of water in the pond we expect over one minute's time.

   (b) Since the amount of water in the pond is decreasing, we expect $\frac{dG}{dt}$ to be negative.

   (c) Since the amount of water in the pond decreases by 8000 gallons each minute, we know that $\frac{dG}{dt}$ at $t = 30$ is -8000 gallons per minute.

   (d) At $t = 30$ there are 2,000,000 gallons in the pond. The pond is losing 8000 gallons each minute, and we want to estimate the amount of water at $t = 35$, or 5 minutes after $t = 30$. Thus we expect $G(35)$ to be $2,000,000 - 5 \times 8000 = 1,960,000$ gallons.

6.4 EQUATIONS OF CHANGE: LINEAR AND EXPONENTIAL FUNCTIONS

S-1. **Technical terms**: An equation of change is known more formally by the common mathematical term "differential equation."

S-3. **Slope**: If f satisfies the equation of change $\dfrac{df}{dx} = 5$, then the rate of change of f is 5, so f is a linear function of slope 5.

S-5. **Solving an equation of change**: If f satisfies the equation of change $\dfrac{df}{dx} = 8f$, then f is an exponential function and hence can be written as $f = Ae^{ct}$. The value of c is 8.

S-7. **A leaky balloon**: If the balloon leaks air at a rate of one third the volume per minute, the rate of change is $-\dfrac{1}{3}V$ (the minus sign because the balloon is leaking and so the volume is decreasing). Thus an equation of change for V is $\dfrac{dV}{dt} = -\dfrac{1}{3}V$.

S-9. **Solving an equation of change**: The equation of change $\dfrac{df}{dx} = 3$ shows that f has a constant rate of change of 3, so f is a linear function with slope 3. Since the initial value of f is 7, we have $f = 3x + 7$.

S-11. **Filling a tank**: The rate of change in the height H of the water level with respect to time t is always 4 feet per minute because the level rises 4 feet every minute. Thus an equation of change is $\dfrac{dH}{dt} = 4$.

1. **Looking up**:

   (a) The exercise tells us that $g = 32$, so $\dfrac{dV}{dt} = -32$. Thus V is a linear function with slope -32. Since the initial velocity is 40, we have $V = -32t + 40$.

   (b) The rock will reach the peak of its flight when the velocity is zero. We must solve the equation $-32t + 40 = 0$ for t. If we do this, we find that the rock will reach its peak when $t = \dfrac{40}{32} = 1.25$ seconds.

   (c) The rock will reach its peak halfway through its flight, and so the rock will strike the ground at $t = 2 \times 1.25 = 2.5$ seconds.

3. **A better investment**:

   (a) The account has continuous compounding with $r = 0.0575$ (the APR in decimal form) and an initial balance of \$250. Let the balance B be measured in dollars. It is an exponential function of time t in years since the initial investment. In standard form, $B = Pa^t$ where P is the initial value and $a = e^r$. Thus $P = 250$ and $a = e^{0.0575} = 1.0592$, and so $B = 250 \times 1.0592^t$. In alternative form, $B = Pe^{rt}$, which is simply $B = 250e^{0.0575t}$.

(b) After 5 years, the balance is expected to be $B = 250 \times 1.0592^5 = 333.30$ dollars using the standard form, and $B = 250e^{0.0575 \times 5} = 333.27$ dollars using the alternative form. The alternative form gives a more accurate answer because it was not rounded until the very end, whereas for the standard form there was early rounding in reporting $e^{0.0575}$ as 1.0592.

5. **Borrowing money**:

(a) We know that B is an exponential function because its equation of change has the form $\dfrac{df}{dx} = rf$ where r is a constant. This is described in Key Idea 6.6.

(b) From the equation of change we have an exponential growth rate of $r = 0.07$ per year. The initial value for B is 10,000 dollars. Thus the formula for B in the alternative form is $B = 10,000e^{0.07t}$.

(c) The yearly growth factor here is $e^{0.07} = 1.073$, and the initial value is 10,000, so the formula in standard form is $B = 10,000 \times 1.073^t$.

(d) The account balance will double when it reaches \$20,000. Thus we need to solve the equation $10,000e^{0.07t} = 20,000$. We do this using the crossing graphs method. Based on the table of values below for B and the constant function 20,000, we use a horizontal span of 0 to 15 years and a vertical span of 0 to 30,000 dollars. In the figure at the right below we have graphed B and the target balance of 20,000 (thick line). We see that the intersection occurs at a time of $t = 9.90$ years. Thus the account will double after 9.90 years.

X	Y1	Y2
0	10000	20000
5	14191	20000
10	20138	20000
15	28577	20000
20	40552	20000
25	57546	20000
30	81662	20000

X=0

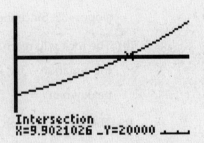

Intersection
X=9.9021026 _Y=20000 ⌐⌐⌐

7. **A population of bighorn sheep**:

(a) As an equation of change, the sentence becomes $\dfrac{dN}{dt} = 0.04N$.

(b) From the equation of change we have that N is an exponential function with an exponential growth rate of $r = 0.04$ per year. The initial value is 30, so the formula in the alternative form is $N = 30e^{0.04t}$. (For the standard form, note that the yearly growth factor is $e^{0.04} = 1.04$, while the initial value is 30, so the formula is $N = 30 \times 1.04^t$.)

(c) We want to find when N is 50, so we want to solve the equation $30e^{0.04t} = 50$. We do this using the crossing graphs method. Based on the table of values below for N and the constant function 50, we use a horizontal span of 0 to 15 years and a vertical span of 0 to 60. In the figure at the right below we have graphed N and the target population of 50. We see that the intersection occurs at a time of $t = 12.77$ years, and this is when the population will have grown to a level of 50. (If we use the standard form for N, we get a time of $t = 13.02$ years.)

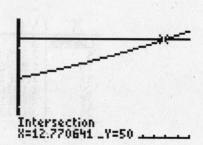

X	Y1	Y2
0	30	50
5	36.642	50
10	44.755	50
15	54.664	50
20	66.766	50
25	81.548	50
30	99.604	50

X=0

Intersection
X=12.770641 Y=50

9. **Growing child**:

(a) Since the girl grew steadily at a rate of 5.5 pounds per year, the weight W has a constant rate of change of 5.5 pounds per year, and thus is a linear function with slope 5.5.

(b) Since the rate of change is the constant 5.5, the equation of change is $\dfrac{dW}{dt} = 5.5$.

(c) From Part (a), W is a linear function of t with slope 5.5, and here we are told that $W = 30$ when $t = 3$. We need to find the initial value b of W. We have $W = 5.5t + b$, and using the fact that $W = 30$ when $t = 3$ gives $30 = 5.5 \times 3 + b$, or $30 = 16.5 + b$. Thus $b = 30 - 16.5 = 13.5$. Hence the formula is $W = 5.5t + 13.5$.

11. **Radioactive decay**:

(a) The equation of change for A has the form $\dfrac{df}{dx} = rf$ where r is a constant, and thus A is an exponential function with exponential growth rate $r = -0.05$ per day.

(b) The initial value is 3 grams, so in the alternative form the formula is $A = 3e^{-0.05t}$. (For the standard form, note that the daily decay factor is $e^{-0.05} = 0.95$, while the initial valuee is 3, so the formula is $B = 3 \times 0.95^t$.)

(c) Half the initial amount is 1.5 grams, so we want to find when $A = 1.5$. That is, we want to solve the equation $3e^{-0.05t} = 1.5$. We do this using the crossing graphs method. Based on the table of values below for A and the constant function 1.5, we use a horizontal span of 0 to 15 days and a vertical span of 0 to 3 grams. In the figure at the right below we have graphed A and the target amount of 1.5. We see that the intersection occurs at a time of $t = 13.86$ days, and this is the half-life. (If we use the standard form for A, we get a time of $t = 13.51$ days.)

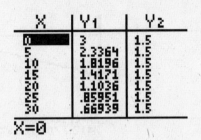
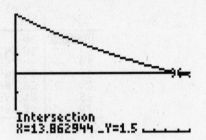

6.5 EQUATIONS OF CHANGE: GRAPHICAL SOLUTIONS

S-1. **Equilibrium solutions**: An equilibrium solution of an equation of change is a solution which does not change, that is, it remains constant.

S-3. **Finding equilibrium solutions**: The equilibrium solutions of an equation of change occur when $\dfrac{df}{dx} = 0$. To find the equilibrium solutions of $\dfrac{df}{dx} = 2f - 6$ is the same as solving $0 = 2f - 6$, so the equilibrium solution is $f = 3$.

S-5. **Water**: If the process of water flowing in and some of it draining out continues for a long time, then we expect the water volume to reach some type of equilibrium. The equilibrium solution to $\dfrac{dv}{dt} = 5 - \dfrac{v}{3}$ occurs when $\dfrac{dv}{dt} = 0$, that is, $0 = 5 - \dfrac{v}{3}$. Solving, we see that $v = 15$ is the equilibrium solution, so there will be 15 cubic feet of water in the tank.

S-7. **Water**: If there are 8 cubic feet of water in the tank, so $v = 8$, then the equation of change gives $\dfrac{dv}{dt} = 5 - \dfrac{v}{3} = 5 - \dfrac{8}{3} = 2.33$. Since $\dfrac{dv}{dt}$ is positive, the volume v is increasing.

S-9. **Population**: If the population is 4238, so $N = 4238$, then the equation of change gives $\dfrac{dN}{dt} = 0.03N \left(1 - \dfrac{N}{6300}\right) = 0.03 \times 4238 \left(1 - \dfrac{4238}{6300}\right) = 41.61$. Since $\dfrac{dN}{dt}$ is positive, the population N is increasing.

S-11. **Equation of change**: When $f = 1$ then $5f - 7 = 5 \times 1 - 7 = -2$, so $\frac{df}{dx}$ is negative. Hence f is decreasing when $f = 1$.

1. **Equation of change for logistic growth:**

   (a) Since Y_1 corresponds to $\dfrac{dN}{dt}$ and X corresponds to N, the function we want to graph is $Y_1 = 0.8X(1 - X/177)$. Based on the table of values at the left below, we use a horizontal span of 0 to 200 and a vertical span of -20 to 40. In the figure at the right below we have made the graph.

   Equilibrium solutions are those for which $\dfrac{dN}{dt} = 0$, or $Y_1 = 0$ in the above correspondence. As shown in the figure at the right below, $Y_1 = 0$ occurs when X is 0 or 177. Thus the equilibrium solutions are $N = 0$ and $N = 177$. The physical significance of the equilibrium solution $N = 0$ is that there will never be any deer if we start with none. The solution $N = 177$ is the environmental carrying capacity for the deer in this reserve. If the number of deer ever reaches 177, then environmental limitations match growth tendencies, and the population stays the same.

   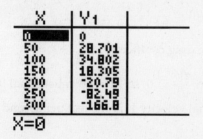   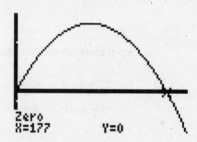

   (b) The relevant information we get from the graph in Part (a) is summarized in the table below.

Range for N	from 0 to 177	greater than 177
Sign of $\dfrac{dN}{dt}$	positive	negative
Effect on N	increasing	decreasing

If $N(0) = 10$ then the population will increase toward the equilibrium solution of $N = 177$. If $N(0) = 225$ then the population will decrease toward the equilibrium solution of $N = 177$. The graphs are shown below.

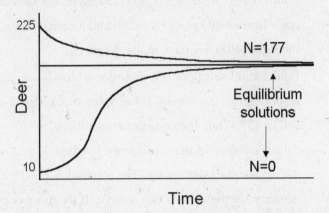

(c) The starting value for N corresponding to the given solution is the initial value of that solution:

$$N(0) = \frac{6.21}{0.035 + 0.45^0} = 6.$$

3. **Experimental determination of the drag coefficient**:

(a) Terminal velocity occurs when $\frac{dV}{dt} = 0$, so $32 - rV$ must equal 0 when V equals terminal velocity. Since terminal velocity V is 176 feet per second, we have

$$32 - r \times 176 = 0.$$

Solving for r yields $r = \frac{32}{176} = 0.1818$.

(b) Reasoning as in Part (a), we put in $V = 4$ for the terminal velocity, and we get $32 - r \times 4 = 0$. Thus the drag coefficient is $r = \frac{32}{4} = 8$.

5. **Fishing for sardines**:

(a) We need to find the value of F such that 1.8 million tons is an equilibrium solution for the equation of change. This means that $\frac{dN}{dt} = 0$ when $N = 1.8$. Using these values in the equation of change, we get

$$0.338 \times 1.8 \left(1 - \frac{1.8}{2.4}\right) - F = 0.$$

We can easily solve for F:

$$F = 0.338 \times 1.8 \left(1 - \frac{1.8}{2.4}\right) = 0.1521.$$

Thus the fishing level for Pacific sardines should be set at about 0.15 million tons per year.

(b) i. First we graph $\dfrac{dN}{dt}$ versus N. Since Y_1 corresponds to $\dfrac{dN}{dt}$ and X corresponds to N, the function we want to graph is $Y_1 = 0.338X(1 - X/2.4) - 0.1$. (Remember that $F = 0.1$.) Based on the table of values at the left below, we use a horizontal span of 0 to 3 and a vertical span of -0.15 to 0.15. In the figure at the right below we have made the graph.

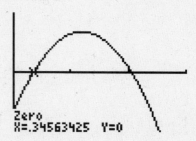

The equilibrium solutions occur where $\dfrac{dN}{dt} = 0$, so in the graph above we want to find where $Y_1 = 0$. There are two such points, and we find them using the single graph method. The graph above shows that one zero occurs when $N = 0.35$, and finding the other gives $N = 2.05$. Thus the two equilibrium solutions are $N = 0.35$ million tons and $N = 2.05$ million tons.

ii. The relevant information from Part i above is summarized in the following table.

Range for N	less than 0.35	from 0.35 to 2.05	greater than 2.05
Sign of $\dfrac{dN}{dt}$	negative	positive	negative
Effect on N	decreasing	increasing	decreasing

From the table we see that the population increases when N is between 0.35 and 2.05 million tons. The population decreases when N is less 0.35 million tons and when N is greater than 2.05 million tons.

iii. If $N(0) = 0.3$ then the population will decrease to 0. Note that 0 is not an equilibrium solution, and in fact $\frac{dN}{dt}$ is negative even at $N = 0$, so with $N(0) = 0.3$ the population will reach 0 quickly. If $N(0) = 1.0$ then the population will increase to the equilibrium solution at 2.05. If $N(0) = 2.3$ then the population will decrease to the equilibrium solution at 2.05.

The graphs are shown below.

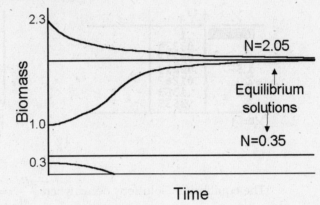

iv. In practical terms, if the initial population is 0.3 million tons, then the population will die out quickly. If the initial population is 1 million tons, then it will increase and slowly approach 2.05 million tons. If the initial population is 2.3 million tons, then it will decrease and slowly approach 2.05 million tons. If the initial population is 0.35 or 2.05 million tons, then the population will remain at that level.

7. **Logistic growth with a threshold:**

(a) Since $r = 0.338$, $K = 2.4$, and $S = 0.8$, the equation of change is
$$\frac{dN}{dt} = -0.338N\left(1 - \frac{N}{0.8}\right)\left(1 - \frac{N}{2.4}\right).$$

(b) Since Y_1 corresponds to $\frac{dN}{dt}$ and X corresponds to N, we want to graph the function $Y_1 = -0.338X(1 - X/0.8)(1 - X/2.4)$. Based on the table of values at the left below, we use a horizontal span of 0 to 3 and a vertical span of -0.25 to 0.25. In the figure at the right below we have made the graph.

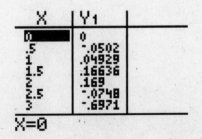

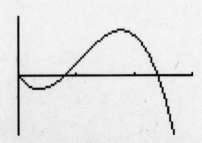

The equilibrium solutions are at the zeros of the function graphed above. Clearly one such zero is at 0, and the graph shows that another is at 0.8. In a similar way, we find that the third zero is at 2.4. Thus the equilibrium solutions are $N = 0$, $N = 0.8$ and $N = 2.4$ (all measured in millions of tons). The solution $N = 0.8$ corresponds to S, the survival threshold, while the solution $N = 2.4$ corresponds to K, the carrying capacity.

(c) The relevant information from Part (b) above is summarized in the following table.

Range for N	from 0 to 0.8	from 0.8 to 2.4	greater than 2.4
Sign of $\dfrac{dN}{dt}$	negative	positive	negative
Effect on N	decreasing	increasing	decreasing

From the table we see that the population increases when N is between 0.8 and 2.4 million tons. The population decreases when N is between 0 and 0.8 million tons and when N is greater than 2.4 million tons.

(d) If the initial population is 0.7 million tons, then, since $N(0)$ is less than 0.8, the population will decrease to 0. This makes sense, because the initial population is less than the survival threshold, and so the population can be expected to dwindle away to nothing.

(e) The population is growing the fastest when $\dfrac{dN}{dt}$ is the greatest. This occurs at the maximum point on the graph in Part (b) above, and the figure below shows that this corresponds to $N = 1.77$ million tons.

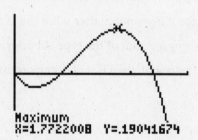

Maximum
X=1.7722008   Y=.19041674

9. **Competition between bacteria:**

(a) Since $a = 2.3$ and $b = 1.7$, we have $a - b = 2.3 - 1.7 = 0.6$, so the equation of change is $\dfrac{dP}{dt} = 0.6P(1 - P)$. This fits the form of a logistic equation of change with $r = 0.6$ and $K = 1$.

(b) We first find the equilibrium solutions, and we do so by graphing $\dfrac{dP}{dt}$ versus P. Since Y_1 corresponds to $\dfrac{dP}{dt}$ and X corresponds to P, we want to graph the function $Y_1 = 0.6X(1 - X)$. Based on the table of values at the left below, we use a horizontal span of 0 to 1 and a vertical span of -0.1 to 0.2. In the figure at the right below we have made the graph.

X	Y1
0	0
.2	.096
.4	.144
.6	.144
.8	.096
1	0
1.2	-.144

X=0

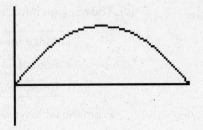

From the table or the graph we see that the equilibrium solutions are $P = 0$ and $P = 1$. (This is what we expect from a logistic equation with carrying capacity $K = 1$.)

If initially P is 0 then, since $P = 0$ is an equilibrium solution, P will stay at 0. This makes sense since $P(0) = 0$ says that there are only type B bacteria and no type A bacteria, so there never will be any type A bacteria.

(c) If $P(0) = 1$ then, since $P = 1$ is an equilibrium solution, P will stay at 1. Also, from the graph in Part (b) above, we see that $\dfrac{dP}{dt}$ is always positive for P strictly between 0 and 1. Thus if $P(0)$ is positive but less than 1 then P will increase to 1. In this case it does not matter what the exact value of $P(0)$ is. This means that as long as some amount of the type A bacteria (the type with the larger growth rate) is initially present, then it will dominate, and eventually there will be only type A bacteria.

11. **Grazing sheep:**

(a) In practical terms, $\dfrac{dC}{dV}$ tells how much more food (in pounds) we can expect a merino sheep to consume in a day if we increase the amount of vegetation available by one pound per acre.

(b) Since Y_1 corresponds to $\dfrac{dC}{dV}$ and X corresponds to C, we want to graph the function $Y_1 = 0.01(2.8 - X)$. We are told to use a horizontal span of 0 to 3, and, based on the table of values at the left below, we use a vertical span of -0.01 to 0.05. In the figure at the right below we have made the graph.

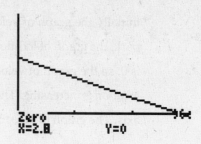

(c) We first find the equilibrium solution. In the graph above we see that it corresponds to $C = 2.8$. (Since the graph is linear, this can also be found by hand.) From that graph we also see that $\dfrac{dC}{dV}$ is positive for C less than 2.8 pounds. Thus the most we would expect the merino sheep to consume in a day is 2.8 pounds.

Chapter 6 Review Exercises

1. **Maximum distance**: At a maximum of directed distance the velocity is 0.

2. **Constant slope**: When the graph of directed distance is a straight line, then the rate of change is constant, so the velocity is constant. The value of that constant is the slope -3. Hence the velocity is a constant -3.

3. **A casual walk**: We measure location as distance east from your home. Your initial location is at home, so the graph starts at the origin. For the first 10 minutes you are walking east at a constant speed, so the graph of location increases in a straight line over that time period. While you rest at the park your location is unchanging, so the graph is a horizontal line segment over that 5-minute period. After that you resume going east, so the graph begins to rise again. The graph of location is below on the left.

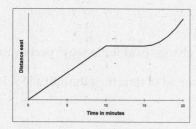

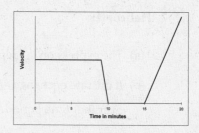

Velocity is the rate of change of location. During the first 10 minutes, the graph of location is a straight line, so velocity is constant and its graph is a horizontal line. Also, during this period the velocity is positive, since the graph of location is increasing. Thus initially the graph of velocity is a horizontal line above the axis. While you are at the park, the graph of location is a horizontal line. Hence over that time period the velocity is 0, so the graph of velocity lies on the horizontal axis. When you resume going east, location is increasing. Then the velocity is positive, so the graph of velocity is above the axis. The graph of velocity is above on the right.

4. **Velocity given**:

  (a) We use a horizontal span of 0 to 3 and a vertical span of -12 to 12. The graph of V is on the left below.

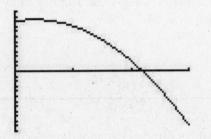
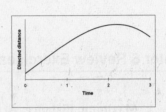

  (b) The velocity V is positive until it reaches a point that is just greater than 2. Using the single-graph method shows that this point is $t = 2.19$. After this point the velocity is negative. Hence the directed distance is increasing from $t = 0$ to $t = 2.19$ and decreasing after that. One possible graph of directed distance is shown on the right above.

5. **Marginal profit**: Because the rate of change $\frac{dP}{dn}$ is negative, the profit P is a decreasing function of n. Hence increasing the number of widgets made decreases the profit.

6. **Water depth**: After time $t = 5$ the depth of water H does not change, so the rate of change $\frac{dH}{dt}$ is 0.

7. **Helicopter**:

  (a) The expression $\frac{dA}{dt}$ represents the change in altitude we expect in 1 unit of time.

  (b) If the rate of change $\frac{dA}{dt}$ is 0 for a period of time then the altitude A is not changing over that time period.

  (c) If the helicopter is coming down for a landing then the altitude A is decreasing, so the rate of change $\frac{dA}{dt}$ is negative.

8. **Account balance:**

    (a) We want to make a graph of $10 + 5t - 3t^2$ versus t. We use a horizontal span of 0 to 3 and a vertical span of -5 to 15. The graph is below.

    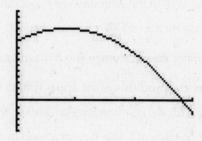

    (b) The balance B is increasing when the rate of change $\frac{dB}{dt}$ is positive. From the graph above the rate of change is positive until it reaches a point that is just less than 3. Using the single-graph method shows that this point is $t = 2.84$. Thus the balance B is increasing from 0 to 2.84 years.

    (c) The balance B is at its maximum when it changes from increasing to decreasing, and that is when the rate of change $\frac{dB}{dt}$ changes from positive to negative. From the graph in Part (a) and the calculation in Part (b) we find that this occurs at $t = 2.84$. Thus the balance B is at its maximum at 2.84 years.

9. **Rate of change from data:** If $f(1) = 3$ and $f(1.002) = 3.012$, then we can estimate the value of $\frac{df}{dx}$ at $x = 1$ using the average rate of change. The average rate of change is

    $$\frac{f(1.002) - f(1)}{1.002 - 1} = \frac{3.012 - 3}{0.002} = 6,$$

    so we estimate the value of $\frac{df}{dx}$ at $x = 1$ as 6.

10. **Estimating rates of change with the calculator:** Using the suggested horizontal and vertical spans, we can see from the figure below that the calculated rate of change at $x = 4$ is 8.31.

    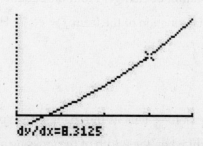

    dy/dx=8.3125

11. **SARS**:

(a) The average rate of change for N from $t = 19$ to $t = 22$ is $\dfrac{3947 - 3547}{22 - 19} = 133.33$ cases per day, and this is the approximate value of $\dfrac{dN}{dt}$ at $t = 19$.

(b) From the 19th to the 20th day we expect the cumulative number of cases of SARS to increase by about 133.

(c) The average rate of change for N from $t = 24$ to $t = 29$ is $\dfrac{5642 - 4439}{29 - 24} = 240.60$ cases per day, and this is the approximate value of $\dfrac{dN}{dt}$ at $t = 24$. Comparing this number with the result in Part (a), we see that the rate of growth in the cumulative number of cases of SARS has increased from day $t = 19$ to day $t = 24$.

12. **Account balance**:

(a) We use a horizontal span of 0 to 15 and a vertical span of 0 to 1500. In the figure on the left below we have made the graph.

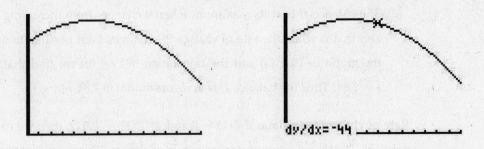

(b) In the graph on the right above we have calculated $\frac{dB}{dt}$ at $t = 8$, and the result is -44 dollars per month.

(c) We expect the balance to decrease \$44 from month 8 to month 9.

13. **Finding an equation of change**: The rate of change of depth H as a function of time t is a constant -0.5 foot per minute. Hence an equation of change satisfied by H is $\dfrac{dH}{dt} = -0.5$.

14. **Solving an equation of change**: From the equation of change $\dfrac{df}{dx} = 6f$ we know that f is an exponential function of the form $f = Pe^{rx}$. Here $r = 6$ and P is the initial value 3. The solution is $f = 3e^{6x}$.

15. **Traffic signals**:

(a) The expression $\frac{dn}{dw}$ tells us the change we expect in the number of seconds for the yellow light when the width of the crossing street is increased by 1 foot.

(b) From the equation of change

$$\frac{dn}{dw} = 0.02$$

we know that n is a linear function of w with slope 0.02. We need to find the initial value b of n. We have $n = 0.02w + b$, and using the fact that $n = 4.7$ when $w = 70$ gives $4.7 = 0.02 \times 70 + b$. Thus $b = 4.7 - 0.02 \times 70 = 3.3$. Hence the formula is $n = 0.02w + 3.3$.

16. **Atmospheric pressure**:

(a) The expression $\frac{dP}{dh}$ tells us the change we expect in the atmospheric pressure when the altitude increases by 1 kilometer.

(b) From the equation of change

$$\frac{dP}{dh} = -0.12P$$

we know that P is an exponential function.

(c) From the equation of change we have that P is an exponential function with an exponential growth rate of $r = -0.12$. The initial value is 1035 because that is the atmospheric pressure when $h = 0$. Thus the formula in the alternative form is $P = 1035e^{-0.12h}$. (For the standard form, note that the decay factor is $e^{-0.12} = 0.89$, while the initial value is 1035, so the formula is $P = 1035 \times 0.89^h$.)

17. **Equilibrium solution**: The equilibrium solutions of such an equation of change occur when $\frac{df}{dx} = 0$. Thus to find an equilibrium solution of $\frac{df}{dx} = 7f - 14$ we solve $7f - 14 = 0$. We obtain the equilibrium solution $f = 2$.

18. **Equation of change**: When $f = 1$ we have $\frac{df}{dx} = 7 \times 1 - 14 = -7$. Thus when $f = 1$ the rate of change is negative, and so f is decreasing.

19. **Sales growth:**

(a) First we make a graph of $\frac{ds}{dt}$ versus s. Since Y_1 corresponds to $\frac{ds}{dt}$ and X corresponds to s, we want to graph the function $Y_1 = 0.3X(4-X)$. We use a horizontal span of 0 to 5 and, based on a table of values, a vertical span of -1.5 to 1.5. In the figure on the left below we have made the graph.

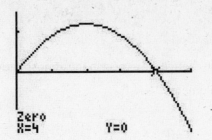   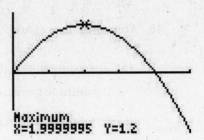

The equilibrium solutions are at the zeros of the function graphed above. Clearly one such zero is at 0, and the graph shows that another is at 4. (This can also be seen just by looking at the equation of change.) The other relevant information from the graph is summarized in the following table.

Range for s	from 0 to 4	greater than 4
Sign of $\frac{ds}{dt}$	positive	negative
Effect on s	increasing	decreasing

From the table we see that sales increase between 0 and 4 thousand dollars and decrease above a level of 4 thousand dollars. Thus a level of 4 thousand dollars will be attained in the long run. Another way to do this is to note that the equation of change can be written in the form

$$\frac{ds}{dt} = 0.3 \times 4 \times s \left(1 - \frac{s}{4}\right),$$

which shows that this equation represents logistic growth with a carrying capacity of 4 thousand dollars.

(b) The largest rate of growth in sales is the maximum value of $\frac{ds}{dt}$. In the figure on the right above we have located the maximum point on the graph, and we see that the largest rate of growth in sales is 1.2 thousand dollars per year.

20. **Critical threshold:**

(a) We substitute $r = 0.5$ and $T = 10$ into the equation and obtain

$$\frac{dN}{dt} = -0.5N \left(1 - \frac{N}{10}\right).$$

(b) Since Y_1 corresponds to $\dfrac{dN}{dt}$ and X corresponds to N, we want to graph the function $Y_1 = -0.5X(1 - X/10)$. We use a horizontal span of 0 to 12 and, based on a table of values, a vertical span of -1.5 to 1.5. In the figure below we have made the graph.

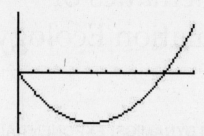

The equilibrium solutions are at the zeros of the function graphed above. Clearly one such zero is at 0, and using the single-graph method (or just looking at the original equation) shows that another is at 10. Note that 10 is the threshold level T. Thus the equilibrium solutions are $N = 0$ and the threshold $N = T = 10$.

(c) The relevant information from the graph is summarized in the following table.

Range for N	from 0 to 10	greater than 10
Sign of $\dfrac{dN}{dt}$	negative	positive
Effect on N	decreasing	increasing

From the table we see that N increases for N greater than 10 and that N decreases for N between 0 and 10.

(d) If initially N is smaller than 10 then N will decrease to 0. If initially N is larger than 10 then N will increase.

Solution Guide for Chapter 7: Mathematics of Population Ecology

7.1 POPULATION DYNAMICS: EXPONENTIAL GROWTH

S-1. **Population growth**: Change in population results from births and deaths. Births add to the population, while deaths diminish the population, so the change in population is births minus deaths. Thus the equation of change for population N as a function of time t is $\dfrac{dN}{dt} = B - D$.

S-3. **An exponential model**: If $B = 0.04N$ and $D = 0.013N$, then, since $B - D = 0.04N - 0.013N = 0.027N$, the equation of change for N is $\dfrac{dN}{dt} = 0.027N$.

S-5. **r value**: If a population has an exponential model $N = N_0 e^{rt}$, then the number r is a measure of the intrinsic per capita growth rate of the population in an environment. It is analogous to the role of APR in continuous compounding.

S-7. **Negative r-value**: If a population has an r-value of -0.03, then its population model is $N_0 e^{-0.03t}$; in particular, the population shows exponential decay, so it is declining.

S-9. **Doubling time**: If the doubling time t_d is 13 years, then the value of r can be found from $t_d = \dfrac{\ln 2}{r}$, so $13 = \dfrac{\ln 2}{r}$. Solving for r, we find that $r = \dfrac{\ln 2}{1.3} = 0.05$ per year.

S-11. **Discrete growth**: The growth factor λ is $\dfrac{100}{20} = 5$. The initial value is 20, so the formula is $N = 20 \times 5^t$, with time measured in years. After 4 years the population is $20 \times 5^4 = 12{,}500$.

1. **Norway rat**:

   (a) Since $r = 3.91$ per year and the initial population is 6, the equation of the exponential function that models the population is $N = 6e^{3.91t}$, where t is measured in years. After 6 months, or 0.5 year, the population will be

   $$N(0.5) = 6e^{3.91 \times 0.5} = 42.38,$$

or about 42 rats. After 2 years the population will be

$$N(2) = 6e^{3.91 \times 2} = 14{,}939.43,$$

or about 14,939 rats.

(b) Using the formula for doubling times $t_d = \dfrac{\ln 2}{r}$, we see that the population will reach 12 in about $\dfrac{\ln 2}{3.91} = 0.177$ year. It will take that long for the population to double again, so the population will reach 24 in 2×0.177, or about 0.35, year.

(c) Since $r = 3.91$, the formula for N is $N = N(0)e^{3.91t}$. When $t = 2$, then $N = 150{,}000$, so we have the equation $N(0)e^{3.91 \times 2} = 150{,}000$. We solve this equation for $N(0)$. Dividing both sides by $e^{3.91 \times 2}$, we get $N(0) = \dfrac{150{,}000}{e^{3.91 \times 2}}$. Evaluating the fraction on the right, we get $N(0) = 60.24$. So the initial population was about 60 rats.

3. **Exponential growth rate from doubling time**: To solve $rt = \ln 2$ for r, divide both sides by t. The result is $r = \dfrac{\ln 2}{t}$. For a population taking 7 years to double, we find $r = \dfrac{\ln 2}{7} = 0.099$ per year. Thus 0.099 is the exponential growth rate per year for a population that takes 7 years to double.

5. **Caribou**: Here we use linear regression to find r. The figure below shows that the regression line for the natural logarithm of the data is $\ln N = 0.35t + 2.91$ if N denotes the population. The slope of the regression line is 0.35 per year, and so this is the value of r.

```
LinReg
y=ax+b
a=.352642991
b=2.91319131
```

Here is another way to estimate r: Notice that the population doubles every two years approximately, and then use the formula in Exercise 3 to estimate $r = \dfrac{\ln 2}{2} = 0.35$ per year.

7. **Field mice:**

   (a) The plot of the natural logarithm of the data is shown in the left-hand figure below. (Here the horizontal axis is the time in years, and the vertical axis is the population.) Since this plot is close to linear, it is reasonable to model the original data with an exponential function.

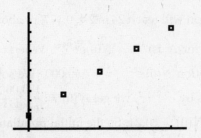

```
LinReg
y=ax+b
a=4.499723736
b=1.386775086
```

   (b) The right-hand figure above shows that the regression line for the logarithm of the data is $\ln N = 4.500t + 1.387$. Thus r is 4.50 per year. The doubling time t_d is then calculated as $t_d = \dfrac{\ln 2}{4.50} = 0.15$ year.

   (c) Since $r = 4.50$ and the initial value is found from the regression line as $e^{1.387} = 4.00$, the exponential model is given by $N = 4e^{4.5t}$.

   (d) Now 9 months is $\dfrac{9}{12} = 0.75$ year, so $t = 0.75$. Thus in 9 months the population should be $N(0.75) = 4e^{4.5 \times 0.75}$, or about 117 mice.

9. **Aphid growth on different plants:** Calculating the r value using the slope of regression line for the logarithm of the data, we find $r = 0.180$ per day for the aphids raised on collard and $r = 0.094$ per day for the aphids raised on yellow rocket. This tells us that the population of aphids raised on collard increased more rapidly than did the population on yellow rocket.

   Another approach is to examine the tables and note that the population on collard doubled within 4 days while the population on yellow rocket did not.

11. **Microbes:** Since the population triples every day and the initial population is 2, then the growth is modeled by $N = 2 \times 3^t$, where t is measured in days. We made a table of values for the function (as shown below) to choose the window. Based on this, we graphed the function using a horizontal span of 0 to 5 and a vertical span of 0 to 500, as shown below. (The horizontal axis corresponds to time in days, and the vertical axis corresponds to population.)

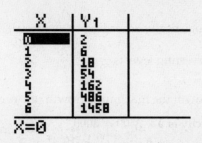
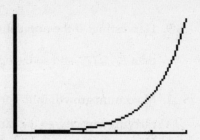

The table shows that after 5 days there are 486 microbes.

13. **How the size of λ affects geometric growth**: If λ is less than 1 then the population will decrease in size. The reason is that λ is by definition the factor by which the population grows from one unit of time to the next; if this factor is less than 1, the population must decrease.

By similar reasoning, if $\lambda = 1$ then the population will not change in size.

7.2 POPULATION DYNAMICS: LOGISTIC GROWTH

S-1. **Logistic growth**: If the notion of environmental carrying capacity is added to the basic assumptions leading to the exponential model, then we get the logistic model.

S-3. **More on the logistic equation of change**: We assume the population grows according to $\dfrac{dN}{dt} = 0.13N\left(1 - \dfrac{N}{378}\right)$.

(a) The r-value for this species in this environment is the leading coefficient, which is 0.13.

(b) The environmental carrying capacity K is the denominator in the fraction with N, so it is 378.

(c) The equilibrium solutions of the equation are found when $\dfrac{dN}{dt} = 0$, so when $0 = 0.13N\left(1 - \dfrac{N}{378}\right)$. Solving for N, we see that the equilibrium solutions are $N = 0$ and $N = K = 378$.

(d) The function N will have the maximum growth rate when $N = \dfrac{K}{2} = \dfrac{378}{2} = 189$.

S-5. **Formula for logistic growth**: The formula for logistic growth corresponding to the equation of change $\dfrac{dN}{dt} = rN\left(1 - \dfrac{N}{K}\right)$ is $N = \dfrac{K}{1 + be^{-rt}}$.

S-7. **More logistic formula**: To find the logistic formula that governs a population which grows logistically with carrying capacity $K = 2390$, $r = 0.015$, and $N_0 = 120$, we first find $b = \dfrac{K}{N_0} - 1 = \dfrac{2390}{120} - 1 = 18.92$. Using the formula for logistic growth $N = \dfrac{K}{1 + be^{-rt}}$, we have $N = \dfrac{2390}{1 + 18.92e^{-0.015t}}$.

S-9. **Harvesting**: If the population grows logistically according to $\frac{dN}{dt} = 0.02N\left(1 - \frac{N}{778}\right)$, then $K = 778$, and so the optimum harvesting level is $\frac{K}{2} = \frac{778}{2} = 389$.

S-11. **Maximum growth rate**: For logistic growth the maximum growth rate occurs at half the carrying capacity, so the carrying capacity is $2 \times 2500 = 5000$.

1. **Estimating optimum yield**: To answer this we locate the inflection point. The optimum yield level is about 200, and this occurs when the time is about 8.

3. **Northern Yellowstone elk**: First we must find the regression line for the data in the table as in Example 7.4. We enter the values for N in the first column and the values for $\frac{1}{N}\frac{dN}{dt}$ in the second column, as shown in the left-hand figure below. The right-hand figure below shows that the slope of the regression line is -2.53×10^{-5}, and the vertical intercept is 0.374. Since r is the vertical intercept, we see that $r = 0.374$ per year. Now $K = -\frac{r}{m}$, so $K = -\frac{0.374}{-2.53 \times 10^{-5}} = 14{,}783$, or about 14,800.

5. **More on the Pacific sardine**:

   (a) Note first that 50,000 tons equals 0.05 million tons. We use the function N from Example 7.5, find t_1 such that $N(t_1) = 0.05$, and find t_2 such that $N(t_2) = 1.2$. Since t_1 is the time representing the current population (50,000 tons), and t_2 is the time at which the population is 1.2 million tons, the difference $t_2 - t_1$ is the time it will take to get from 50,000 tons to 1.2 million tons.

   To find t_1 we need to solve the equation $\frac{2.4}{1 + 239e^{-0.338t}} = 0.05$. We do this using the crossing graphs method. Based on the table below, we used a horizontal span of 0 to 7 and a vertical span of 0 to 0.1 in the graph below. (The horizontal axis is time in years, and the vertical axis is population in millions of tons.) We see that the desired time t_1 is 4.812 years.

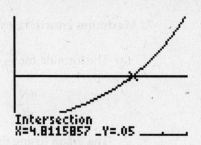

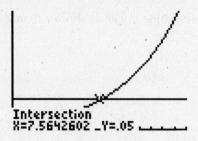

The time t_2 at which the population reaches the optimum growth level of 1.2 million tons was found in Figure 7.21 of Example 7.5 to be 16.203 years. Then the difference is $t_2 - t_1 = 16.203 - 4.812$, or about 11.39 years. This is the time required to recover from a level of 50,000 tons to 1.2 million tons.

(b) Here we need to adjust the r value in the formula found in Example 7.5 to get $N = \dfrac{2.4}{1 + 239e^{-0.215t}}$. We then proceed as in the solution to Part (a) above by entering this new formula for N and finding t_1 and t_2.

To find t_1 we make the graph using a horizontal span of 0 to 15 and a vertical span of 0 to 0.2. The left-hand figure below shows that, with the reduced growth rate, the level of 50,000 tons is reached at $t_1 = 7.564$ years.

To find t_2 we put 1.2 in place of 0.05 on the function entry screen and make the graph using a horizontal span of 0 to 40 and a vertical span of 0 to 2.5 as in Example 7.5. The right-hand figure below shows that, with the reduced growth rate, the level of 1.2 million tons is reached at $t_2 = 25.472$ years. Then the difference is $t_2 - t_1 = 25.472 - 7.564$, or about 17.91 years. This is the time required to recover from a level of 50,000 tons to 1.2 million tons with the reduced growth rate. Note that this is considerably longer than the time of 11.39 years found in Part (a).

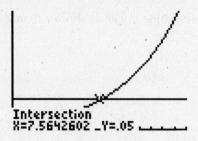

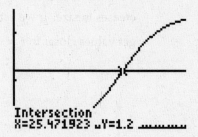

7. **Maximum growth rate for tuna:**

(a) The formula for $\dfrac{dN}{dt}$ is

$$\frac{dN}{dt} = 2.61 \times \frac{148}{1 + 3.6e^{-2.61t}} \left(1 - \frac{148/(1 + 3.6e^{-2.61t})}{148}\right).$$

The graph (using the same window as in Exercise 6) is shown below at the left.

(b) We find the maximum of $\dfrac{dN}{dt}$ on the graph as shown below at the right. The growth rate is largest at $t = 0.49$ year. (Note that this agrees with our answer to Part (e) of Exercise 6.)

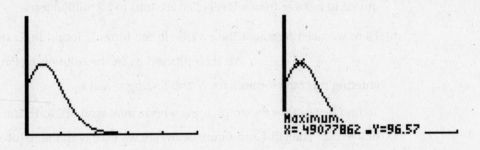

```
Maximum.
X=.49077862 ▪Y=96.57
```

(c) If we evaluate the function N using the graph we had in Exercise 6, we find that $N(0.49) = 73.93$, or about 74 thousand tons. This is the amount we found in Part (c) of Exercise 6, since the optimum yield level is the population at the time of fastest growth.

9. **Yeast growth rate:** The solution to this exercise can be found in the discussion surrounding Equation 7.7. The r value from the logistic model gives the per capita growth rate in the absence of constraints. The r value obtained in Example 7.2 gives a lower estimated per capita growth rate because growth is constrained as the population increases in size. If we estimated r using time intervals shorter than 7 hours, we would get values closer to $r = 0.54$.

11. **Logarithm of the logistic curve**: A sketch of the graph of $\ln N$ against t is shown below. Since the first part of the graph in Figure 7.15 looks like an exponential curve, when we take the logarithm this part should look like a linear function. The last part of the graph in Figure 7.15 is nearly constant, and so that part of the graph of $\ln N$ should be nearly constant.

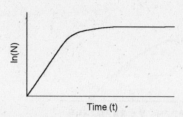

13. **Gompertz model**:

(a) We want to graph $\dfrac{dN}{dt} = N \ln \left(\dfrac{10}{N} \right)$ as a function of N. Based on the table of values below, we chose a horizontal span of 0 to 12 and a vertical span of -1 to 6 to get the graph below.

(b) From the graph, we see that the maximum value for $\dfrac{dN}{dt}$ occurs at about $N = 3.68$.

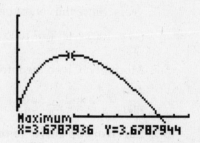

(c) From the graph below at the left, we see that when $K = 1$ the maximum for $\dfrac{dN}{dt}$ occurs at about $N = 0.368$. From the graph below at the right, we see that when $K = 100$ the maximum for $\dfrac{dN}{dt}$ occurs at about $N = 36.8$. Hence the optimum yield level seems to be about $0.368K$ for the Gompertz model, as opposed to $\frac{1}{2}K = 0.5K$ for the logistic model.

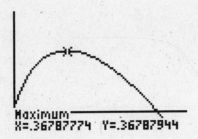
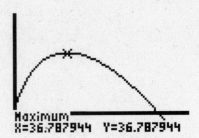

7.3 POPULATION STRUCTURE: SURVIVORSHIP CURVES

S-1. **Groups**: A "cohort" is a group of individuals in the same age range.

S-3. **Calculation of mortality statistics**: We can calculate q_x from l_x and d_x using $q_x = \dfrac{d_x}{l_x}$.

S-5. **Number of survivors**: The value of l_{31} is the number of survivors at the start of age 31 years. Since the cohort begins at 1000 individuals and, by age 31 years, 60% have died, $l_{31} = 1000 - 60\%$ of $1000 = 1000 - 600 = 400$.

S-7. **Mortality**: Given the information from Exercises S-5 and S-6, we calculate $q_{31} = \dfrac{d_{31}}{l_{31}} = \dfrac{120}{400} = 0.3$ and $l_{32} = l_{31} - d_{31} = 400 - 120 = 280$ individuals.

S-9. **Survivorship curve**: In practical terms a survivorship curve shows how many individuals survive to a given age.

S-11. **Type II**: The shape of a type II survivorship curve on a semi-logarithmic scale is a straight line.

1. **Classifying survivorship curves**:

   (a) Figure 7.29 appears to show a Type I survivorship curve. The first part of the curve indicates that many juvenile animals survive to adulthood, and the last part of the curve indicates that the older animals die off quickly

   (b) Figure 7.30 appears to show a Type III survivorship curve. The first part of the graph indicates a high mortality rate among very young seedlings. This is followed by a lower rate in the last part of the curve.

(c) Figure 7.31 appears to show a Type II survivorship curve. The graph has many dips, but it appears to indicate a relatively constant mortality rate (the juvenile and older animals have similar mortality rates). There is, however, a dramatic drop at the end.

3. **Graphing and classifying survivorship curves**:

   (a) The graph of $\log l_x$ against x for the desert night lizard is shown below at the left. (The horizontal span is -0.8 to 8.8, and the vertical span is 1.77 to 3.18.) This appears to be a Type I survivorship curve: The first part of the graph indicates a low mortality rate among the juvenile lizards, and the last part of the graph indicates a high mortality rate among the older lizards.

   (b) The graph of $\log l_x$ against x for the American robin is shown below at the right. (The horizontal span is -0.6 to 6.6, and the vertical span is 0.4 to 3.38.) This appears to be a Type II curve: It is close to being linear, so the mortality is nearly the same among all the age groups in the population.

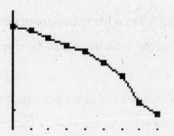
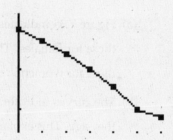

5. **Sagebrush lizard mortality**:

   (a) The suggestion given in the exercise shows that $d_0 = 770$ and $l_1 = 230$. Then

   $$d_1 = q_1 \times l_1 = 0.378 \times 230 = 86.94,$$

   which we round to 87. Thus

   $$l_2 = l_1 - d_1 = 230 - 87 = 143.$$

   Continuing in this way gives the following table.

x	l_x	d_x	q_x
0	1000	770	0.770
1	230	87	0.378
2	143	67	0.469
3	76	42	0.553
4	34	21	0.618
5	13	6	0.462

(b) Here is the graph of $\log l_x$ against x. (The horizontal span is -0.5 to 5.5, and the vertical span is 0.79 to 3.32.)

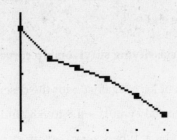

There is a relatively steep drop at the first, but after that the curve is nearly linear. This is a Type II curve.

7. **Experience of no use**: For this animal we expect that all age groups will have the same mortality rate. So this animal has a Type II survivorship curve.

9. **Life insurance rates**: The general principle here is that we expect higher insurance rates for higher mortality rates.

(a) Figure 7.35 indicates that the mortality rates for males are generally higher than those for females. That explains why life insurance rates for men are higher than those for women.

(b) The curves indicate that the mortality rates for older persons increase greatly as they age. This explains why insurance rates for the elderly increase so rapidly.

(c) If the curves were of Type II then the insurance rates should be about the same for all ages, since for Type II curves all ages have about the same mortality rate.

If the curves were of Type III then the insurance rates would be higher for children, since their mortality rates would be higher than those of older persons.

Chapter 7 Review Exercises

1. **Exponential growth rate**: Using the formula for doubling times $t_d = \dfrac{\ln 2}{r}$, we see that the population will double in size in about $\dfrac{\ln 2}{0.02} = 34.66$ years.

2. **Doubling time**: The formula for doubling times is $t_d = \dfrac{\ln 2}{r}$. If we put in $t_d = 5$ we get the equation $5 = \dfrac{\ln 2}{r}$. Solving for r gives $r = \dfrac{\ln 2}{5} = 0.14$. Thus the exponential growth rate r is about 0.14 per month.

3. **Population data**: Here we use linear regression to find r. The figure below shows that the regression line for the natural logarithm of the data is $\ln N = 0.20t + 5.86$. The slope of the regression line is 0.20 per month, and so this is the value of r.

```
LinReg
y=ax+b
a=.1999616002
b=5.858131219
```

4. **Doubling time**: According to the table the population doubles every 6 years, so $t_d = 6$ years. Proceeding as in Review Exercise 2 above, we get the equation $r = \dfrac{\ln 2}{6} = 0.12$. Thus the exponential growth rate r is about 0.12 per year.

5. **Logistic model**: We assume the population grows according to $\dfrac{dN}{dt} = 0.01N\left(1 - \dfrac{N}{200}\right)$. The environmental carrying capacity K is the denominator in the fraction with N, so it is 200.

6. **Logistic model**: If the r value is 0.02 and the environmental carrying capacity is 2500 then the equation of change is $\dfrac{dN}{dt} = 0.02N\left(1 - \dfrac{N}{2500}\right)$.

7. **Logistic formula**:

    (a) The equation given in the exercise matches the integral form of the logistic equation, so we read the r value from the formula: The r value for this population is 1.4 per year.

    (b) The constant in the numerator is 25, so the carrying capacity for this population is 25 thousand.

    (c) The optimum yield level is half of the carrying capacity, so it is 12.5 thousand.

8. **Inflection point**: For logistic growth the inflection point corresponds to the maximum growth rate, and this occurs at half the carrying capacity. Thus the carrying capacity is $2 \times 320 = 640$.

9. **Calculation of mortality statistics**: Now d_3 is the number dying during the given period, so $d_3 = l_3 - l_4 = 460 - 420 = 40$. The proportion q_3 is calculated as $q_3 = \dfrac{d_3}{l_3} = \dfrac{40}{460} = 0.087$.

10. **Calculation of mortality statistics**: Since $q_7 = \dfrac{d_7}{l_7}$, we have $0.08 = \dfrac{d_7}{500}$. Solving for d_7 gives $d_7 = 0.08 \times 500 = 40$. Now $l_8 = l_7 - d_7 = 500 - 40 - 460$.

11. **Low juvenile mortality**: If a species has low juvenile mortality, but mortality increases rapidly for older individuals, then it has a Type I survivorship curve.

12. **Life table**: The following is the graph of $\log l_x$ against x. (The horizontal span is -0.8 to 8.8, and the vertical span is 0.93 to 3.30.)

Clearly this shows a constant mortality rate, so it is a Type II survivorship curve.

PREFACE TO TECHNOLOGY GUIDE

This guide is written as a technology supplement for *Functions and Change: A Modeling Approach to College Algebra* by Bruce Crauder, Benny Evans, and Alan Noell. It is intended to provide basic instruction in the use of graphing calculators and spreadsheets. Specific instructions are provided for the Texas Instruments TI-82, TI-83, TI-83 Plus, and TI-84 Plus, and for Microsoft Excel. Many other graphing calculators and spreadsheets operate in fashions similar to these. All of the operations necessary for the text are included, and they are presented in the order in which they are needed for the text.

The text *Functions and Change* is a technology-dependent work, and effective calculator or computer use is absolutely essential for success. A concerted effort is made to keep the instructions here basic and straightforward and to make this manual as brief as is practical. This is made easy by the technology itself, which is designed to provide a maximum of power with a minimum learning curve. As you become familiar with the operation of your calculator or spreadsheet program, you will find that there are many shortcuts available which are not covered here. Most often we show only what we consider to be the easiest method of producing a desired output. The TI-82, TI-83, and TI-84 are very flexible instruments which may offer a number of options for getting the same result, and Excel, which may on the surface appear to be a simple spreadsheet program, is in fact a very complex and powerful piece of software. You will discover some shortcuts yourself and others by interacting with your colleagues, but you should also look at the extensive manuals which come with the calculator or with Excel. You may find methods not presented here that you prefer, and you are encouraged to use them.

This guide is organized into three major parts. Part I consists of calculator Quick Reference pages cross-referenced with the *Functions and Change* text. This seems practical only for calculators, and so quick reference pages for Excel are not included. Part II is the Calculator Keystroke Guide. For each topic from *Functions and Change* which requires technology, instructions are given for the graphing calculator. In Part III are the corresponding instructions for Excel. Instructions for calculators and Excel are quite different, and so are kept separate. Excel users can employ this guide without encountering calculator instructions and vice versa. These three divisions of this guide are described in more detail below.

In Part I of this guide are the Quick Reference pages. These pages provide the TI-83, TI-83 Plus, and TI-84 Plus keystrokes for creating tables and graphs, entering expressions, and making the other calculator operations needed in using the text. These Quick Reference pages provide two types of calculator information: instructions for calculator operations and a list of cross-referenced keystroke instructions. The Quick Reference pages are organized section-by-section with the *Functions and Change* text for those sections which require new calculator skills. For each section, succinct instructions are given for the needed calculator operations, followed by a list of explicit keystrokes cross-referenced to the text. These keystrokes are cross-referenced using footnote boxes in the *Functions and Change* text. For example, [2.5] on page 98 of the text indicates that the exact keystrokes needed to create the table there are found as item number **2.5**, the fifth such item for Chapter 2, in the Quick Reference pages (in this case, page 315) of this guide. You may find it convenient to select the reference pages that go with your calculator, remove them from this manual, and keep them with your copy of *Functions and Change*.

Part II of this guide is an extensive reference guide to the use of the graphing calculator. This part goes far beyond the keystrokes themselves and carefully details the use of the calculator to perform specific procedures. Not every section of *Functions and Change* involves new technology procedures, so Part II is not a section-by-section guide, but rather is more general. Three topics are covered in depth: arithmetic operations, making tables and graphs, and the treatment of discrete data, particularly the use of regression to find best-fitting functions.

Part III is a repeat of much of Part II for Excel. Special thanks to Deborah Benton of Wake Technical Community College for her help with this part. Whatever errors or shortcomings may be here are the result of our poor execution of her excellent advice.

If the manual is to be brief, the Preface should be so too.

Bruce Crauder, Benny Evans, Alan Noell

2005

PART I *Quick Reference Pages*

QUICK REFERENCE FOR THE PROLOGUE

> **Instructions for the Prologue**
> **Keyboard use**
>
> - For the TI-83 and TI-83 Plus, the gold $\boxed{\text{2nd}}$ key is used to access special gold symbols on the keyboard face. For the TI-84 Plus, the blue $\boxed{\text{2nd}}$ key is used to access special blue symbols on the keyboard face.
>
> - $\boxed{\text{2nd}}$ [$\sqrt{\ }$] displays the square root symbol and an open parenthesis.
>
> - $\boxed{\text{2nd}}$ [π] displays the special number π.
>
> - $\boxed{\text{2nd}}$ [ANS] recalls the answer from the previous calculation.
>
> - $\boxed{\text{2nd}}$ [e] displays the special number e.
>
> - $\boxed{-}$ denotes subtraction between two numbers while $\boxed{(-)}$ denotes a negative sign.

Keystroke cross reference

0.1: 71 $\boxed{\div}$ 7 $\boxed{+}$ 3 $\boxed{\wedge}$ 2 $\boxed{\times}$ 5

0.2: 4 $\boxed{(}$ 2 $\boxed{+}$ 1 $\boxed{)}$

0.3: 17 $\boxed{\div}$ $\boxed{(}$ 5 $\boxed{+}$ 3 $\boxed{)}$

0.4: $\boxed{(}$ 8 $\boxed{+}$ 9 $\boxed{)}$ $\boxed{\div}$ $\boxed{(}$ 7 $\boxed{+}$ 2 $\boxed{)}$

0.5: 3 $\boxed{\wedge}$ $\boxed{(}$ 2.7 $\boxed{\times}$ 1.8 $\boxed{)}$

0.6: $\boxed{-}$ is used for subtraction as in 9 $\boxed{-}$ 6 while $\boxed{(-)}$ is used to denote a negative number as in $\boxed{(-)}$ 3 .

0.7: `(−)` 8 `−` 4

0.8: `(` 3 `−` 7 `)` `÷` `(` `(−)` 2 `×` 3 `)`

0.9: 2 `∧` `(−)` 3

0.10: 2 `∧` `−` 3 gives a syntax error

0.11: `2nd` `[ √ ]` 11.4 `−` 3.5 `)` `÷` 26.5

0.12: `(` 7 `×` 3 `∧` `(−)` 2 `+` 1 `)` `÷` `(` 3 `−` 2 `∧` `(−)` 3 `)`

0.13: `2nd` `[ π ]`

0.14: `ENTER`

0.15: `2nd` `[ e ]`

0.16: `2nd` `[ eˣ ]` 1.02 `)`

0.17: `2nd` `[ ANS ]`

0.18: `(` `2nd` `[ √ ]` 13 `)` `−` `2nd` `[ √ ]` 2 `)` `)` `∧` 3

0.19: `2nd` `[ ANS ]` `+` 17 `÷` `(` 2 `+` `2nd` `[ π ]` `)`

0.20: `(` 3 `∧` `2nd` `[ ANS ]` `+` 2 `∧` `2nd` `[ ANS ]` `)` `÷`
`(` 5 `∧` `2nd` `[ ANS ]` `−` 4 `∧` `2nd` `[ ANS ]` `)`

0.21: 5000 `(` 1 `+` .07 `×` 10 `)`
   or 5000 `×` `(` 1 `+` .07 `×` 10 `)`

0.22: 5000 `×` 1.07 `∧` 10

0.23: 5000 `(` 1 `+` `2nd` `[ ANS ]` `)` `∧` 120

QUICK REFERENCE FOR CHAPTER 1

Keystroke cross reference

1.1: `(` 3 `∧` 2 `+` 1 `)` `÷` `2nd` `[ √ ]` 3 `)`

1.2: 1.0067 `∧` 48 `−` 1

1.3: 7800 `×` .0067 `×` 1.0067 `∧` 48 `÷` `2nd` `[ ANS ]`

1.4: 1.0058 $\boxed{\wedge}$ 36 $\boxed{-}$ 1

1.5: 5000 $\boxed{\times}$ .0058 $\boxed{\times}$ 1.0058 $\boxed{\wedge}$ 36 $\boxed{\div}$ $\boxed{\text{2nd}}$ [ANS]

QUICK REFERENCE FOR CHAPTER 2

$\boxed{\text{Instructions for Section 2.1}}$
How to enter functions

Step 1: Press the $\boxed{Y=}$ key to get to the function screen.

Step 2: Use $\boxed{\text{CLEAR}}$ to delete any unwanted functions that appear.

Step 3: Enter your function using the $\boxed{X, T, \Theta, n}$ key for the variable.

$\boxed{\text{Instructions for Section 2.1}}$
How to make a table of values

Step 1: Enter your function as described above.

Step 2: Press $\boxed{\text{2nd}}$ [TBLSET] to get to the TABLE SETUP menu.

Step 3: Set TblStart= to the starting place for the table.

Step 4: Set △Tbl= to the increment value for the table.

Step 5: Press $\boxed{\text{2nd}}$ [TABLE] to view the table. The arrow keys $\boxed{\triangle}$ and $\boxed{\nabla}$ will let you see more of the table.

Keystroke cross reference

2.1: $\boxed{Y=}$ 6.21 $\boxed{\div}$ $\boxed{(}$ 0.035 $\boxed{+}$ 0.45 $\boxed{\wedge}$ $\boxed{X, T, \Theta, n}$ $\boxed{)}$

2.2: $\boxed{\text{2nd}}$ [TBLSET] TblStart=0 △Tbl=5

2.3: $\boxed{Y=}$ 176 $\boxed{(}$ 1 $\boxed{-}$ .834 $\boxed{\wedge}$ $\boxed{X, T, \Theta, n}$ $\boxed{)}$

2.4: $\boxed{\text{2nd}}$ [TBLSET] TblStart=0 △Tbl=5

2.5: $\boxed{\text{2nd}}$ [TABLE]

2.6: $\boxed{\nabla}$

2.7: $\boxed{Y=}$ $\boxed{X,T,\Theta,n}$ $\boxed{(}$ $\boxed{X,T,\Theta,n}$ $\boxed{-}$ $\boxed{1}$ $\boxed{)}$ $\boxed{(}$ $\boxed{X,T,\Theta,n}$ $\boxed{-}$ $\boxed{2}$ $\boxed{)}$
$\boxed{\div}$ $\boxed{750}$ $\boxed{\times}$ $\boxed{(}$ $\boxed{5}$ $\boxed{\div}$ $\boxed{6}$ $\boxed{)}$ $\boxed{\wedge}$ $\boxed{X,T,\Theta,n}$

2.8: $\boxed{\text{2nd}}$ $\boxed{[\ \text{TBLSET}\ ]}$ TblStart=1 ΔTbl=1

2.9: $\boxed{\nabla}$

2.10: $\boxed{Y=}$ $\boxed{X,T,\Theta,n}$ $\boxed{(}$ $\boxed{36}$ $\boxed{-}$ $\boxed{X,T,\Theta,n}$ $\boxed{)}$

2.11: $\boxed{\text{2nd}}$ $\boxed{[\ \text{TBLSET}\ ]}$ TblStart=1 ΔTbl=1

Instructions for Section 2.2
How to make graphs

Step 1: Use the $\boxed{Y=}$ key to open the function input window.

Step 2: Enter the function using $\boxed{X,T,\Theta,n}$ for the variable.

Step 3: Determine an appropriate $\boxed{\text{WINDOW}}$ setup by using practical information you have about the function. A table of values can be helpful in choosing the vertical span.

Alternative Step 3: For some functions it is appropriate to bypass Step 3 and go directly to the ZStandard view using $\boxed{\text{ZOOM}}$ 6 .

Instructions for Section 2.2
How to adjust graphs and get function values

1. For some graphs it is best to look first at the **ZStandard** viewing screen using ZOOM 6 .

2. The TRACE key will make the cursor follow the graph, and some function values can be read from the bottom of the screen.

3. To get more precise function values, first be sure TRACE is on. Then type in the number you want and press ENTER . The cursor will move to the point you want and give the function value at the Y= prompt.

4. If a graph does not appear in the **ZStandard** screen, try TRACE ENTER to find and show the graph.

5. If you want to get a closer look near a point (zoom in), first move the cursor there, and then use ZOOM 2 ENTER .

6. If you want to see a larger view (zoom out), use ZOOM 3 ENTER .

7. Manual adjustment of the viewing screen can be made using WINDOW . The **Xmin** and **Xmax** values determine the horizontal span of the viewing screen. The **Ymin** and **Ymax** values determine the vertical span of the viewing screen. Proper values for **Ymin** and **Ymax** can often be found by looking at a table of values.

2.12: $\boxed{Y=}$ $\boxed{X,T,\Theta,n}$ $\boxed{\wedge}$ 2 $\boxed{-}$ 1

2.13: $\boxed{\text{GRAPH}}$

2.14: $\boxed{\text{ZOOM}}$ and then select **6:ZStandard** from the menu.

Note that the **ZStandard** view can be (and may have been) changed on your calculator.

If **ZStandard** does not show the picture we made, see p. 346 in Part II of this Guide.

2.15: $\boxed{\text{TRACE}}$

2.16: $\boxed{\text{TRACE}}$ 3 $\boxed{\text{ENTER}}$

2.17: $\boxed{Y=}$ 22.75 $\boxed{X,T,\Theta,n}$ $\boxed{-}$ 300

2.18: $\boxed{\text{ZOOM}}$ 6

2.19: $\boxed{\text{WINDOW}}$ **Xmin**=0 **Xmax**=25 **Ymin**= $\boxed{(-)}$ 325 **Ymax**=300

2.20: $\boxed{\text{GRAPH}}$

2.21: $\boxed{\text{TRACE}}$ 10 $\boxed{\text{ENTER}}$

2.22: $\boxed{Y=}$ 750 $\boxed{\div}$ $\boxed{(}$ 8 $\boxed{X,T,\Theta,n}$ $\boxed{)}$

2.23: $\boxed{\text{2nd}}$ [TBLSET] TblStart=1 ΔTbl=2

2.24: $\boxed{\text{WINDOW}}$ Xmin=1 Xmax=11 Ymin=0 Ymax=100

2.25: $\boxed{\text{GRAPH}}$

2.26: $\boxed{\text{TRACE}}$ 7 $\boxed{\text{ENTER}}$

2.27: $\boxed{Y=}$ $\boxed{X,T,\Theta,n}$ $\boxed{\div}$ 40

2.28: $\boxed{\text{WINDOW}}$ Xmin=500 Xmax=1000 Ymin=0 Ymax=30

2.29: $\boxed{Y=}$ 13 $\boxed{\div}$ $\boxed{(}$ .93 $\boxed{\wedge}$ $\boxed{X,T,\Theta,n}$ $\boxed{+}$ .05 $\boxed{)}$

2.30: $\boxed{Y=}$ Type 260 on the $Y_2=$ line. $\boxed{\text{GRAPH}}$

2.31: $\boxed{Y=}$ 235 $\boxed{-}$ 105 $\boxed{\text{2nd}}$ [e^x] $\boxed{(-)}$ .3 $\boxed{X,T,\Theta,n}$ $\boxed{)}$

2.32: $\boxed{\text{WINDOW}}$ Xmin=0 Xmax=36 Ymin=100 Ymax=250

2.33: $\boxed{Y=}$ Y_2=200 $\boxed{\text{GRAPH}}$

2.34: $\boxed{Y=}$ Y_3=235 $\boxed{\text{GRAPH}}$

2.35: $\boxed{Y=}$ Y_1= 40 $\boxed{+}$ .29 $\boxed{X,T,\Theta,n}$ Y_2= 50 $\boxed{+}$ .14 $\boxed{X,T,\Theta,n}$

2.36: $\boxed{\text{WINDOW}}$ Xmin=0 Xmax=100 Ymin=20 Ymax=90

2.37: $\boxed{Y=}$ Move the cursor to the left of Y_2 and press $\boxed{\text{ENTER}}$

Instructions for Section 2.4
How to solve an equation of the form *Left Side=Right Side*

Crossing-Graphs Method

Step 1: Use $\boxed{Y=}$ to enter *Left Side* on the $Y_1=$ line and *Right Side* on the $Y_2=$ line.

Step 2: $\boxed{\text{GRAPH}}$ and make necessary adjustments so that the crossing point shows on the graphing screen.

Step 3: Press $\boxed{\text{2nd}}$ [CALC] and select 5:intersect from the menu.

Step 4: Press $\boxed{\text{ENTER}}$ at each of the prompts First curve? and Second curve?.

Step 5: If there is only one crossing point on the screen, press $\boxed{\text{ENTER}}$ at the Guess? prompt. If there is more than one crossing point on the screen, move the cursor near the one you want before pressing $\boxed{\text{ENTER}}$.

Step 6: At the Intersection prompt read the solution from the X= display.

Single-Graph Method

Step 1: Write the equation as *Left Side−Right Side*= 0.

Step 2: Use $\boxed{Y=}$ to enter *Left Side−Right Side* as a single function.

Step 3: $\boxed{\text{GRAPH}}$ and make necessary adjustments so that the zero is shown.

Step 4: Press $\boxed{\text{2nd}}$ [CALC] and select 2:zero from the menu.

Step 5: At the Left Bound? prompt move the cursor to the left of the zero and press $\boxed{\text{ENTER}}$.

Step 6: At the Right Bound? prompt move the cursor to the right of the zero and press $\boxed{\text{ENTER}}$.

Step 7: At the Guess? prompt press $\boxed{\text{ENTER}}$.

Step 8: At the Zero prompt read the solution from the X= display.

2.38: $\boxed{Y=}$ $Y_1=$ 800 $\boxed{-}$ 730 $\boxed{\text{2nd}}$ [e^x] $\boxed{(-)}$ .06 $\boxed{X,T,\Theta,n}$ $\boxed{)}$
$Y_2=$ 600

2.39: $\boxed{\text{WINDOW}}$ Xmin=0 Xmax=30 Ymin=0 Ymax=700

2.40: $\boxed{\text{2nd}}$ [CALC] 5:intersect $\boxed{\text{ENTER}}$ $\boxed{\text{ENTER}}$ $\boxed{\text{ENTER}}$

2.41: $\boxed{Y=}$ 102 $\boxed{+}$ 12 $\boxed{X,T,\Theta,n}$ $\boxed{-}$ 100 $\boxed{\text{2nd}}$ [e^x] .1 $\boxed{X,T,\Theta,n}$ $\boxed{)}$

2.42: $\boxed{\text{WINDOW}}$ Xmin=0 Xmax=5 Ymin= $\boxed{(-)}$ 5 Ymax=6

2.43: $\boxed{\text{2nd}}$ [CALC] 2:zero Move cursor left of zero $\boxed{\text{ENTER}}$.
Move cursor right of zero $\boxed{\text{ENTER}}$ $\boxed{\text{ENTER}}$

2.44: $\boxed{Y=}$ Y$_1$= 6.21 $\boxed{\div}$ $\boxed{(}$ .035 $\boxed{+}$ .45 $\boxed{\wedge}$ $\boxed{X,T,\Theta,n}$ $\boxed{)}$

2.45: $\boxed{Y=}$ Y$_2$= 85

2.46: $\boxed{\text{2nd}}$ [CALC] 5:intersect $\boxed{\text{ENTER}}$ $\boxed{\text{ENTER}}$ $\boxed{\text{ENTER}}$

2.47: $\boxed{Y=}$ Y$_1$= 62.4 $\boxed{\text{2nd}}$ [π] $\boxed{X,T,\Theta,n}$ $\boxed{\wedge}$ 2 $\boxed{(}$ 2 $\boxed{-}$ $\boxed{X,T,\Theta,n}$ $\boxed{\div}$ 3 $\boxed{)}$
Y$_2$= 436

2.48: $\boxed{\text{2nd}}$ [CALC] 5:intersect $\boxed{\text{ENTER}}$ $\boxed{\text{ENTER}}$ $\boxed{\text{ENTER}}$

Instructions for Section 2.5
How to locate maxima and minima on a graph

First graph the function and adjust the window as necessary so that the maxima and minima are clearly shown. Once this is done proceed as follows.

Step 1: Press $\boxed{\text{2nd}}$ [CALC] and select 3:minimum or 4:maximum from the menu.

Step 2: At the Left Bound? prompt move the cursor to the left of the maximum or minimum and then press $\boxed{\text{ENTER}}$.

Step 3: At the Right Bound? prompt move the cursor to the right of the maximum or minimum and then press $\boxed{\text{ENTER}}$.

Step 4: At the Guess? prompt press $\boxed{\text{ENTER}}$.

Step 5: At the Minimum or Maximum prompt read the location of the maximum or minimum from the bottom of the display.

2.49: $\boxed{Y=}$ $\boxed{X,T,\Theta,n}$ $\boxed{-}$ 32 $\boxed{(}$ $\boxed{X,T,\Theta,n}$ $\boxed{\div}$ 250 $\boxed{)}$ $\boxed{\wedge}$ 2

2.50: $\boxed{\text{2nd}}$ [TBLSET] TblStart=0 ΔTbl=500 $\boxed{\text{2nd}}$ [TABLE]

2.51: $\boxed{\text{2nd}}$ [CALC] 2:zero Move cursor left of zero $\boxed{\text{ENTER}}$.
Move cursor right of zero $\boxed{\text{ENTER}}$ $\boxed{\text{ENTER}}$

2.52: $\boxed{\text{2nd}}$ [CALC] 4:maximum Move cursor left of peak $\boxed{\text{ENTER}}$. Move cursor right of peak $\boxed{\text{ENTER}}$ $\boxed{\text{ENTER}}$.

2.53: $\boxed{Y=}$ 32 $\boxed{X,T,\Theta,n}$ $\boxed{\wedge}$ $\boxed{(-)}$ 2 $\boxed{\text{2nd}}$ [e^x] 10 $\boxed{-}$ 32 $\boxed{X,T,\Theta,n}$ $\boxed{\wedge}$ $\boxed{(-)}$ 1 $\boxed{)}$

2.54: $\boxed{\text{2nd}}$ [TBLSET] TblStart=0 ΔTbl=10 $\boxed{\text{2nd}}$ [TABLE]

2.55: $\boxed{\text{WINDOW}}$ Xmin=0 Xmax=60 Ymin=0 Ymax=500

2.56: $\boxed{\text{2nd}}$ [CALC] 4:maximum Move cursor left of maximum $\boxed{\text{ENTER}}$.
Move cursor right of maximum $\boxed{\text{ENTER}}$ $\boxed{\text{ENTER}}$.

2.57: $\boxed{Y=}$ Y$_2$= $\boxed{\text{2nd}}$ [e^x] 10 $\boxed{-}$ 32 $\boxed{X,T,\Theta,n}$ $\boxed{\wedge}$ $\boxed{(-)}$ 1 $\boxed{)}$

2.58: $\boxed{Y=}$ Put the cursor on the = sign on the Y$_1$ line and press $\boxed{\text{ENTER}}$.

2.59: $\boxed{Y=}$ Highlight = on Y$_1$ line and press $\boxed{\text{ENTER}}$.

Highlight = on Y$_2$ line and press $\boxed{\text{ENTER}}$.

2.60: $\boxed{\text{WINDOW}}$ Xmin= $\boxed{(-)}$ 3750 **Xmax**=6250 Ymin= $\boxed{(-)}$ 1150 **Ymax**=1650

2.61: $\boxed{Y=}$ $\boxed{(}$ $\boxed{X,T,\Theta,n}$ $\boxed{\div}$ 4 $\boxed{)}$ $\boxed{\wedge}$ 2 $\boxed{+}$ $\boxed{(}$ 100 $\boxed{-}$ $\boxed{X,T,\Theta,n}$ $\boxed{)}$ $\boxed{\wedge}$ 2
$\boxed{\div}$ $\boxed{(}$ 4 $\boxed{\text{2nd}}$ [π] $\boxed{)}$

2.62: $\boxed{\text{2nd}}$ [CALC] **3:minimum** Move cursor left of minimum $\boxed{\text{ENTER}}$.

Move cursor right of minimum $\boxed{\text{ENTER}}$ $\boxed{\text{ENTER}}$.

QUICK REFERENCE FOR CHAPTER 3

Keystroke cross reference

3.1: $\boxed{Y=}$ 32 $\boxed{X,T,\Theta,n}$ $\boxed{-}$ 192

Instructions for Section 3.3
How to set up statistical plots

Step 1: Press $\boxed{\text{2nd}}$ [STAT PLOT] .

Step 2: Select 1:Plot 1 from the menu.

Step 3: Use the arrow keys to highlight On and press $\boxed{\text{ENTER}}$.

Step 4: Check the other lines in this menu to see that they are correct. Usually you will want
Xlist set to $\boxed{\text{2nd}}$ [L1] , indicating that the L1 column goes on the horizontal axis, and
Ylist set to $\boxed{\text{2nd}}$ [L2] , indicating that the L2 list goes on the vertical axis.

<u>Note</u>: You do not have to turn on statistical plots each time you want to graph data. This setting remains in effect until you specifically turn it Off. To do so, proceed as above, but in Step 3, highlight Off. A plot can also be turned off from the function entry window: Press $\boxed{Y=}$ and then $\boxed{\triangle}$ to move to the top line. When a plot is on, its name is highlighted there.

To change the status, place the cursor on the name and then press $\boxed{\text{ENTER}}$.

Instructions for Section 3.3
How to enter data

Step 1: Press ⃞ STAT .

Step 2: Select 1:**Edit** from the menu.

Step 3: If it is necessary to clear old data, highlight **L1** and then press ⃞ CLEAR   ⃞ ENTER .
Repeat this for each column as necessary.

Step 4: Locate the cursor in the column where you want to enter data and type each data entry
followed by either ⃞ ENTER or ⃞ ∇ .

Instructions for Section 3.3
How to plot data points

Step 1: Be sure that statistical plotting is turned on and the **Plot 1** screen properly configured
as described above. Also you should use ⃞ $Y=$ to check if old functions are stored and
if necessary ⃞ CLEAR them.

Step 2: Enter your data points as described above. In the statistical plot setup outlined above,
the **L1** column corresponds to the horizontal axis, and the **L2** column corresponds to the
vertical axis.

Step 3: When data is properly entered use ⃞ ZOOM 9 to display automatically all data points
on the screen.

3.2:   ⃞ STAT 1:**Edit** Enter values for d in **L1** column. Enter values for N in **L2** column.

3.3:   Turn statistical plots on using ⃞ 2nd [STAT PLOT] 1:**Plot1** Highlight **On** ⃞ ENTER .
In **Xlist**: type ⃞ 2nd [L1], and in **Ylist**: type ⃞ 2nd [L2]. (You only need to do
this the first time you plot. Statistical plotting will remain on until you turn it off.)
⃞ ZOOM 9 to make the graph.

3.4:   ⃞ $Y=$ 462 ⃞ X,T,Θ,n ⃞ + 28321 ⃞ GRAPH

3.5:   ⃞ STAT 1:**Edit** Enter data for m in **L1** column and data for F in **L2** column.

3.6: Be sure statistical plotting is turned on and that the function list is clear. Now ⃞ ZOOM 9

3.7:   ⃞ $Y=$ 5 ⃞ X,T,Θ,n

3.8:   ⃞ GRAPH

Instructions for Section 3.4
How to get a regression line

Step 1: Be sure your data points are properly entered in the L1 and L2 lists.

Step 2: Press [STAT] and use [▷] to highlight CALC.

Step 3: Select 4:LinReg(ax+b) from the menu and press [ENTER] .

Step 4: You will see a screen that gives the equation of the regression line $y = ax + b$ and values for a and b. The given a value is the slope, and the given b value is the vertical intercept.

3.9: [STAT] 1:Edit Enter t data in L1. Enter E data in L2.

(To make the graph, be sure [2nd] [STAT PLOT] Plot1 is properly configured and [ZOOM] 9 .)

3.10: With the data entered, press [STAT] . Move the highlight at the top of the screen to CALC. Select 4:LinReg(ax+b). Press [ENTER] when returned to the calculation screen.

3.11: [Y =] 1.21 [X,T,Θ,n] [+] 27.46 [GRAPH]

3.12: [STAT] 1:Edit Enter t values in L1 column and M values in the L2 column.

3.13: Use [Y =] and be sure the function list is [CLEAR] . Use [2nd] [STAT PLOT] and be sure Plot1 is On and that Xlist: is set to L1 and that Ylist: is set to L2. [ZOOM] 9

3.14: [STAT] CALC 4:LinReg(ax+b) [ENTER]

3.15: [Y =] 12.89 [X,T,Θ,n] [+] 252.62 [GRAPH]

3.16: [Y =] Y₁= [(] 900 [−] 28 [X,T,Θ,n] [)] [÷] 32

Y₂= 3 [X,T,Θ,n]

3.17: [WINDOW] Xmin=0 Xmax=30 Ymin=0 Ymax=30

3.18: If you get extra dots on the screen, you may need to turn statistical plotting off using [2nd] [STAT PLOT] 4:PlotsOff [ENTER] .

3.19: [2nd] [CALC] 5:intersect [ENTER] [ENTER] [ENTER]

3.20: [Y =] Y₁= [(] 56 [−] 12 [X,T,Θ,n] [)] [÷] .5

Y₂=4 [X,T,Θ,n]

3.21: [WINDOW] Xmin=0 Xmax=5 Ymin=0 Ymax=20

3.22: | 2nd | [CALC] 5:intersect | ENTER | | ENTER | | ENTER |

QUICK REFERENCE FOR CHAPTER 4

Keystroke cross reference

4.1: | WINDOW | Xmin=0 Xmax=5 Ymin=0 Ymax=100000

4.2: | WINDOW | Xmin=0 Xmax=5 Ymin=0 Ymax=3500

4.3: | 2nd | [CALC] 5:intersect | ENTER | | ENTER | | ENTER |

4.4: | $Y=$ | $Y_1=$ 64 | $-$ | 64 | $\times$ | 7.1 | $\wedge$ | | X, T, Θ, n | $Y_2=$ 57

4.5: | 2nd | [CALC] 5:intersect | ENTER | | ENTER | | ENTER |

4.6: | $Y=$ | $Y_1=$ 6 | $+$ | 69 | $\times$ | .96 | $\wedge$ | | X, T, Θ, n | $Y_2=$ 32

4.7: | 2nd | [CALC] 5:intersect | ENTER | | ENTER | | ENTER |

4.8: | STAT | 1:Edit Enter time in **L1** column and population in **L2** column.

4.9: | 2nd | [STAT PLOT] and be sure Plot1 is **On**. | ZOOM | 9 .

4.10: | STAT | CALC 0:ExpReg | ENTER |

4.11: | $Y=$ |

5.3372 | $\times$ | 1.0298 | $\wedge$ | | X, T, Θ, n | | GRAPH |

4.12: | STAT | 1:Edit Enter t data in **L1** column and C data in **L2** column.

4.13: (You may need to use | $Y=$ | and | CLEAR | out old functions.) | 2nd | [STAT PLOT] and be sure Plot1 is **On**. | ZOOM | 9 .

4.14: | $Y=$ | .704 | $\times$ | 1.1 | $\wedge$ | | X, T, Θ, n | | GRAPH |

4.15: | LOG | 5 |) |

4.16: | LN | 7 |) |

4.17: | STAT | 1:Edit Enter x values in **L1** column and y values in **L2** column.

4.18: | 2nd | [STAT PLOT] and be sure Plot1 is **On**. Then | ZOOM | 9 .

4.19: | STAT | 1:Edit Highlight L3 | 2nd | [e^x] | 2nd | [L2] |) | | ENTER | .

4.20: 2nd [STAT PLOT] 1:Plot1 Change Ylist to L3, then ZOOM 9 .

4.21: STAT 1:Edit Highlight L2 LN 2nd [L3]) ENTER .

4.22: STAT 1:Edit Enter time in L1 column and population in L3 column.

4.23: STAT 1:Edit Highlight L2 and LN 2nd [L3]) ENTER .

4.24: 2nd [STAT PLOT] Plot1 set Ylist to L2. ZOOM 9

4.25: STAT CALC 4:LinReg(ax+b) ENTER

4.26: $Y=$ .0294 X, T, Θ, n + 1.6747 GRAPH

4.27: STAT 1:Edit Enter t data in L1 column and C data in L3 column.

4.28: STAT 1:Edit Highlight L2. LN 2nd [L3]) ENTER

4.29: $Y=$ .095 X, T, Θ, n − .351 GRAPH

QUICK REFERENCE FOR CHAPTER 5

Keystroke cross reference

5.1: $Y=$ 16 X, T, Θ, n ∧ 2

5.2: WINDOW Xmin=0 Xmax=5 Ymin=0 Ymax=100

5.3: 2nd [CALC] 5:intersect ENTER ENTER ENTER

5.4: $Y=$ 21.83 X, T, Θ, n ∧ 0.18

5.5: $Y=$ (on Y_1 line) 5.34 × 1.03 ∧ X, T, Θ, n
     (on Y_2 line) 0.04 X, T, Θ, n ∧ 1.5 + 5.34

5.6: 2nd [TBLSET] TblStart=0 ΔTbl=5

5.7: STAT 1:Edit Enter x data in L3 column. Highlight L4 and then 3 2nd [L3] ∧ 2

5.8: Highlight L1 LN 2nd [L3])

5.9: Highlight L2 LN 2nd [L4])

5.10: 2nd [STAT PLOT] 1:Plot1 On. Set Xlist: to L1 and Ylist: to L2. ZOOM 9

5.11: $\boxed{\text{STAT}}$ CALC   4:LinReg(ax+b) $\boxed{\text{ENTER}}$

$\boxed{\text{Instructions for Section 5.2}}$
How to fit power functions to data

The procedure is similar to the one used for exponential regression except we take the logarithm of the variable values in addition to the function values.

Step 1: Press $\boxed{\text{STAT}}$, select 1:Edit, and $\boxed{\text{CLEAR}}$ out old lists as necessary. In the L3 column enter the values for the variable, and in the L4 column enter the corresponding values for the function. Now highlight L1 and type $\boxed{\text{LN}}$ $\boxed{\text{2nd}}$ [L3] $\boxed{)}$ and then $\boxed{\text{ENTER}}$. Then highlight L2 and type $\boxed{\text{LN}}$ $\boxed{\text{2nd}}$ [L4] $\boxed{)}$ and then $\boxed{\text{ENTER}}$.

Step 2: To plot the data on a logarithmic scale, press $\boxed{\text{2nd}}$ [STAT PLOT] and select 1:Plot 1 from the menu. Highlight On and press $\boxed{\text{ENTER}}$. Make sure that Xlist is set to L1 and Ylist is set to L2. Use $\boxed{Y=}$ and $\boxed{\text{CLEAR}}$ old functions if necessary, then $\boxed{\text{ZOOM}}$ 9 .

Step 3: Perform linear regression as usual: Press $\boxed{\text{STAT}}$ and use $\boxed{\triangleright}$ to move the highlight at the top of the screen to CALC. Select 4:LinReg(ax+b), then press $\boxed{\text{ENTER}}$.

5.12: $\boxed{\text{STAT}}$ 1:Edit and enter r values in L3 and V values in L4.

5.13: Highlight L1 and $\boxed{\text{LN}}$ $\boxed{\text{2nd}}$ [L3] $\boxed{)}$ $\boxed{\text{ENTER}}$. Highlight L2 and $\boxed{\text{LN}}$ $\boxed{\text{2nd}}$ [L4] $\boxed{)}$ $\boxed{\text{ENTER}}$.

5.14: $\boxed{\text{2nd}}$ [STAT PLOT] 1:Plot 1 On. Set Xlist: to L1 and Ylist: to L2. $\boxed{\text{ZOOM}}$ 9

5.15: $\boxed{\text{STAT}}$ CALC   4:LinReg(ax+b) $\boxed{\text{ENTER}}$.

5.16: $\boxed{\text{2nd}}$ [STAT PLOT] and set Xlist: to L3 and Ylist: to L4. $\boxed{Y=}$ and enter 4.18 $\boxed{X, T, \Theta, n}$ $\boxed{\wedge}$ 3 . Then $\boxed{\text{ZOOM}}$ 9 .

5.17: $\boxed{\text{STAT}}$ 1:Edit enter D values in L3 and P values in L4.

5.18: Highlight L1 $\boxed{\text{LN}}$ $\boxed{\text{2nd}}$ [L3] $\boxed{)}$. Highlight L2 $\boxed{\text{LN}}$ $\boxed{\text{2nd}}$ [L4] $\boxed{)}$.

5.19: $\boxed{\text{STAT}}$ 1:Edit and enter data for L in L3 column

5.20: $\boxed{\text{STAT}}$ 1:Edit and enter data for T in L4 column

5.21: Highlight L1. $\boxed{\text{LN}}$ $\boxed{\text{2nd}}$ [L3] $\boxed{)}$

5.22: Highlight L2. $\boxed{\text{LN}}$ $\boxed{\text{2nd}}$ [L4] $\boxed{)}$

5.23: $\boxed{\text{2nd}}$ [STAT PLOT] Set Xlist: to L1 and Ylist: to L2 $\boxed{\text{ZOOM}}$ 9

5.24: [STAT] CALC 4:LinReg(ax+b)   [ENTER] .

5.25: Highlight L1   [LOG]   [2nd] [L3]   [)] . Highlight L2   [LOG]   [2nd] [L4]   [)] .

Instructions for Section 5.4
How to get a quadratic regression equation

Step 1: Be sure your data points are properly entered in the L1 and L2 lists.

Step 2: Press [STAT] and use [▷] to highlight CALC.

Step 3: Select 5:QuadReg from the menu and press [ENTER] .

Step 4: You will see a screen that gives the quadratic regression equation $y = ax^2 + bx + c$ and values for a, b, and c.

Instructions for Section 5.5
How to get a cubic (or quartic) regression equation

Step 1: Be sure your data points are properly entered in the L1 and L2 lists.

Step 2: Press [STAT] and use [▷] to highlight CALC.

Step 3: Select 6:CubicReg for cubic regression (or 7:QuartReg for quartic regression) from the menu and press [ENTER] .

Step 4: You will see a screen that gives the cubic regression equation $y = ax^3 + bx^2 + cx + d$ and values for a, b, c, and d (or the quartic regression equation $y = ax^4 + bx^3 + cx^2 + dx + e$ and values for a, b, c, d, and e).

QUICK REFERENCE FOR CHAPTER 6

Keystroke cross reference

6.1: (Be sure [2nd] [STAT PLOT] Plot1 is set to Off.) [Y =] 30 [+] 18 [X, T, Θ, n]
[−] 16 [X, T, Θ, n] [∧] 2 . [WINDOW] Xmin=0 Xmax=5 Ymin=0 Ymax=50 [GRAPH] .

6.2: [2nd] [CALC] 2:zero Move cursor left of zero [ENTER] . Move cursor right of zero
[ENTER]   [ENTER] .

6.3: [2nd] [CALC] 4:maximum Move cursor left of maximum [ENTER] . Move cursor
right of maximum [ENTER]   [ENTER] .

Instructions for Section 6.3
How to find rates of change for functions given by formulas

Step 1: Graph the function and make sure the desired point shows on the graphing screen.

Step 2: Use $\boxed{\text{2nd}}$ [CALC] and choose **6:dy/dx** . Type in the desired x value, then press $\boxed{\text{ENTER}}$.

6.4: $\boxed{Y=}$ 16 $\boxed{X,T,\Theta,n}$ $\boxed{\wedge}$ 2 $\boxed{\text{WINDOW}}$ **Xmin**=0 **Xmax**=5 **Ymin**=0 **Ymax**=300 $\boxed{\text{GRAPH}}$

6.5: $\boxed{\text{2nd}}$ [CALC] **6:dy/dx** Type 2.5 and $\boxed{\text{ENTER}}$

6.6: $\boxed{Y=}$ $\boxed{X,T,\Theta,n}$ $\boxed{-}$ 32 $\boxed{(}$ $\boxed{X,T,\Theta,n}$ $\boxed{\div}$ 300 $\boxed{)}$ $\boxed{\wedge}$ 2

6.7: $\boxed{\text{TRACE}}$ 734 $\boxed{\text{ENTER}}$

6.8: $\boxed{\text{2nd}}$ [CALC] **6:dy/dx** 734 $\boxed{\text{ENTER}}$

6.9: $\boxed{Y=}$ Y_1= 72 $\boxed{+}$ 118 $\boxed{\text{2nd}}$ [e^x] $\boxed{(-)}$ .06 $\boxed{X,T,\Theta,n}$ $\boxed{)}$
Y_2= 130

6.10: $\boxed{\text{2nd}}$ [CALC] **5:intersect** $\boxed{\text{ENTER}}$ $\boxed{\text{ENTER}}$ $\boxed{\text{ENTER}}$

6.11: $\boxed{Y=}$ .338 $\boxed{X,T,\Theta,n}$ $\boxed{(}$ 1 $\boxed{-}$ $\boxed{X,T,\Theta,n}$ $\boxed{\div}$ 2.4 $\boxed{)}$

6.12: $\boxed{\text{2nd}}$ [CALC] **2:zero**
Move cursor left of zero $\boxed{\text{ENTER}}$. Move cursor right of zero $\boxed{\text{ENTER}}$ $\boxed{\text{ENTER}}$.

6.13: $\boxed{Y=}$ 32 $\boxed{-}$ .1818 $\boxed{X,T,\Theta,n}$

6.14: $\boxed{\text{WINDOW}}$ **Xmin**=0 **Xmax**=250 **Ymin**= $\boxed{(-)}$ 20 **Ymax**=35

QUICK REFERENCE FOR CHAPTER 7

Keystroke cross reference

7.1: $\boxed{\text{STAT}}$ **1:Edit** and $\boxed{\text{CLEAR}}$ out old lists as necessary. Enter t values in the L1 column and N values in the L3 column. Highlight L2 and type $\boxed{\text{LN}}$ $\boxed{\text{2nd}}$ [L3] $\boxed{)}$ $\boxed{\text{ENTER}}$.

7.2: Set Plot1 to On and $\boxed{\text{ZOOM}}$ 9 .

7.3: $\boxed{\text{STAT}}$ CALC 4:LinReg(ax+b) $\boxed{\text{ENTER}}$

7.4: $\boxed{\text{STAT}}$ 1:Edit and $\boxed{\text{CLEAR}}$ out old lists as necessary. Enter N values in the L1 column and values of $\dfrac{1}{N}\dfrac{dN}{dt}$ in the L2 column. Set Plot1 to On and $\boxed{\text{ZOOM}}$ 9 .

7.5: $\boxed{\text{STAT}}$ CALC 4:LinReg(ax+b) $\boxed{\text{ENTER}}$

Instructions for Section 7.3
How to generate survivorship curves

The procedure is similar to the one used for plotting with exponential regression. The two differences are the use of the common logarithm instead of the natural logarithm, and connecting the data points with line segments.

Step 1: Press $\boxed{\text{STAT}}$, select 1:Edit, and $\boxed{\text{CLEAR}}$ out old lists as necessary. In the L1 column enter the x values from the life table, and in the L3 column enter the corresponding l_x values. Now highlight L2 and type $\boxed{\text{LOG}}$ $\boxed{\text{2nd}}$ [L3] $\boxed{)}$ and then $\boxed{\text{ENTER}}$.

Step 2: Press $\boxed{\text{2nd}}$ [STAT PLOT] and select 1:Plot 1 from the menu. Highlight On and press $\boxed{\text{ENTER}}$. To connect the data points on the graph, from the Type: list select the second type of plot (connected icon). Make sure that Xlist is set to L1 and Ylist is set to L2.

Step 3: Use $\boxed{Y=}$ and $\boxed{\text{CLEAR}}$ old functions if necessary, then $\boxed{\text{ZOOM}}$ 9 .

Arithmetic Operations

The TI-82, TI-83, TI-83 Plus, and TI-84 Plus are designed to make arithmetic calculations easy, and many will find that little if any instruction is really necessary. Often all that is needed is a bit of practice to gain familiarity with the operation of the keyboard, and when difficulties do occur, they may be traceable to problems with parentheses and grouping. These are at least partly mathematical in nature, and you are encouraged to read the Prologue of *Functions and Change*. The keystrokes for arithmetic operations on the TI-82 and TI-83/84 are virtually identical, but where they differ we will provide separate instructions. There are no differences in our operation of the TI-83, TI-83 Plus, and TI-84 Plus. All but a few of the figures in this text were made using the TI-83, and the output from the TI-82 is sometimes slightly different. The first time this is encountered, special pictures are made from the TI-82 screen, but in general the output of the two calculators is very similar. Where the difference may cause confusion we have pointed out the differences that are to be expected.

PROLOGUE: BASIC CALCULATIONS

To perform a simple calculation such as $\frac{72}{9} + 3 \times 5$ on the TI-82/83/84, we type in the expression just as we might write it on paper: 72 $\boxed{\div}$ 9 $\boxed{+}$ 3 $\boxed{\times}$ 5 . This is displayed in Figure 2.1. To get the answer, we press $\boxed{\text{ENTER}}$, and we see 23 as expected in Figure 2.2. For some calculations you need to access special keys.

Figure 2.1: *Entering a simple calculation*

```
72/9+3*5
```

Figure 2.2: *Using* ENTER *to complete the calculation*

```
72/9+3*5
              23
■
```

The exponent key: To enter an exponent like 2^3 you need to use the black ∧ key located just below the CLEAR key. Typing 2 ∧ 3 ENTER will produce the correct result, 8.

The square root key is at the top-left of the x^2 key.

TI-82

To access these *alternate keys*, you first press the blue 2nd key and the cursor will be replaced by a flashing, upward-pointing, arrow. Now press the x^2 key, and the square root symbol will appear on the display. In what follows we will use 2nd [√] to indicate this sequence of keystrokes. So for example, if you want to get $\sqrt{7}$, you type 2nd [√] 7 ENTER . The answer, 2.645751311, is shown in the figure below.

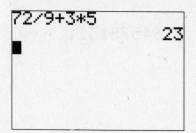

```
√7
        2.645751311
```

TI-83/84

To access these *alternate keys*, you first press the gold (blue for the TI-84) 2nd key and the cursor will be replaced by a flashing, upward pointing arrow. Now press the x^2 key, and the square root symbol will appear on the display. The TI-83/84 automatically adds a left parenthesis following the square root symbol in anticipation of a longer expression. To complete your entry, you should remember to finish with) . In what follows we will use 2nd [√] to indicate this sequence of keystrokes. So for example, if you want to get $\sqrt{7}$, you type 2nd [√] 7) ENTER . The answer is shown in Figure 2.3.

The number π: The special number π is printed above the ∧ key. To access it you use 2nd [π] . When you do this, the calculator will display the symbol for π as shown in the first line of Figure 2.4. When you press ENTER , the calculator will display the numerical approximation shown in the second line of Figure 2.4.

Figure 2.3: *Using the square root key*

Figure 2.4: *Accessing* π

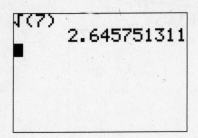

The number e:

TI-82

The special number e is most often used in the context of an *exponential function*. For this reason, when you press [2nd] [e^x] (above the black [LN] key) the calculator automatically adds the $\wedge$ in anticipation of an exponent. To get the number $e = e^1$ we provide an exponent of 1. That is, we type [2nd] [e^x] 1 and [ENTER]. The result, 2.718281828, is the numerical approximation of e shown below.

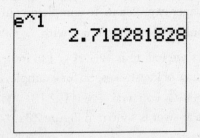

TI-83/84

The special number e appears in gold (blue for the TI-84) above the division key [÷]. You access it using [2nd] [e] and [ENTER]. The result, 2.718281828, is a numerical approximation of e which is shown in Figure 2.6.

We should note however that this number is most often used in the context of an *exponential function*, and this provides an alternative way of getting the number e. While this method takes more keystrokes, it is the way e will be most commonly accessed in what follows. There is a second key [2nd] [e^x] (above the black [LN] key) which automatically adds the $\wedge$ and a left parenthesis in anticipation of an exponent. To get the number $e = e^1$ using this key we use [2nd] [e^x] 1 [ )] and [ENTER]. The result, which matches our first approximation of e in Figure 2.6, is in Figure 2.5.

Figure 2.5: ∧ *(and (on the TI-83/84) automatically displayed when you use* 2nd *[e^x]*

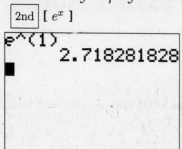

Figure 2.6: *Using* 2nd *[e] to access e directly (not available on the TI-82)*

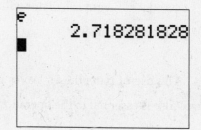

PROLOGUE: PARENTHESES

Two of the most important keys on your calculator are the black parentheses keys (and). When parentheses are indicated in a calculation, you should use them, and sometimes it is necessary to add additional parentheses. To make a calculation such as 2.7(3.6+1.8) you type 2.7 (3.6 + 1.8) and then ENTER . The result is shown in Figure 2.7. To calculate $\frac{2.7 - 3.3}{6.1 + 4.7}$, we must use parentheses around 2.7 − 3.3 to indicate that the whole expression goes in the numerator, and we must use parentheses around 6.1 + 4.7 to indicate that the whole expression goes in the denominator. Thus we type

(2.7 − 3.3) ÷ (6.1 + 4.7) and ENTER .

The result is in Figure 2.8.

Figure 2.7: *A calculation using parentheses*

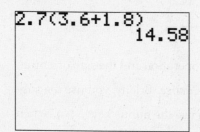

Figure 2.8: *A calculation where additional parentheses must be supplied*

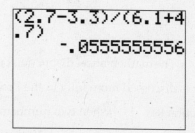

When you are in doubt about whether parentheses are necessary or not, it is good practice to use them. For example, if you calculate $\frac{3 \times 4}{7}$ using 3 × 4 ÷ 7 , you will get the right answer, 1.71. You will get the same answer if you use parentheses to emphasize to the calculator that the entire expression 3 × 4 goes in the numerator: (3 × 4) ÷ 7 .

If on the other hand you calculate $\dfrac{7}{3 \times 4}$, you must use parentheses to get the correct answer, 0.58. The correct entry is 7 $\boxed{\div}$ $\boxed{(}$ 3 $\boxed{\times}$ 4 $\boxed{)}$. If you leave out the parentheses, using 7 $\boxed{\div}$ 3 $\boxed{\times}$ 4 , your calculator will think that only the 3 goes in the denominator, and you will get the wrong answer.

The use of parentheses leaves no room for doubt about what goes where, and their correct use is essential to the operation of the calculator.

PROLOGUE: MINUS SIGNS

If you look on your TI-82/83/84 keyboard, you will find two minus signs. One is the key $\boxed{-}$ located just above the $\boxed{+}$ key. The other is the $\boxed{(-)}$ key that is just to the left of the $\boxed{\text{ENTER}}$ key. The $\boxed{-}$ key is used for subtraction as in 9 $\boxed{-}$ 4 while the $\boxed{(-)}$ key is used to denote a negative number. For example if you want to get 2^{-3} you must use   2 $\boxed{\wedge}$ $\boxed{(-)}$ 3 as shown in Figure 2.9. If you try to use 2 $\boxed{\wedge}$ $\boxed{-}$ 3 you will get a *syntax error* as seen in Figure 2.10.

Figure 2.9: *Using* $\boxed{(-)}$ *to calculate* 2^{-3}

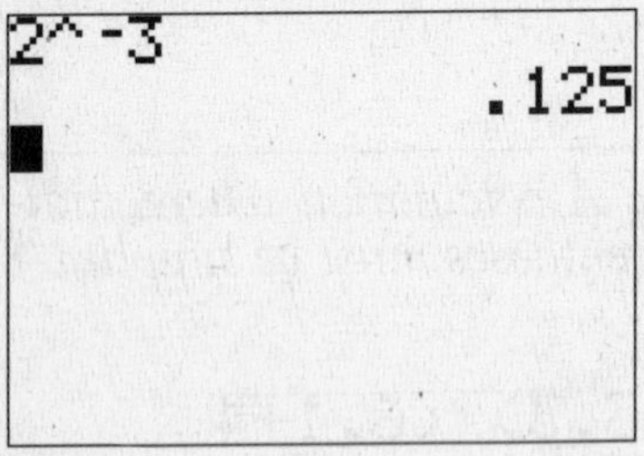

```
2^-3
              .125
■
```

Figure 2.10: *Syntax error displayed when* $\boxed{-}$ *is used in* 2^{-3}

```
ERR:SYNTAX
1■Quit
2:Goto
```

The mathematical distinction between the subtraction operation and the sign of a number is discussed more fully in the Prologue of *Functions and Change*. Briefly, you use the subtraction key $\boxed{-}$ when two numbers are involved, and you use the minus key $\boxed{(-)}$ when only one number is involved. With a little practice, the distinction is easy.

PROLOGUE: CHAIN CALCULATIONS

Some calculations are most naturally done in stages. The TI-82/83/84 [2nd] [ANS] key is helpful here. This key refers to the results of the last calculation. To show its use, let's look at

$$\left(\frac{2+\pi}{7}\right)^{\left(3+\frac{7}{9}\right)}.$$

This can be done in a single calculation, but we will show how to do it in two steps. First we calculate the exponent $3 + \frac{7}{9}$ as shown in Figure 2.11. To finish, we need

$$\left(\frac{2+\pi}{7}\right)^{\text{Answer from the last calculation}}$$

The answer from the last calculation is obtained using [2nd] [ANS] . Thus to complete the calculation, we use [( ] [( ] 2 + [2nd] [π] [ )] ÷ 7 [ )] [∧] [2nd] [ANS] .

The completed calculation is in Figure 2.12.

Figure 2.11: *The first step in a chain calculation*

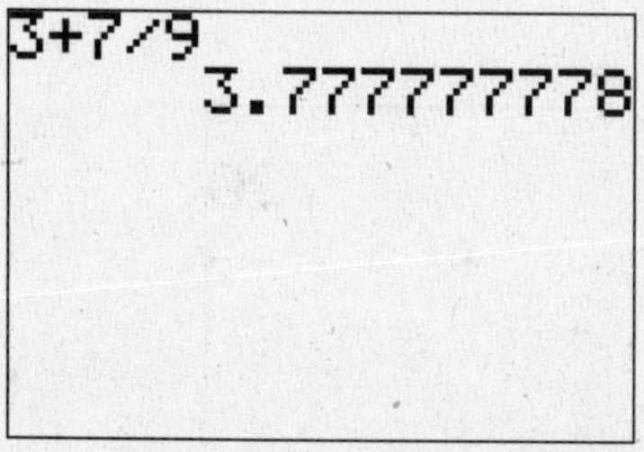

```
3+7/9
        3.777777778
```

Figure 2.12: *Completing a chain calculation using* [2nd] [ANS]

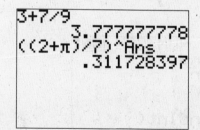

```
3+7/9
        3.777777778
((2+π)/7)^Ans
        .311728397
```

CHAPTER 2: THE FACTORIAL FUNCTION

The *factorial* function occurs most often in probabilistic and statistical applications of mathematics. It applies only to non-negative integers and is denoted by $n!$. It is defined as follows:

$$0! = 1$$

$$\text{For } n > 0, \; n! = n(n-1)(n-2)\ldots3 \times 2 \times 1.$$

Thus for example $3! = 3 \times 2 \times 1 = 6$ and $5! = 5 \times 4 \times 3 \times 2 \times 1 = 120$. Both the TI-82 and TI-83/84 can calculate the factorial function. Let's see how to use the calculator to get $6!$. First enter 6. Next press the $\boxed{\text{MATH}}$ key. Then use the right arrow key to highlight **PRB** as shown in Figure 2.13. Finally use the down arrow to highlight the exclamation point shown in Figure 2.13 and press $\boxed{\text{ENTER}}$. The calculator will display $6!$. Press $\boxed{\text{ENTER}}$ once more to get the answer shown in Figure 2.14.

Figure 2.13: *Accessing the factorial symbol*

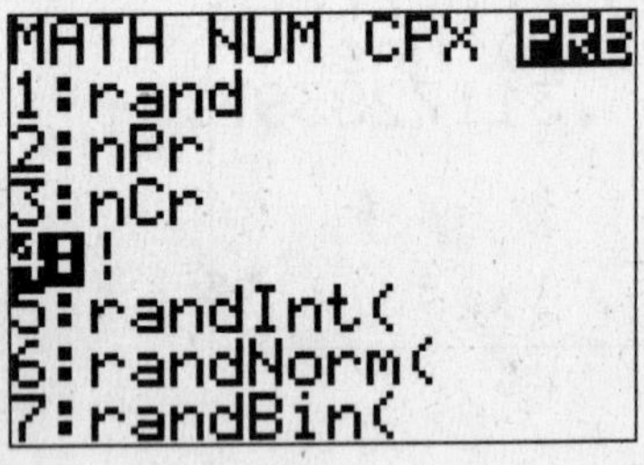

Figure 2.14: *Calculating* $6!$

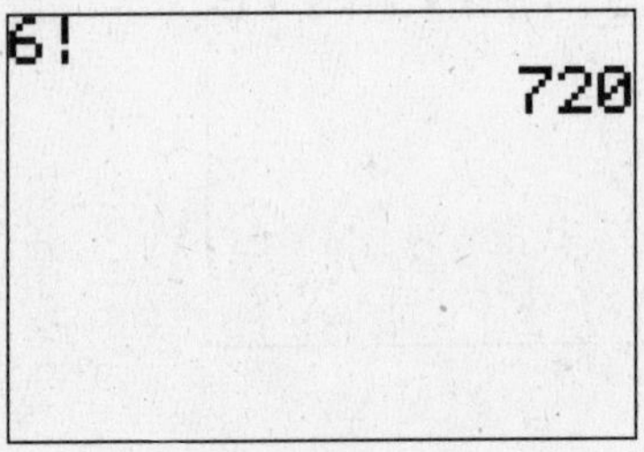

```
MATH NUM CPX PRB
1:rand
2:nPr
3:nCr
4:!
5:randInt(
6:randNorm(
7:randBin(
```

```
6!
              720
```

Tables and Graphs

The feature of modern calculators which makes them different from their forerunners is their ability to make tables and produce graphs. This is a significant technological step, but more importantly it enhances the calculator's utility as a mathematical and scientific problem-solving tool.

CHAPTER 2: TABLES OF VALUES

For a function given by a formula, we can always produce a hand-generated table of values by calculating many individual function values, but the TI-82/83/84 acts as a significant time saver by making such tables automatically.

Three steps are required to make tables of values on a calculator. We will show them in an example. Let's make a table of values for $f = 3x + 1$.

Step 1, Entering the function: The first step is to tell the calculator which function we are using. To do this we use $\boxed{Y=}$ to show the *function entry* window. The TI-82/83/84 remembers functions that have been entered before, and so when you do this, you may find leftover formulas already on the screen as shown in Figure 2.15. If so, move the cursor to each formula that appears and press $\boxed{\text{CLEAR}}$. After the clean-up, you should have a clear function entry window like the one in Figure 2.16. (The TI-82 will not show Plot1 Plot2 Plot3 at the top of the screen, nor will it show the marks to the left of the Y's.) The presence of the lines $Y_1 =$ through $Y_7 =$ in Figure 2.16 indicates that

Figure 2.15: Clutter from previous work left on the function entry screen

Figure 2.16: Using $\boxed{\text{CLEAR}}$ *to clean up the function entry window*

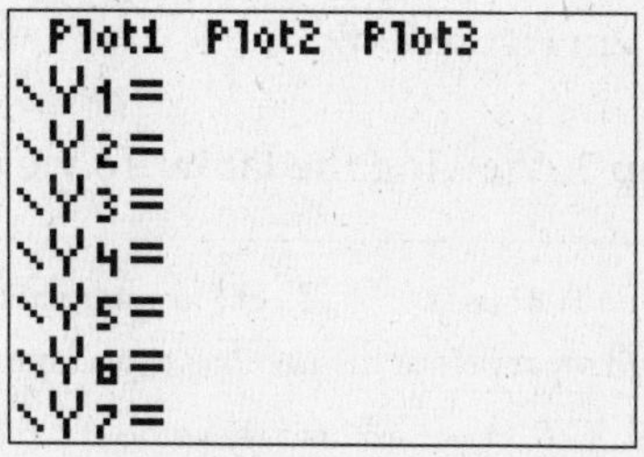

many functions can be entered. For the moment, we will only use $Y_1 =$. Now we enter the function.[1]

TI-82	**TI-83/84**

To enter the function, we use $\boxed{X, T, \Theta}$ for the variable x. (We would use the $\boxed{X, T, \Theta}$ key no matter what the name of the variable in our function.) Thus on the $Y_1 =$ line we type 3 $\boxed{X, T, \Theta}$ $\boxed{+}$ 1. The properly entered function is in Figure 2.17. Notice that the calculator writes the variable x as X. Once again, this is the symbol that the calculator will display no matter what the name of our variable.

To enter the function, we use $\boxed{X, T, \Theta, n}$ for the variable x. (We would use the $\boxed{X, T, \Theta, n}$ key no matter what the name of the variable in our function.) Thus on the $Y_1 =$ line we type 3 $\boxed{X, T, \Theta, n}$ $\boxed{+}$ 1. The properly entered function is in Figure 2.17. Notice that the calculator writes the variable x as X. Once again, this is the symbol that the calculator will display no matter what the name of our variable.

Step 2, Setting up the table: The next step is to tell the calculator how we want the table to look. We do this using $\boxed{\text{2nd}}$ [TBLSET] . When you press $\boxed{\text{2nd}}$ [TBLSET] you should see the **TABLE SETUP** menu shown in Figure 2.18. The first option on the

Figure **2.17**: *The completed entry of* $3x + 1$

Figure **2.18**: *The completed* **TABLE SETUP** *menu*

```
Plot1 Plot2 Plot3
\Y1■3X+1
\Y2=
\Y3=
\Y4=
\Y5=
\Y6=
\Y7=
```

```
TABLE SETUP
 TblStart=0
 △Tbl=1
Indpnt: Auto Ask
Depend: Auto Ask
```

screen.[2] is TblStart=. This is where the table is to start. We want this table to start at $x = 0$, and so we enter 0 there. The next option is △Tbl=. This gives the increment for the table. For the first view, we will look at the table for $x = 1, 2, 3, \ldots$. That is, we want to increment x by 1 at each step. (We will see later what happens if a different value for △Tbl= is used.) Thus we type 1 here. The properly completed **TABLE SETUP** menu is in Figure 2.18.

Step 3, Viewing the table: To view the table we press $\boxed{\text{2nd}}$ [TABLE] , and the calculator

---

[1]The TI-82 uses $\boxed{X, T, \Theta}$ to enter variables, while the TI-83/84 uses $\boxed{X, T, \Theta, n}$. On this first occurrence, we will separate our instructions to accommodate this, but since these keys are so similar, in what follows we will use $\boxed{X, T, \Theta, n}$ exclusively without direct reference to the TI-82 $\boxed{X, T, \Theta}$ key.

[2]On the TI-82 this appears as TblMin= As with $\boxed{X, T, \Theta}$ and $\boxed{X, T, \Theta, n}$, we will not distinguish between them in what follows.

will present the display shown in Figure 2.19. To get information from the table, it is important to remember that the X column corresponds to the variable x, and the Y_1 column gives the corresponding function value $f(x)$. Thus the first line in Figure 2.19 tells us that $f(0) = 1$, the second that $f(1) = 4$, and so on. There is more to the table than is shown on the screen. If you use $\boxed{\nabla}$ or $\boxed{\triangle}$, you will see additional entries. In Figure 2.20 we have used $\boxed{\nabla}$ to view further function values.

Figure 2.19: *A table of values for* $3x + 1$

X	Y₁	
0	1	
1	4	
2	7	
3	10	
4	13	
5	16	
6	19	
X=0		

Figure 2.20: *Extending the table using* $\boxed{\nabla}$

X	Y₁	
6	19	
7	22	
8	25	
9	28	
10	31	
11	34	
12	37	
X=12		

Adjusting the table: Many times we can see what we want from a table by making adjustments with $\boxed{\triangle}$ or $\boxed{\nabla}$, but for some alterations, this is impractical or impossible. Suppose for example that we wanted to see what happens for $x = 300, 301, 302, \ldots$. It not very efficient to do this using the $\boxed{\nabla}$ key. A better strategy is to use $\boxed{\text{2nd}}$ [TBLSET] to return to the TABLE SETUP menu and set TblStart= 300 . This is shown in Figure 2.21. Now when we use $\boxed{\text{2nd}}$ [TABLE] to go back to the table, the new TblStart= value causes the entries to start at $x = 300$ as shown in Figure 2.22.

Figure 2.21: *Changing* TblStart= *to* 300

```
TABLE SETUP
 TblStart=300
 △Tbl=1
Indpnt: Auto Ask
Depend: Auto Ask
```

Figure 2.22: *The table with a new starting value*

X	Y₁	
300	901	
301	904	
302	907	
303	910	
304	913	
305	916	
306	919	
X=300		

Let's make another adjustment which will allow us to view the table for $x = 0, 5, 10, 15, \ldots$. That is, we want to view a table for f which starts at $x = 0$ and has an increment of 5. To do this, we use $\boxed{\text{2nd}}$ [TBLSET] to return to the TABLE SETUP menu and then use TblStart=0 and △Tbl=5 . The correctly configured TABLE SETUP menu is in Figure 2.23, and when we press ⌷2nd⌷ [TABLE], we see the new table in Figure 2.24.

Figure 2.23: *Changing the value of* ΔTbl=

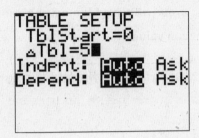

Figure 2.24: *The table with an increment of 5*

X	Y₁	
0	1	
5	16	
10	31	
15	46	
20	61	
25	76	
30	91	

X=0

Comparing functions: The calculator's ability to deal with more than one function allows us to use tables to make comparisons. For example, let's show how to compare the values of $f = 3x + 1$ with those of $g = 4x - 2$. First we need to use ⌷ Y = ⌷ to enter the formula for g. We want to keep f as entered on the Y₁= line, so we don't clear it but move the cursor directly to the Y₂= line. There we enter 4 ⌷X, T, Θ, n⌷ ⌷ − ⌷ 2 as shown in Figure 2.25. Use **TABLE SETUP** values of **TblStart**=0 and **ΔTbl**=1 . When we press ⌷2nd⌷ [TABLE], we see the table in Figure 2.26, which shows function values for f in the Y₁ column and values for g in the Y₂ column. Note that f starts out larger than g. They have the same value when $x = 3$, and after that g has the larger values.

Figure 2.25: *Entering a second function*

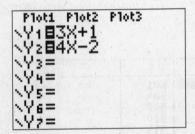

Figure 2.26: *Comparing function values*

X	Y₁	Y₂
0	1	-2
1	4	2
2	7	6
3	10	10
4	13	14
5	16	18
6	19	22

X=0

CHAPTER 2: GRAPHS

The TI-82/83/84 can generate the graph of a function as easily as it makes tables. We will illustrate using $f = \dfrac{x^4}{50} - x$. The first step is to tell the calculator which function we want to graph. We do this using the $\boxed{Y =}$ key exactly as we did in the previous section. Use $\boxed{\text{CLEAR}}$ to delete old functions and enter the new one on the Y_1 line using $\boxed{X, T, \Theta, n}$ $\boxed{\wedge}$ 4 $\boxed{\div}$ 50 $\boxed{-}$ $\boxed{X, T, \Theta, n}$. As with tables, the TI chooses its own names for functions and variables; in this case Y_1 , which will appear on the vertical axis for f, and X, which will appear on the horizontal axis for x.

Once the function is properly entered as in Figure 2.27, we press $\boxed{\text{ZOOM}}$ and the menu shown in Figure 2.28 is shown. From this menu, select 6:ZStandard and the graph in Figure 2.29 will be shown.[3]

Figure 2.27: *Using* $\boxed{Y =}$ *to enter* $\dfrac{x^4}{50} - x$

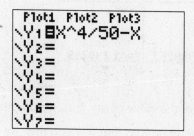

Figure 2.28: *Using* $\boxed{\text{ZOOM}}$ *to get to the ZOOM menu*

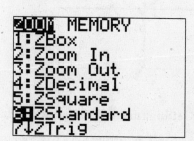

There are various ways of adjusting the picture shown by the calculator, and the TI-82/83/84 remembers the settings it used the last time it made a graph. The instructions we gave made what is called the ZStandard view. In this view, the horizontal span is from -10 to 10, and so is the vertical span. If your picture does not match the one in Figure 2.29, use $\boxed{Y =}$ to check that your function is properly entered. If this fails, refer to the *Graphics configuration notes* in the section entitled Chapter 2: Graphics configuration notes below. *It is important that you make your picture match the one in Figure 2.29; otherwise, you will have difficulty following the examples and explanations in what follows.*

---

[3]This assumes the calculator is in its default graphics display mode. As is explained in what follows, this mode may have been altered by any number of calculator operations. Instructions for returning the calculator to its default display mode are given in the section entitled Chapter 2: Graphics configuration notes below.

Once we have a graph on the screen there are several ways to adjust the view or to get information from it.

Tracing the graph: The [TRACE] key places a cursor on the graph and enables the left and right arrow keys, [◁] and [▷] , to move the cursor along the graph. (If there is more than one graph on the screen then [△] or [▽] moves the cursor to other graphs.) As the cursor moves, its location is recorded at the bottom of the screen. The X= prompt shows where we are on the horizontal axis, and the Y= prompt shows where we are on the vertical axis. In Figure 2.30 we have used [TRACE] and moved the cursor to the lowest point of the graph, and we see X=2.3404255 and Y=−1.740345 at the bottom of the screen. This tells us that the cursor is located at the point (2.3404255, −1.740345). Since this point lies on the graph it also tells us that, rounded to two decimal places,

$$f(2.34) = -1.74.$$

Figure 2.29: *The graph of* $\dfrac{x^4}{50} - x$

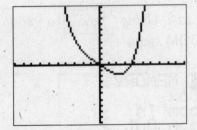

Figure 2.30: *Using* [TRACE] *to locate the cursor at the bottom of the graph*

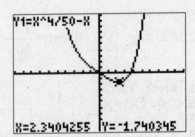

Getting function values: If we want to use [TRACE] to get the value of say $f(3)$, we might use the arrow keys to try to make the cursor land exactly on X=3. But if we do this, we find that the arrow keys make the X= prompt skip 3, going from 2.9787234 to 3.1914894. Both the TI-82 and the TI-83/84 offer us a way to correct this and make the cursor go exactly to X=3, but the keystrokes to accomplish this are different.

TI-82	TI-83/84

We first press [2nd] [CALC]. You will see the **CALCULATE** menu shown in Figure 2.31. We want a function value, so we choose 1:value from the menu. We will be returned to the graph with the prompt **Eval X=** at the bottom of the screen as shown in Figure 2.32. We are interested in $x = 3$, so we type in 3 as seen in Figure 2.33. Now when we press [ENTER] we see that the cursor has indeed moved to X=3, and read from Figure 2.34 that $f(3) = -1.38$.

Be sure the [TRACE] option is on. Now type 3, and X=3 will appear at the bottom of the screen as shown in Figure 2.35. When we press enter, the cursor will move to X=3 as shown in Figure 2.36, and we read from the bottom of the screen that $f(3) = -1.38$.

Figure 2.31: The TI-82 CALCULATE menu

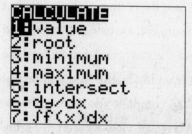

Figure 2.32: The TI-82 Eval X= prompt

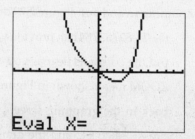

Figure 2.33: Entering X=3 on the TI-82

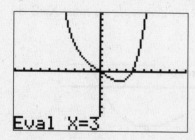

Figure 2.34: Evaluating f at $x = 3$ on the TI-82

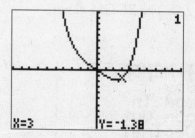

Figure 2.35: Entering X=3 on the TI-83/84

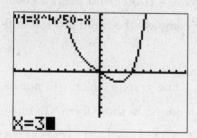

Figure 2.36: Evaluating f at $x = 3$ on the TI-83/84

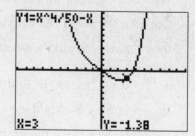

We should note that this only works if the X value you are looking for is actually on the viewing screen. If the X value you want is not far to the right or left of the viewing screen, you can use TRACE and move toward it. When the tracing cursor gets to the edge of the viewing area, the screen will change to follow it. If the X value is far to the right or left of the viewing area, you will need first to adjust the screen manually as described in **Manual window adjustments** below.

Zooming in and out: Figure 2.29 doesn't show much detail near the bottom of the graph, but the TI-82/83/84 can provide you with a closer look. Be sure you are using TRACE and put the cursor as near as you can to X=2.34. Now press ZOOM , and you will see the ZOOM menu shown in Figure 2.37. Select **2:Zoom In** from the menu. You will be put back in the graphing screen with nothing changed. To complete the zoom process, press ENTER , and the graph will be redrawn with a closer view as seen in Figure 2.38. Zooming turns off the trace, and so you will need to TRACE again to put the cursor back on the graph. If you ZOOM 2 ENTER again you will see a still closer

---

Figure 2.37: The **ZOOM** *menu*

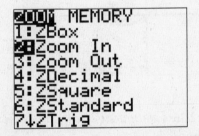

Figure 2.38: Zooming in near X=2.34

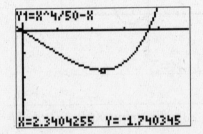

---

view of the graph. To back away from the graph, press ZOOM and select **3:Zoom Out** from the menu. As before, we press ENTER to complete the zoom process, and then TRACE to put the cursor back on the graph.

Using TRACE   ENTER **to find lost graphs**: There are some graphs which will not be shown in the ZStandard view, and adjustments are required to show them. To illustrate this, let's make the graph of $f(x) = x^2 + 1000$. Use Y = and CLEAR out the old functions. On the Y$_1$= line type X, T, Θ, n $\wedge$ 2 + 1000 . Now if we ZOOM 6 , we will see the axes but no graph at all as shown in Figure 2.39. The difficulty is that this graph sits high above the horizontal axis and is out of the viewing area. Press TRACE to put the cursor onto the graph. Now if you press ENTER , the viewing area will move to find the cursor and hence the graph. The resulting graph is shown in Figure 2.40.

Figure 2.39: *A lost graph*

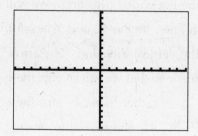

Figure 2.40: *Finding a lost graph with* TRACE ENTER

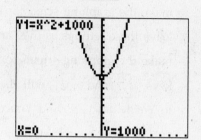

Manual window adjustments: You can get satisfactory views of many graphs by tracing and zooming, but some need manual adjustments. To illustrate this, let's look at the graph of $f(x) = \dfrac{x}{50}$. Use $\boxed{Y=}$ and $\boxed{\text{CLEAR}}$ out the old functions. Now type $\boxed{X, T, \dot{\Theta}, n}$ $\boxed{\div}$ 50 and $\boxed{\text{ZOOM}}$ 6 to get the **ZStandard** view shown in Figure 2.41. (We have also pressed $\boxed{\text{TRACE}}$ so that the TI-83/84 will identify the graph in the upper-left corner of the screen. The TI-82 identifies graphs by placing a single number n corresponding to Y_n in the upper-right corner.) This is not a very good picture of the graph. The difficulty is that the graph is a straight line that in the **ZStandard** view lies so close to the horizontal axis that the calculator has difficulty distinguishing them.

This is a problem that is not easily fixed by zooming. We want to adjust the vertical span of the window to get a better view. We will make a table of values to help us choose an appropriate window size. Since the **ZStandard** window has a horizontal span of -10 to 10, we make the table start at -10 with an increment of 4. The result is shown in Figure 2.42, and it shows that as x ranges from -10 to 10, $f(x)$ ranges from -0.2 to 0.2.

Figure 2.41: *An unsatisfactory view of a line*

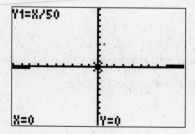

Figure 2.42: *A table of values for* $\dfrac{x}{50}$

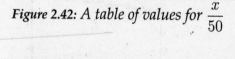

Allowing a little extra room, we want to change the vertical span so that it goes from

−1 to 1. To accomplish this, we first press the WINDOW key to get the WINDOW menu shown in Figure 2.43. The first two lines of this screen, Xmin=-10 and Xmax=10, make the graphing screen span from −10 to 10 in the horizontal direction. We will leave these settings as they are. The fourth and fifth lines, Ymin=-10 and Ymax=10, make the graphing screen span from −10 to 10 in the vertical direction. The arrow keys, △ and ▽ , will allow us to position the cursor so that we can change these to Ymin= (−) 1 , and Ymax=1 . (Be sure to use (−) rather than − for these entries.) These new settings, which are shown in Figure 2.44, will make the viewing screen span from −10 to 10 in the horizontal direction and from −1 to 1 in the vertical direction. Now GRAPH to get the better view of the line shown in Figure 2.45.

Figure 2.43: *The* WINDOW *menu*

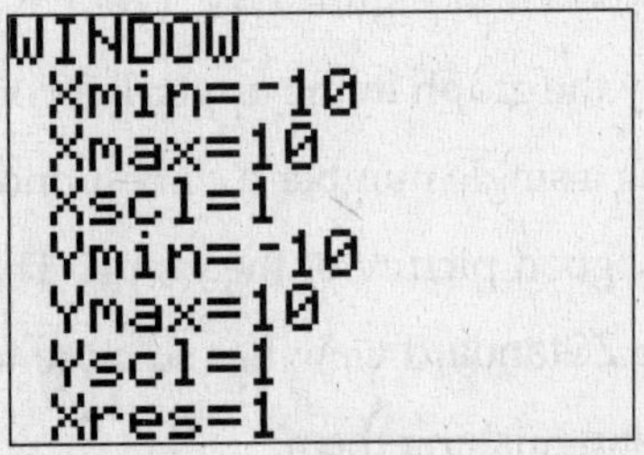

Figure 2.44: *Changing the vertical span*

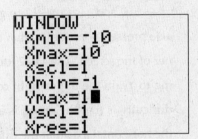

Figure 2.45: *A better view of the line*

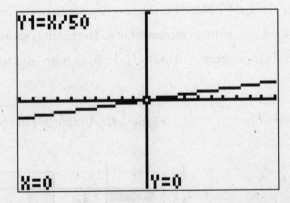

The text gives instructions on how to use a table of values to choose the graphing window in practical settings. See Section 2 in Chapter 2 of *Functions and Change*.

Showing more than one graph: To show multiple graphs, all you need to do is to enter each function you want to show on the function entry screen. For example, if we want to see the graphs of $f = x^2$ and $g = 5 - x^2$ at the same time, we first use $\boxed{Y =}$ and type $\boxed{X, T, \Theta, n}$ $\boxed{\wedge}$ 2 on the $Y_1 =$ line and 5 $\boxed{-}$ $\boxed{X, T, \Theta, n}$ $\boxed{\wedge}$ 2 on the $Y_2 =$ line as shown in Figure 2.46. Now if we use $\boxed{\text{ZOOM}}$ 6 we get the **ZStandard** view of both graphs shown in Figure 2.47.

Figure 2.46: *Entering two functions*

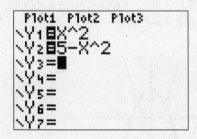

Figure 2.47: *Displaying the graphs of two functions*

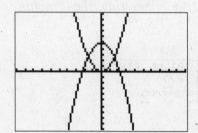

Which graph is which?: We can use $\boxed{\text{TRACE}}$ to help us keep track of which graph goes with which function. When we do this with the TI-83/84, we see in Figure 2.48 that the formula for the graph the cursor is on shows in the upper-left corner of the screen. (On the TI-82 a **1** will appear in the upper-right corner indicating that this is the graph of Y_1.) If we use $\boxed{\triangle}$ or $\boxed{\nabla}$, the cursor switches to the other graph, and as we see in Figure 2.49, the corresponding function is shown in the upper left corner of the screen. (On the TI-82 a **2** will appear in the upper right corner indicating that this is the graph of Y_2.)

Figure 2.48: *How* $\boxed{\text{TRACE}}$ *labels graphs on the TI-83/84*

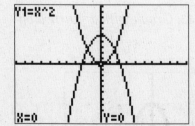

Figure 2.49: *Using* $\boxed{\triangle}$ *to move to the next graph*

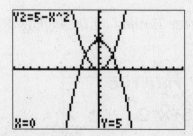

Changing the look of the graph on the TI-83/84: The TI-83/84 allows the display to show different graphs in different styles, providing additional help in keeping track of which graph is which. (This feature is not available on the TI-82.) Let's change the look of the graph of $5 - x^2$. Use $\boxed{Y=}$ and move the highlight to the left of $Y_2=$ and press $\boxed{\text{ENTER}}$. This will cause the slash displayed there to start blinking. Now if we press $\boxed{\text{ENTER}}$, its shape will change as shown in Figure 2.50. (If you press $\boxed{\text{ENTER}}$ more times, you will cycle through several display options.) Now when we $\boxed{\text{GRAPH}}$, the graph of $5 - x^2$ is shown with a heavy line as in Figure 2.51.

Figure 2.50: Changing the graphing tag on the TI-83/84

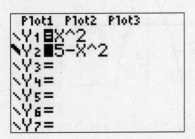

Figure 2.51: $5 - x^2$ shown with a heavy line on the TI-83/84

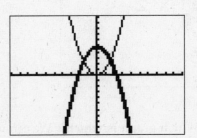

Turning off graph displays: There may be occasions when we want to show one graph but not erase the other function from the function entry screen. Let's show how to turn off the display of the graph of x^2 without erasing its formula. Use $\boxed{Y=}$ to return to the function entry screen. Notice that the equal sign on both the Y_1 and Y_2 lines are highlighted. Move the cursor to the equal sign on the Y_1 line and press $\boxed{\text{ENTER}}$. This will turn off the highlight as shown in Figure 2.52. Now when we $\boxed{\text{GRAPH}}$ as in Figure 2.53, only the graphs of functions with highlighted equal signs will be displayed. To turn the display back on, use $\boxed{Y=}$, move the highlight to the equal sign, and press $\boxed{\text{ENTER}}$ again.

Figure 2.52: Turning off the highlight on =

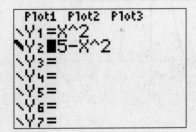

Figure 2.53: Graph with x^2 turned off

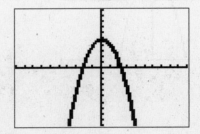

CHAPTER 2: SOLVING EQUATIONS USING THE CROSSING-GRAPHS METHOD

Let's show how to use the graphing capabilities of the calculator to solve

$$\frac{x^3}{10} + x = 5 - \frac{x^2}{2} \tag{2.1}$$

using what the text refers to as *the crossing-graphs method*. We want to graph each side of Equation (2.1) and see where they are the same; that is where the graphs cross. The first step is to enter the left-hand side of the equation as one function and the right-hand side as a second function. Use $\boxed{Y=}$ and $\boxed{\text{CLEAR}}$ out old functions. On the $Y_1 =$ line enter $\boxed{X,T,\Theta,n}$ $\boxed{\wedge}$ 3 $\boxed{\div}$ 10 $\boxed{+}$ $\boxed{X,T,\Theta,n}$, and on the $Y_2 =$ line enter 5 $\boxed{-}$ $\boxed{X,T,\Theta,n}$ $\boxed{\wedge}$ 2 $\boxed{\div}$ 2 . The correctly entered functions are in Figure 2.54. Use $\boxed{\text{ZOOM}}$ 6 to get the ZStandard view of the graphs shown in the Figure 2.55. Note that we have used $\boxed{\text{TRACE}}$ to show which graph goes with which function. The point in Figure 2.55 where the graphs cross is where the left-hand and right-hand sides of the equation are the same. To solve Equation (2.1), we want the X-value of that crossing point.

To locate the crossing point we first press $\boxed{\text{2nd}}$ [CALC] to get the CALCULATE menu shown in Figure 2.56. We want to know where the graphs cross, so we choose 5:intersect from the menu. The calculator will return you to the graphing screen with the prompt First curve? at the bottom of the screen. Press $\boxed{\text{ENTER}}$ and you will see the prompt Second curve?. Press $\boxed{\text{ENTER}}$ again. You will now see the graph with Guess? at the bottom. Press $\boxed{\text{ENTER}}$ a final time, and Intersection will appear at the bottom of the screen together with the coordinates of the crossing point. From Figure 2.57 we see that the graphs cross when $x = 2.0470371$, and so, rounding to two decimal places, we report the solution to Equation (2.1) as $x = 2.05$.

Briefly, to solve with the crossing-graphs method, you use $\boxed{Y=}$ and enter each side of the equation on a separate line. Then $\boxed{\text{GRAPH}}$ and make necessary adjustments so that the crossing point is shown. Press $\boxed{\text{2nd}}$ [CALC] and select 5:intersect from the menu. In general you can get the intersection point from here by pressing $\boxed{\text{ENTER}}$ three times. The intermediate prompts First curve? and Second curve? are there in case you have more than two graphs on the screen. The prompt Guess? is there in case more than one crossing point is on the screen. In that case, move the cursor near the crossing point you want before pressing $\boxed{\text{ENTER}}$. You may also have to move the cursor near the crossing point you want on the

Figure 2.54: Entering $\dfrac{x^3}{10} + x$ *and* $5 - \dfrac{x^2}{2}$

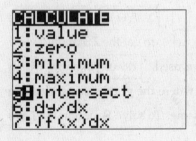

Figure 2.55: The graphs of $\dfrac{x^3}{10} + x$ *and* $5 - \dfrac{x^2}{2}$

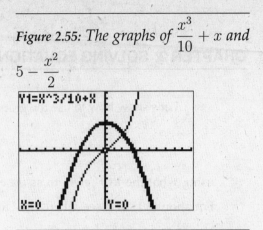

Figure 2.56: The **CALCULATE** *menu*

Figure 2.57: Solving with the crossing-graphs method

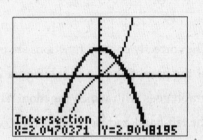

rare occasion that the sequence 5:intersect $\boxed{\text{ENTER}}$ $\boxed{\text{ENTER}}$ $\boxed{\text{ENTER}}$ fails to produce an answer. Finally, you should be aware that for this method to work the crossing point you want must appear in the graphing window.

CHAPTER 2: SOLVING EQUATIONS USING THE SINGLE-GRAPH METHOD

We will show an alternative way, which we will refer to as *the single-graph method*, for solving Equation (2.1). First we move everything from the right-hand side of the equation over to the left-hand side, remembering to change the sign of each term:

$$\frac{x^3}{10} + x = 5 - \frac{x^2}{2} \qquad (2.2)$$

$$\frac{x^3}{10} + x - 5 + \frac{x^2}{2} = 0. \qquad (2.3)$$

Now use $\boxed{Y=}$ and $\boxed{\text{CLEAR}}$ the old functions before entering

$\boxed{X,T,\Theta,n}$ $\boxed{\wedge}$ 3 $\boxed{\div}$ 10 $\boxed{+}$ $\boxed{X,T,\Theta,n}$

$\boxed{-}$ 5 $\boxed{+}$ $\boxed{X,T,\Theta,n}$ $\boxed{\wedge}$ 2 $\boxed{\div}$ 2 as shown in Figure 2.58. Use $\boxed{\text{ZOOM}}$ 6 to get the **ZStandard** view shown Figure 2.59. The solution of Equation (2.3) is where the graph crosses the horizontal axis. This point is referred to as a *root* or *zero*. The TI-82 uses *root* while the TI-83/84 uses *zero*. We will adhere to the TI-83/84 convention, and refer to it as a zero.

Figure 2.58: *Entering the function in preparation for the single-graph method*

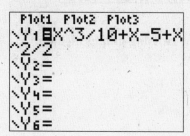

Figure 2.59: *The root or zero*

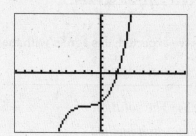

To find this point press $\boxed{\text{2nd}}$ [CALC], but this time from the **CALCULATE** menu, we select **2:zero** (or **2:root** on the TI-82) as shown in Figure 2.60. The graphing screen will appear once more with the prompt **Left Bound?** (or **Lower Bound?** on the TI-82) at the bottom as shown in Figure 2.61.

Figure 2.60: *The* **CALCULATE** *menu with* **2:zero** *highlighted*

Figure 2.61: *Prompting for the* **Left Bound**

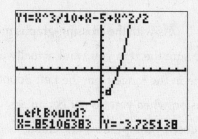

The calculator is asking for help to obtain the solution. We provide it by using $\boxed{\triangleleft}$ to move the cursor to any point to the left of the zero and pressing $\boxed{\text{ENTER}}$. We are presented with Figure 2.62, where our selection is marked at the top of the screen, and at the bottom we see the new prompt **Right Bound?** (or **Upper Bound?** on the TI-82). This time we use $\boxed{\triangleright}$ to move the cursor to the right of the zero and press $\boxed{\text{ENTER}}$ again. We see in Figure 2.63 that there are now two marks at the top of the screen and a **Guess?** prompt at the bottom. The crossing point must be between the displayed marks. We respond to the **Guess?** prompt by pressing $\boxed{\text{ENTER}}$, and we see the solution $x = 2.0470371$ to Equation (2.3) in Figure 2.64.

Figure 2.62: The left bound marked and the Right Bound? *prompt*

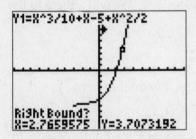

Figure 2.63: Both bounds marked and the Guess? *prompt*

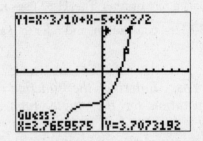

As we expected, this agrees with the answer we got using the crossing-graphs method.

Figure 2.64: The solution of $\dfrac{x^3}{11} + x - 5 + \dfrac{x^2}{2} = 0$ using the single-graph method

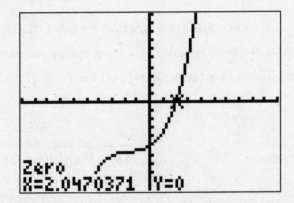

As with the crossing-graphs method, there are a couple of things that should be noted. Be sure the zero you want actually appears in the graphing window, and that there is only one in the range from the Left Bound to the Right Bound that you provide. On the rare occasion when you do not get an answer, you may need to move the bounds closer, or at the Guess? prompt move the cursor as near the root as you can.

CHAPTER 2: OPTIMIZATION

Maxima and minima or peaks and valleys of a graph can be located on the TI-82/83/84 in much the same way that roots or zeros are found.

To illustrate the method, we will look at the graph of

$$f = 2^x - x^2 .$$

Use $\boxed{Y=}$, $\boxed{\text{CLEAR}}$ out any old functions from the list, and enter the new one using 2 $\boxed{\wedge}$ $\boxed{X,T,\Theta,n}$ $\boxed{-}$ $\boxed{X,T,\Theta,n}$ $\boxed{\wedge}$ 2 . The graph of the function shown in Figure 2.65 uses Xmin= $\boxed{(-)}$ 1 , Xmax=4 , Ymin= $\boxed{(-)}$ 2 , and Ymax=2 .

Figure 2.65: *The graph of* $2^x - x^2$

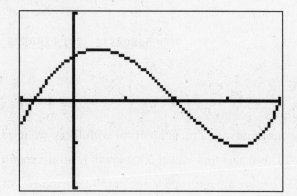

The graph in Figure 2.65 shows a maximum, or a peak, and a minimum, or a valley. The keystrokes required to find them are very similar, but we will carefully find both. Let's first find the maximum. Press $\boxed{\text{2nd}}$ [CALC] to get to the CALCULATE menu shown in Figure 2.66. We want to find the maximum, so we select 4:maximum from the menu. We are returned to the graphing window with the trace option turned on, and the prompt **Left Bound?** (or **Lower bound?** on the TI-82) appears at the bottom of the screen as in Figure 2.67. The information wanted now by the calculator is similar to what is wanted when we search for zeros or roots. We use the arrow keys to move the cursor to the left of the maximum and press $\boxed{\text{ENTER}}$. A tick mark is placed at the top of the screen showing our selection, and we see the **Right Bound?** (or **Upper bound?** on the TI-82) prompt in Figure 2.68. We move the cursor to the right of the maximum and press $\boxed{\text{ENTER}}$ again. Now we see two tick marks at the top of the screen *which must enclose the maximum we want* and the **Guess?** prompt in Figure 2.69. Rarely is any response other than $\boxed{\text{ENTER}}$ required here, and the maximum

Figure 2.66: Selecting 4:maximum *from the* CALCULATE *menu*

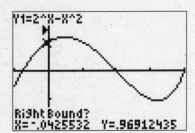

Figure 2.67: The Left Bound? *prompt*

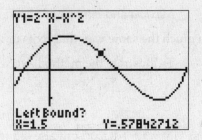

Figure 2.68: The Right Bound? *prompt*

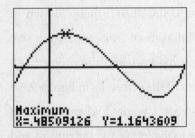

Figure 2.69: The Guess? *prompt*

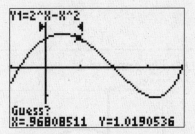

will be located as in Figure 2.70.

The steps for finding the minimum are almost identical with those we used to get the maximum. Press ⟨2nd⟩ [CALC], but this time select 3:minimum from the menu as shown in Figure 2.71. As with the maximum we are returned to the graphing window with the trace

Figure 2.70: Locating a maximum for $2^x - x^2$

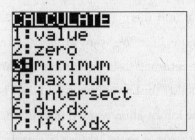

Figure 2.71: Selecting 3:minimum *from the* CALCULATE *menu*

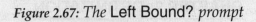

option on and the Left Bound? prompt showing in Figure 2.72. We locate the cursor left of the minimum and press ⟨ENTER⟩ , move the cursor to the right of the minimum and press ⟨ENTER⟩ , and finally press ⟨ENTER⟩ at the Guess? prompt. The minimum will be found as in Figure 2.73.

Figure 2.72: The Left Bound? *prompt for the minimum*

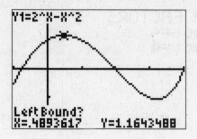

Figure 2.73: A minimum value for $2^x - x^2$

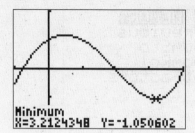

We should note that on the rare occasion when the method described here fails to produce the desired result, you should repeat the procedure, but at the Guess? prompt, move the cursor as near the maximum or minimum as you can and press ENTER .

CHAPTER 2: GRAPHICS CONFIGURATION NOTES

There are a number of ways your calculator can be configured to display graphics. If you were able to follow the examples in the earlier section entitled Chapter 2: Graphs, reproducing the screens we made, then you won't need to use these notes. If not, the difficulty may be that your calculator graphics configuration has been altered from the factory set defaults. These notes address only the two most common graphics settings which may have been changed on your calculator. If this does not solve your graphics configuration difficulties, consult your calculator manual.

Problem: *My graphs match those in the examples until I zoom in or out. Then I get a different picture.* The probable cause is that the default zoom factors have been changed. To find out and fix the problem, press ZOOM and use ▷ to highlight MEMORY as shown in Figure 2.74. Select 4:Set Factors from the menu. This will take you to the ZOOM FACTORS menu shown in Figure 2.75. Your screen should show XFact=4 and YFact=4. If necessary, change these numbers to match Figure 2.75.

Problem: *When I* GRAPH *, I get an error as in Figure 2.76, or I get the graphs, but I also get some extra dots as in Figure 2.77 or some dots connected by lines as in Figure 2.78.* The probable

Figure 2.74: *The* ZOOM MEMORY *menu*

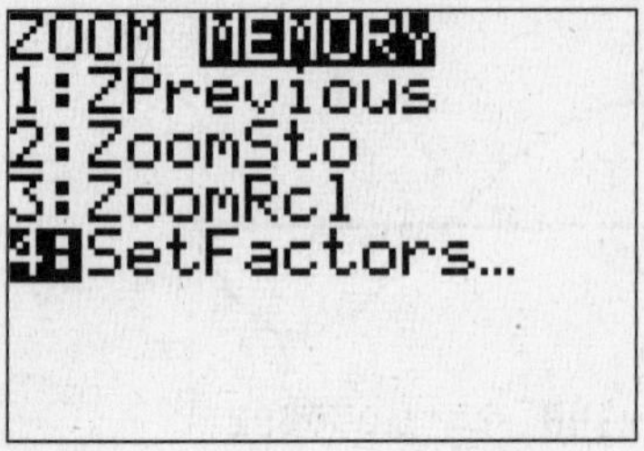

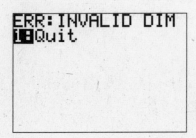

Figure 2.75: *The* ZOOM FACTORS *menu*

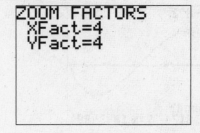

Figure 2.76: *Error reported when you* GRAPH

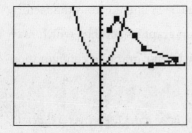

Figure 2.77: *Discrete statistical plot turned on*

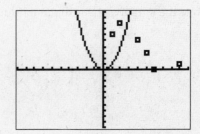

cause is that you have one form or another of statistical plotting turned on. To correct this problem press 2nd [STAT PLOT] to get to the STAT PLOTS menu shown in Figure 2.79. We want to turn statistical plotting Off, so we press 4 and then ENTER .

Figure 2.78: *Connected statistical plot turned on*

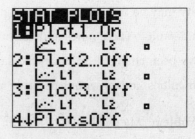

Figure 2.79: *The* STAT PLOTS *menu*

CHAPTER 5: PIECEWISE-DEFINED FUNCTIONS

Sometimes functions are defined by different formulas over different intervals. Such functions are known as *piecewise-defined* functions. For example, we define

$$f(x) = \begin{cases} x+1, & \text{if } x < 1 \\ x-1, & \text{if } x \geq 1. \end{cases}$$

This means that $f(x)$ is $x+1$ if x is less than 1 but it is $x-1$ when x is greater than or equal to 1. Thus $f(0) = 0 + 1 = 1$ since $0 < 1$. But $f(3) = 3 - 1 = 2$ since $3 \geq 1$.

The TI-82 and TI-83/84 will graph piecewise-defined functions. Let's see how to graph the function f defined above. The first step is to use the $\boxed{\text{MODE}}$ key and change from Connected to Dot as shown in Figure 2.80. On the function entry screen we want to enter keystrokes that produce the following:

$$(x+1)(x<1) + (x-1)(x \geq 1).$$

We do this because the calculator evaluates an expression like $x < 1$ as a function of x that has the value 1 if the relation is true and the value 0 if the relation is false. The only difficulty in entering this expression is locating the proper keystrokes for $<$ and $\geq$. We begin using

$$\boxed{(} \quad \boxed{X,T,\Theta,n} \quad \boxed{+} \quad 1 \quad \boxed{)} \quad \boxed{(} \quad \boxed{X,T,\Theta,n}$$

as shown in Figure 2.81.

Figure 2.80: *Setting to* Dot *mode*   **Figure 2.81:** *Beginning the entry*

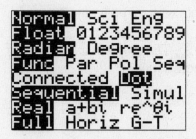

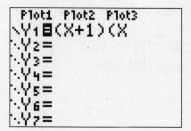

Now press $\boxed{\text{2nd}}$ [TEST] and then highlight the $<$ symbol as shown in Figure 2.82. Press $\boxed{\text{ENTER}}$, and the $<$ symbol will appear as shown in Figure 2.83.

We make the remaining entry up to the $\geq$ symbol using

$$1 \quad \boxed{)} \quad \boxed{+} \quad \boxed{(} \quad \boxed{X,T,\Theta,n} \quad \boxed{-} \quad 1 \quad \boxed{)} \quad \boxed{(} \quad \boxed{X,T,\Theta,n} .$$

Figure 2.82: Using [2nd] [TEST] *to locate the < symbol*

Figure 2.83: The < symbol entered

Figure 2.84: Continuing the entry

Figure 2.85: Using [2nd] [TEST] *to enter the ≥ symbol*

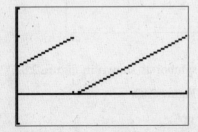

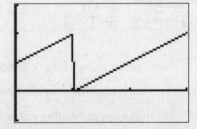

This is shown in Figure 2.84. To enter the ≥ symbol we proceed as before, using [2nd] [TEST] and highlighting the appropriate symbol. Press [ENTER] to display the symbol as shown in Figure 2.85.

Complete the entry using 1 [ )]. In Figure 2.86 we have used [GRAPH] with a horizontal span of 0 to 3 and a vertical span of −1 to 3.

In Figure 2.87 we have changed back to Connected mode. This makes the calculator add an extraneous vertical line to the picture, and it illustrates why we prefer the Dot mode.

Figure 2.86: The correct graph displayed using Dot *mode*

Figure 2.87: The graph displayed using Connected *mode*

Discrete Data

The TI-82/83/84 offers many features for handling discrete data sets. This chapter shows how to enter data and perform some basic analysis.

CHAPTER 3: GRAPHING DISCRETE DATA

Many times information about physical or social phenomena is obtained by gathering individual bits of data and recording it in a table. For example, the following table taken from the *Information Please Almanac* shows median American family income I by year in terms of 1996 dollars. That means the dollar amounts shown have been adjusted to account for inflation. The variable d in the table is years since 1980. Thus, for example, the $d = 2$ column corresponds to 1982.

d = years since 1980	0	1	2	3	4	5
I = median income	21,023	22,388	23,433	24,580	26,433	27,735

Let's show how to enter this data into the calculator and display it graphically.

Entering data: To enter the values from the table above, we press $\boxed{\text{STAT}}$ to get the EDIT CALC TESTS menu shown in Figure 2.88. Be sure EDIT is highlighted. If it is not, use $\boxed{\triangleleft}$ to change that. Choose 1:Edit, and you will see a screen similar to the one in Figure 2.89. Figure 2.89 shows that our calculator has some old data already entered, and

Figure 2.88: *Using* $\boxed{\text{STAT}}$ *to get the* EDIT CALC TESTS *menu*

Figure 2.89: *The data entry screen with old data*

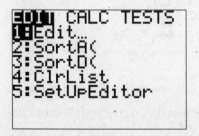

we need to clear it out before entering new data. Use $\boxed{\triangle}$ to highlight L1 as shown in Figure 2.89 and then press $\boxed{\text{CLEAR}}$ $\boxed{\text{ENTER}}$. Do the same with each column that has unwanted entries. You should now have the clear data entry screen shown in Figure 2.90. The L1 column corresponds to d, and we type in the d values 0 through 5, pressing either $\boxed{\text{ENTER}}$ or $\boxed{\triangledown}$ after each entry. Now use $\boxed{\triangleright}$ to move to the top of the L2 column and enter the median incomes there. The completed data entry is in Figure 2.91.

Figure 2.90: A fresh data entry screen

Figure 2.91: Completed data entry

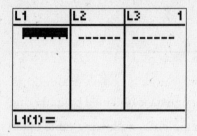

Turning on (or off) statistical plots: Before we can make of graphical display of data, we need to set up the calculator so that it can display discrete data. That is, we must turn on statistical plotting. This only has to be done the first time we want to make such a picture (or if we ourselves have disabled this feature). Once turned on, the statistical plotting will stay on until we turn it off. To set up statistical plotting, press 2nd [STAT PLOT] to get the menu shown in Figure 2.92. Select 1:Plot 1 from that menu, and the TI-83/84 will present you with the menu in Figure 2.93. The menu for the TI-82 is similar.

Figure 2.92: The STAT PLOTS *menu*

Figure 2.93: The Plot 1 *menu*

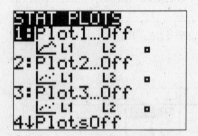

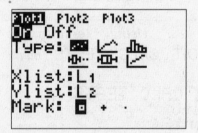

Highlight On and press ENTER . Your screen should now match the one in Figure 2.93. Check the Type:, Xlist, Ylist, and Mark lines to see that they agree with Figure 2.93. If they do not, use the arrow keys to move the highlight where you want it, change the entry, and press ENTER . When you no longer want data points to appear in your graphs, you can repeat this procedure, turning Plot 1 Off.

It is particularly important to note the role of the Xlist and Ylist lines in Figure 2.93. The Xlist line determines which of the data columns will be shown on the horizontal axis.

Usually, this is L1. To set it on the TI-83/84, use ⌈2nd⌉ [L1] (above the 1 key); on the TI-82, highlight L1 and press ⌈ENTER⌉. The Ylist line determines which of the data columns will be shown on the vertical axis. Usually, this is L2, which you set on the TI-83/84 by using ⌈2nd⌉ [L2] (above the 2 key) and on the TI-82 by highlighting L2 and pressing ⌈ENTER⌉.

Plotting the data: Now we are ready to view the graph. Generally with discrete data, we don't have to worry with how to set up the window, because the TI-82/83/84 has a special feature which does the job automatically. This feature is accessed as follows. Press ⌈ZOOM⌉, and move down the ZOOM menu to 9:ZoomStat as shown in Figure 2.94. When you press 9 (or ⌈ENTER⌉ with 9:ZoomStat highlighted) you will see the graphical display in Figure 2.95. If you got the picture in Figure 2.95 with one or more extra graphs added, you need to use ⌈Y =⌉ and clear out any old functions that are on the function entry screen.

Figure 2.94: The ZoomStat *option*

Figure 2.95: Displaying median income data

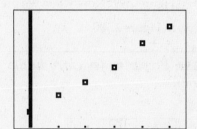

When you use ZoomStat to display data, the calculator decides the window size to use. You can see what its choices were by pressing ⌈WINDOW⌉. We see in Figure 2.96 that the horizontal span is −0.5 to 5.5 and the vertical span is from 19,881.96 to 28,876.04.

Adding regular graphs to data plots: You can add any graph you wish to a data plot. For example, using a technique known as *linear regression*, which is discussed in the text *Functions and Change* and is reviewed in the next section of this guide, we find that the linear function $y = 1338x + 20919$ closely approximates the median income data. To show its graph, use ⌈Y =⌉, type 1338 ⌈X, T, Θ, n⌉ ⌈+⌉ 20919 , and then ⌈GRAPH⌉. This line added to the data plot is in Figure 2.97.

Figure 2.96: *The window size selected* by ZoomStat

```
WINDOW
 Xmin=■.5
 Xmax=5.5
 Xscl=1
 Ymin=19881.96
 Ymax=28876.04
 Yscl=1
 Xres=1
```

Figure 2.97: *Adding an additional graph*

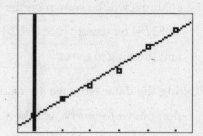

Editing columns: There are shortcuts that make it easy to do certain kinds of data edits. For example, suppose we wish to show median income remaining after each family pays a 17% tax. That is, we want to replace each entry in the L2 column by 83% of its current value. Use STAT and choose 1:Edit from the menu to get to the data we have already entered. Use the arrow keys to highlight L3 as shown in Figure 2.98. We want to put in the L3 column the entries in the L2 column times 0.83. We could calculate these values one at a time and enter each individually, but the TI-82/83/84 offers a much easier way to accomplish this. Just type . 83 2nd [L2] as seen at the bottom of the screen in Figure 2.99. Now when we press ENTER , the L3 column is filled in automatically as in Figure 2.100.

Figure 2.98: *Preparing for entry in the* L3 *column*

L1	L2	**L3** 3
0	21023	------
1	22388	
2	23433	
3	24580	
4	26433	
5	27735	
------	------	

L3 =

Figure 2.99: *The formula entered*

L1	L2	**L3** 3
0	21023	------
1	22388	
2	23433	
3	24580	
4	26433	
5	27735	
------	------	

L3 =.83L2

Let's show how to plot this new data. Now we want the L1 column on the horizontal axis, but the L3 column rather than the L2 column for the vertical axis. To make this happen, use 2nd [STAT PLOT] and select 1:Plot1 from the menu. Go to the Ylist: line and type 2nd [L3] (on the TI-82, highlight L3 and press ENTER) as shown in Figure 2.101. Now when we ZOOM 9 we see the new data displayed in Figure 2.102. (Don't forget to use Y = and clear out the Y_1 line, which we don't want to see anymore.)

Figure 2.100: *The* L3 *column filled automatically*

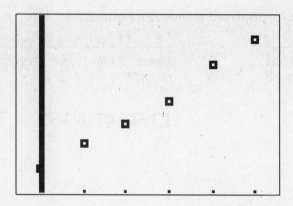

L3(1)=17449.09

Figure 2.101: *Changing the vertical axis from* L2 *to* L3

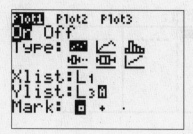

Figure 2.102: *Data after 17% tax*

CHAPTER 3: LINEAR REGRESSION

Linear regression is a method of getting a linear function which approximates almost linear data. The TI-82/83/84 has built-in features which will do this automatically. Let's show how to do it for the following data table.

x	0	1	2	3	4
y	17	19.3	20.9	22.7	26.1

The first step is to enter the data. This is done exactly as described in the previous section. Use STAT and select 1:Edit from the menu. CLEAR out old data as necessary to get the clear data entry screen in Figure 2.103. Next enter the x data in the L1 column and the y data in the L2 column as shown in Figure 2.104.

To get the regression line press STAT and use ▷ to move the highlight at the top

Figure 2.103: *A clear data entry screen*

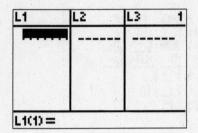

Figure 2.104: *The correctly entered data*

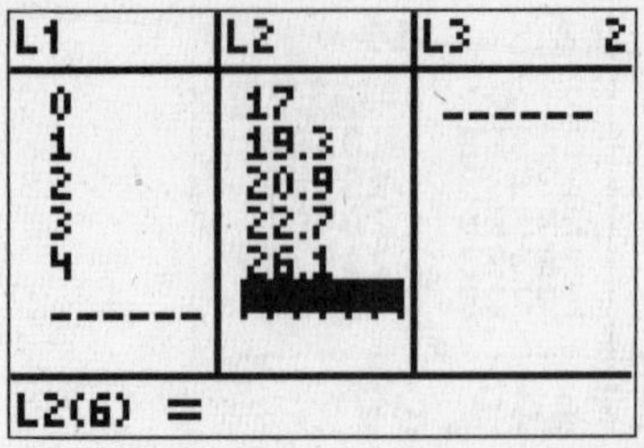

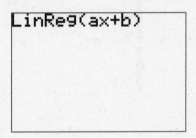

of the screen to CALC. This will change the menu to the one shown in Figure 2.105. From this menu select 4:LinReg(ax+b) (this is 5:LinReg(ax+b) on the TI-82). You will be taken to the calculation screen with LinReg(ax+b) displayed as shown in Figure 2.106. Press ENTER , and

Figure 2.105: *The statistical calculation menu (*LinReg(ax+b) *is menu item* 5 *on the TI-82.)*

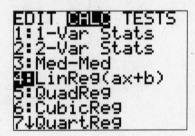

Figure 2.106: *The regression line prompt*

you will see the information in Figure 2.107. (The TI-82 displays an additional line, which we will not use.)

This screen tells us that the equation of the regression line is $y = ax + b$, where the slope is $a = 2.16$, and the vertical intercept is $b = 16.88$. (Note here that the TI-82/83/84 uses a for the slope rather than the more familiar m.) Thus the equation of the regression line is

$$y = 2.16x + 16.88 \, .$$

We have in fact completed the calculation of the regression line, but it is almost always important to continue and make a display of the data and the regression line together. Among other things, such a display provides a valuable check for our work. To graph the line, we must get the regression line formula into the function entry screen. We will first show how to do this manually and then later show a method for having it automatically entered in the function list at the time it is calculated. For manual entry use $Y =$, CLEAR out old functions, and type the equation of the regression line as 2.16 X, T, Θ, n + 16.88 .

The correctly entered function is in Figure 2.108.

Figure 2.107: *Regression line parameters*

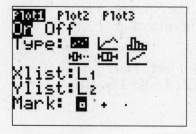

Figure 2.108: *Manual entry of the regression line formula*

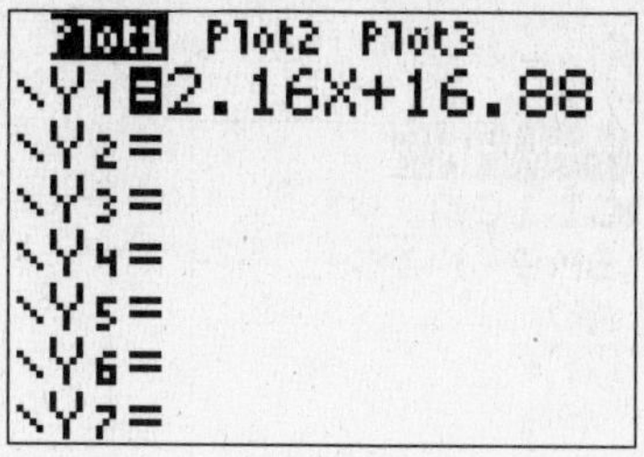

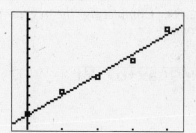

Before we make the graph, we use $\boxed{\text{2nd}}$ [STAT PLOT] and check to see that Plot1 is configured as in Figure 2.109. Finally $\boxed{\text{ZOOM}}$ 9 produces the picture in Figure 2.110.

Figure 2.109: *The properly configured* Plot1 *window*

Figure 2.110: *Data and regression line*

The important feature to note in Figure 2.110 is that the regression line closely approximates the data points. If this does not happen, we know either that we have made a mistake, or that the data is not properly modeled using a linear function.

Automatic entry of the regression line on the function list: Let's calculate the regression line again, this time showing how to make it automatically appear on the function list. This method is available only on the TI-83/84.[4]

Just as before, we use $\boxed{\text{STAT}}$, select the CALC menu, and choose 4: LinReg(ax+b). Now at the regression line prompt shown in Figure 2.106 above, press $\boxed{\text{VARS}}$ and highlight Y-VARS at the top of the screen to see the y-variables menu. Be sure that 1:Function is high-

---

[4]There is another method for transferring the equation of the regression line to the function entry screen without typing it, a method that is available on both calculators. Use $\boxed{Y=}$, $\boxed{\text{CLEAR}}$ the Y_1= line, and then press $\boxed{\text{VARS}}$. Select 5:Statistics. Move the highlight at the top of the screen to EQ and then choose 1:RegEQ (on the TI-82 this is 7:RegEQ) from the menu.

lighted in that menu as shown in Figure 2.111. Now when you press ENTER , you will be presented with the list in Figure 2.112. Be sure 1:Y1 is highlighted and press ENTER . You

---

Figure 2.111: *The y-variables menu*

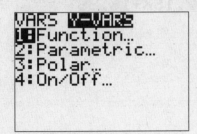

Figure 2.112: *Function list*

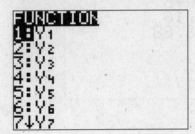

will be returned to the home screen with Y1 added to the regression prompt, as seen in Figure 2.113. Now when you press ENTER , the regression line parameters will be calculated as usual, but also the regression line formula will appear on the function list. In Figure 2.114 we have pressed Y = to verify that this has occurred.

Figure 2.113: *Completing the regression prompt*

Figure 2.114: *Regression formula automatically entered*

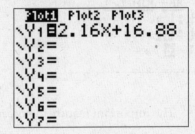

We should note that this is a case where the explanation is so long that it may appear to make the procedure more complicated than it really is. If you go through the keystrokes a few times, you may find this quick and easy indeed. This procedure is available to you, and manual function entry is also available. You should choose the method that seems best to you.

CHAPTER 3: DIAGNOSTICS

Sometimes it is important to get a statistical measure of how well a regression line fits data. One such measure is the *correlation coefficient*, commonly denoted by r. The number r is always between -1 and 1. Values of r near 1 indicate a good fit with a positive slope. Values of r near -1 indicate a good fit with a negative slope. Values of r near 0 indicate a poor fit.

The value of r is automatically displayed by the TI-82 when linear regression is performed. To get the TI-83/84 to display the correlation coefficient, press [2nd] [CATALOG] and select DiagnosticOn as shown in Figure 2.115. Press [ENTER] [ENTER], and the diagnostic feature is turned on as shown in Figure 2.116.

Figure 2.115: *Selecting* Diagnos-ticOn *using* [2nd] [CATALOG]

Figure 2.116: *Enabling the diagnostic feature*

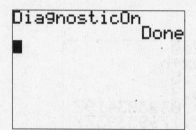

Now if we perform linear regression the value of r will be displayed when regression coefficients are calculated. This is shown in Figure 2.117. (A second statistical measure, denoted by r^2, is also shown on the TI-83/84. We will not discuss this measure here.) Note that the value of r is near 1, indicating a good fit with a positive slope. The data and regression line are plotted in Figure 2.118, and the quality of the fit is apparent.

Figure 2.117: *A correlation coefficient near 1*

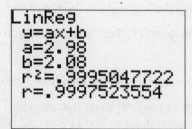

Figure 2.118: *Good fit indicated by r near 1*

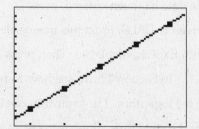

In Figure 2.119 note that the correlation coefficient is near −1, and the good fit with negative slope is shown in Figure 2.120.

Figure 2.119: A correlation coefficient near −1

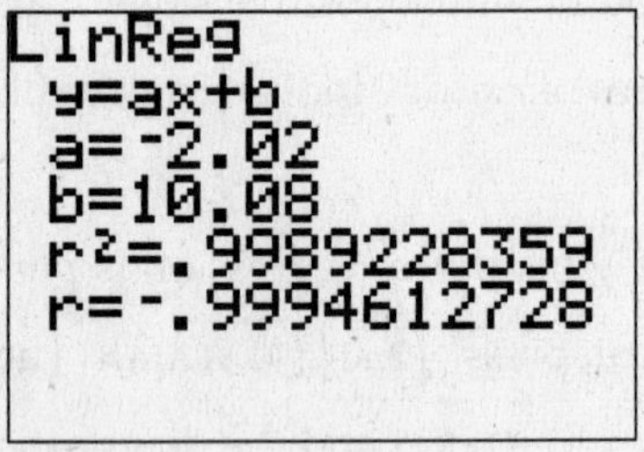

Figure 2.120: Good fit indicated by r near −1

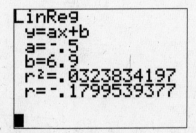

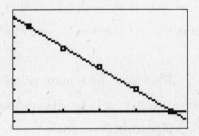

In Figure 2.121 the correlation coefficient is near 0, and the corresponding poor data fit is shown in Figure 2.122.

Figure 2.121: A correlation coefficient near 0

Figure 2.122: Poor fit indicated by r near 0

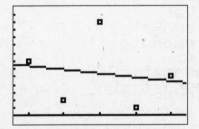

CHAPTER 4: EXPONENTIAL REGRESSION

Exponential regression can be performed directly. The steps are the same as for linear regression except that a different regression method is selected from the CALC menu: After the data has been entered, press $\boxed{\text{STAT}}$ and use $\boxed{\triangleright}$ to move the highlight at the top of the screen to CALC. From this menu select 0:ExpReg. You will be taken to the calculation screen with ExpReg displayed. Then press $\boxed{\text{ENTER}}$.

In Section 4.5 we see how to perform exponential regression indirectly, using the link to the logarithm. The logarithm can be used to transform exponential or power data into linear data and thus allow us to get an appropriate model using linear regression. To transform exponential data, we take the logarithm of function values, leaving input values as they are.

To make an exponential model for data such as

input values	a	b	...
function values	z	w	...

we apply the logarithm to the function values to produce new data

input values	a	b	...
function values	$\ln z$	$\ln w$	...

Next we get the regression line $y = mx + b$ for this transformed data. The exponential model we want is then $e^b \times (e^m)^x$. The alternative form for the model (described in Section 6.4 of *Functions and Change*) is $e^b e^{mx}$.

We can illustrate the method by getting an exponential model for the data in the following table:

x	1	2	3	4	5
y	0.55	0.83	1.5	2.7	4.4

The first step is to use $\boxed{\text{STAT}}$ and 1:Edit to enter the data as shown in Figure 2.123. It will prove convenient if we enter the data in columns 1 and 3, leaving column 2 blank for now. To transform the data, we highlight the L2 at the top of the second column and enter $\boxed{\text{LN}}$ $\boxed{\text{2nd}}$ [L3] (on the TI-83/84 this is followed by $\boxed{)}$) as shown in Figure 2.124. Now

---

Figure 2.123: Correctly entered data

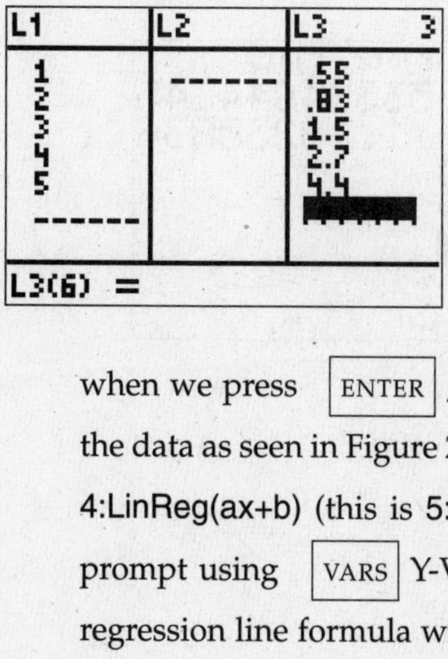

Figure 2.124: Preparing to transform the data

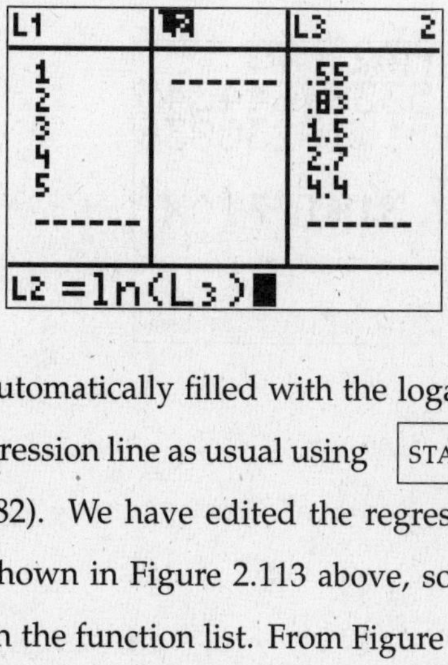

when we press $\boxed{\text{ENTER}}$, the L2 column will be automatically filled with the logarithm of the data as seen in Figure 2.125. Now we get the regression line as usual using $\boxed{\text{STAT}}$ CALC 4:LinReg(ax+b) (this is 5:LinReg(ax+b) on the TI-82). We have edited the regression line prompt using $\boxed{\text{VARS}}$ Y-VARS 1:Function Y1, as shown in Figure 2.113 above, so that the regression line formula will be automatically put on the function list. From Figure 2.126 we see that the regression line formula, with rounding to three decimal places, is $0.534x - 1.182$.

Figure 2.125: The L2 *column filled automatically*

Figure 2.126: Regression line parameters

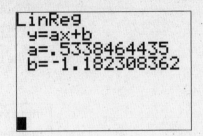

The needed exponential model in the standard form is then 0.31×1.71^x because $e^{-1.182} = 0.31$ and $e^{0.534} = 1.71$. (In the alternative form it is $e^{-1.182}e^{0.534x} = 0.31e^{0.534x}$.) We want now to use the calculator to display the original data along with the exponential model we have made. We use $\boxed{Y=}$ and manually enter the exponential model in the standard form on the Y2= line as shown in Figure 2.127. (As with the regression line, there is a different way of entering the exponential model in the alternative form. You may or may not find it more convenient than manual entry. For the TI-83/84, on the Y2= line type $\boxed{\text{2nd}}$ [e^x] $\boxed{\text{VARS}}$ Y-VARS 1:Function 1:Y1, followed by $\boxed{)}$. For the TI-82, on the Y2= line type $\boxed{\text{2nd}}$ [e^x] $\boxed{\text{2nd}}$ [Y-VARS] 1:Function 1:Y1. This different entry method is shown in Figure 2.128.)

Note that both in Figure 2.127 and in Figure 2.128 we have removed the highlight from the = sign on the Y1 line (by moving the cursor to the equal sign and pressing $\boxed{\text{ENTER}}$) so that function will not be graphed.

Figure 2.127: Manual entry of the exponential model

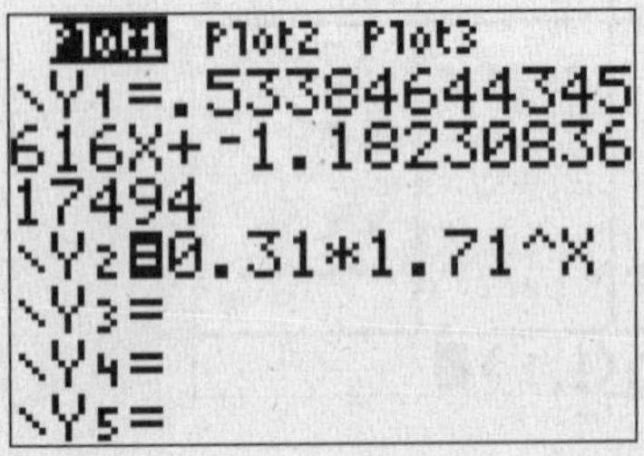

Figure 2.128: Alternative method of entering exponential model

Before graphing, it is important that we graph the original data that we put in columns 1 and 3 of the data list. To make sure, we use ⌐2nd⌐ [STAT PLOT] 1:Plot 1, and make sure Xlist is set to L1 and Ylist is set to L3. The properly configured menu is in Figure 2.129. Now we ⌐ZOOM⌐ 9 , and the data along with the exponential model appear as in Figure 2.130.

Figure 2.129: *The properly configured* Stat Plot *menu*

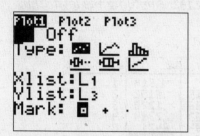

Figure 2.130: *The data along with the exponential model*

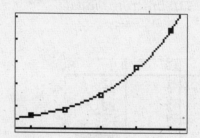

CHAPTER 5: POWER REGRESSION

Power models can be constructed in much the same way as exponential models. The key difference is that we apply the logarithm to both input values and function values.

To make a power model for data such as

input values	a	b	...
function values	z	w	...

we apply the logarithm to produce new data

input values	$\ln a$	$\ln b$	...
function values	$\ln z$	$\ln w$	...

Next we get the regression line $y = mx + b$ for this transformed data. The power model we want is then $e^b x^m$.

We illustrate the method by getting a power model for the following data:

x	1	2	3	4	5
y	0.22	0.54	0.88	1.25	1.91

We use $\boxed{\text{STAT}}$ 1:Edit and enter the data in columns L3 and L4 as shown in Figure 2.131. We want the logarithm of the L3 data in column L1, so we highlight L1 and $\boxed{\text{LN}}$ $\boxed{\text{2nd}}$ [L3] (followed by $\boxed{)}$ on the TI-83/84). Next highlight L2 and type $\boxed{\text{LN}}$ $\boxed{\text{2nd}}$ [L4] (followed by $\boxed{)}$ on the TI-83/84). The completed data entry is shown in Figure 2.132.

Figure 2.131: *Entering the data*

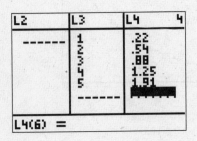

Figure 2.132: *Applying the logarithm*

Now we can get the regression line parameters using $\boxed{\text{STAT}}$ CALC 4:LinReg(ax+b) (this is 5:LinReg(ax+b) on the TI-82). We see from Figure 2.133 that the regression line formula is $1.31x - 1.529$. Since $e^{-1.529} = 0.22$, the appropriate power model is $0.22x^{1.31}$. In Figure 2.134 we have entered this on the function list.

Figure 2.133: *Regression line parameters*

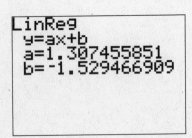

Figure 2.134: *Entering the power model*

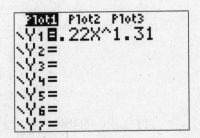

Our original data is in columns L3 and L4, so we use $\boxed{\text{2nd}}$ [STAT PLOT] and check that Xlist is set to L3 and Ylist is set to L4, as shown in Figure 2.135. Now if we $\boxed{\text{ZOOM}}$ 9 , the original data along with the power model will appear as in Figure 2.136.

Figure 2.135: *Properly configured* Xlist *and* Ylist

Figure 2.136: *Data with power model*

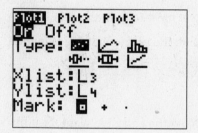

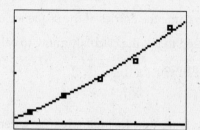

CHAPTER 5: SEQUENCE GRAPHS

Sometimes we are asked to graph data in a sequence where the next term is calculated from the one before. Often this is given as a starting point and an iterative formula. Let's look for example at the sequence we get starting at 1 and using the iterative formula $3x - 1$. Here is how we make the sequence.

The first term is 1.

Apply the formula to 1 to get the second term $3 \times 1 - 1 = 2$.

Apply the formula to 2 to get the third term $3 \times 2 - 1 = 5$.

Apply the formula to 5 to get the fourth term $3 \times 5 - 1 = 14$.

The process continues by applying the formula to the last term calculated to get the next. You should verify that the next term in the sequence is 41. We want to see how to get the calculator to calculate the terms of such a sequence automatically and plot the points $(1, 1), (2, 2), (3, 5), (4, 14), \ldots$

We used the sequence above because the first few terms are easy to calculate, making it a simple illustration of what we are trying to do. But it is clear that, after a few more terms, the sequence will get quite large. For an example to illustrate the use of the calculator we will look at a more interesting sequence. Let's start at 0 and use the formula $x^2 - 0.7$. It is not at all clear from looking at the formula how the sequence behaves. The first step is to press MODE and select Seq (for sequence) in the fourth line of the display as shown in Figure 2.137. Now press $Y =$ to get to the function entry screen. Note that the calculator wants to name the entries of the sequence $u(1), u(2), u(3), \ldots$. We set nMin to 1. In the next line we need to tell the calculator how to calculate the next value from the previous value using the formula

$$u(n) = u^2(n - 1) - 0.7.$$

To enter this, first note that the letter u is above the 7 key. Now the entry we want for u(n)= is as follows.

2nd [u]   (X, T, Θ, n – 1   )   ∧ 2   – 0.7

Also, u(nMin) is our starting value, 0. The correctly entered screen is in Figure 2.138.

Figure 2.137: *Choosing the sequence option*

Figure 2.138: *The function entry screen*

The next step is to look at a table of values, as shown in Figure 2.139, to determine how to set up the graphing window. It appears that the function values are between -1 and 0. Now we select $\boxed{\text{WINDOW}}$. Let's plot the first 30 points. That means we should set nMin to 1 and nMax to 30. This part of the WINDOW menu is shown in Figure 2.140.

We set Xmin to 0 and Xmax to 30. Finally we set Ymin to -1 and Ymax to 0. This part of the WINDOW menu is shown in Figure 2.141. Finally press $\boxed{\text{GRAPH}}$ to get the picture in Figure 2.142.

Figure 2.139: *A table of values for the sequence*

n	u(n)
1	0
2	-.7
3	-.21
4	-.6559
5	-.2698
6	-.6272
7	-.3066

$n=1$

Figure 2.140: *The top of the* WINDOW *menu*

```
WINDOW
 nMin=1
 nMax=30
 PlotStart=1
 PlotStep=1
 Xmin=0
 Xmax=30
↓Xscl=1
```

Figure 2.141: *The completed* WINDOW *menu*

```
WINDOW
↑PlotStep=1
 Xmin=0
 Xmax=30
 Xscl=1
 Ymin=-1
 Ymax=0
 Yscl=1
```

Figure 2.142: *The sequence graph*

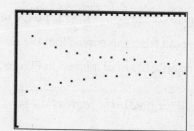

The graph shows that the sequence is in two branches which appear to come together. You may wish to plot the first 100 points of the sequence to see if this trend continues.

CHAPTER 5: QUADRATIC, CUBIC, AND QUARTIC REGRESSION

Quadratic regression is a method of fitting data with a quadratic polynomial, that is, a function of the form $ax^2 + bx + c$. On the TI calculators, the process is quite similar to linear regression. We will show how to fit the following data set with a quadratic.

x	0	1	2	3	4	5
y	2.5	1.1	4.3	16.2	47.4	95.3

The first step is to use $\boxed{\text{STAT}}$ and EDIT to enter the data as usual. The properly entered data is in Figure 2.143. Next we use $\boxed{\text{STAT}}$ and CALC and highlight 5:QuadReg (6:QuadReg on the TI-82) as shown in Figure 2.144.

Figure 2.143: *Properly entered data*

Figure 2.144: *Selecting quadratic regression*

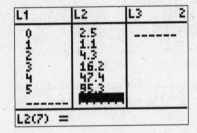

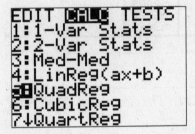

Now press $\boxed{\text{ENTER}}$ and the regression coefficients will be displayed as in Figure 2.145. We read from this screen that the function we want is $6.4x^2 - 14.44x + 5.23$. The data and the quadratic fit are displayed in Figure 2.146.

Figure 2.145: *Quadratic regression parameters*

Figure 2.146: *Data and the quadratic fit*

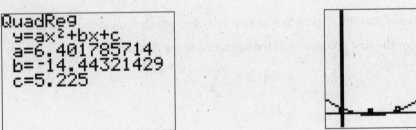

Cubic regression fits data with a function of the form $ax^3 + bx^2 + cx + d$, and quartic regression fits data with a function of the form $ax^4 + bx^3 + cx^2 + dx + e$. The procedure is the same as for quadratic regression except that for cubic regression we use 6:CubicReg (7:Cubic-cReg on the TI-82) as shown in Figure 2.147. The cubic regression coefficients for the data in the table above are displayed in Figure 2.148. This tells us that if we wish to fit the data with a cubic, we should use $0.85x^3 - 0.01x^2 - 2.73x + 2.66$.

Figure 2.147: *Selecting cubic regression*

Figure 2.148: *Cubic regression parameters*

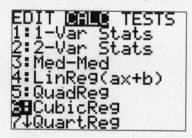

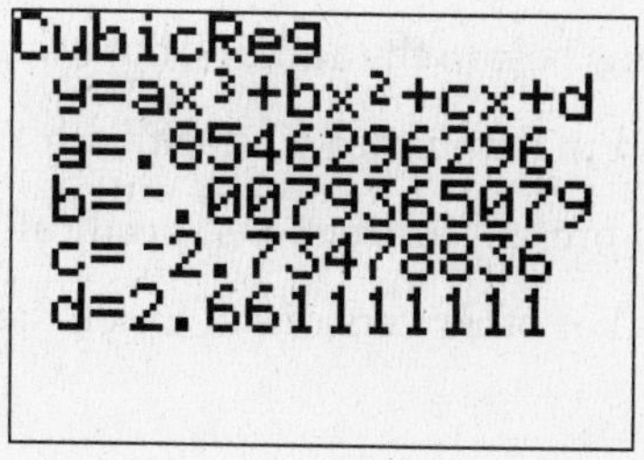

If we wish to use quartic regression, we select 7:QuartReg (8:QuartReg on the TI-82) as shown in Figure 2.149. The quartic regression parameters in Figure 2.150 tell us that the proper quartic to use for the data above is $-0.14x^4 + 2.25x^3 - 4.3x^2 + 1.25x + 2.42$.

Figure 2.149: *Selecting quartic regression*

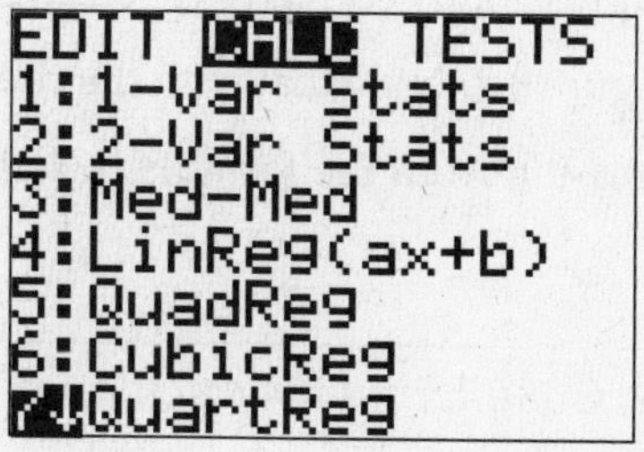

Figure 2.150: *Quartic regression parameters*

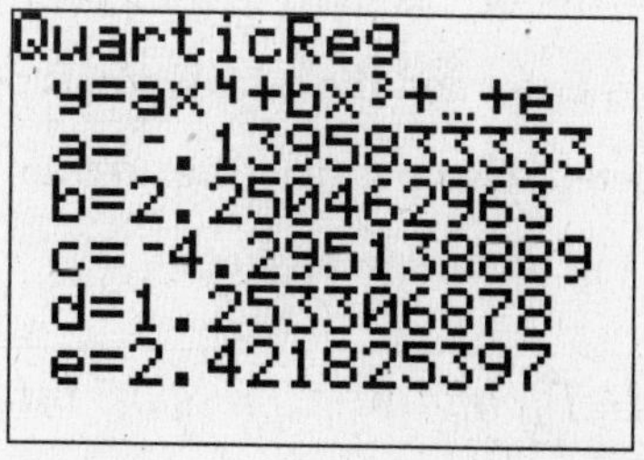

Arithmetic Operations

Excel is a powerful and complex piece of software. It will do many things, and there is often more than one way to get them done. In general we will present only the way which seems best to the authors, but other methods may be more natural to some. You are encouraged to familiarize yourself with Excel and to choose methods which fit your own style. We will present instructions specifically for the version of Excel which is part of Microsoft Office XP. For other versions of Excel, the individual steps may vary slightly.

PROLOGUE: BASIC CALCULATIONS

To perform a simple calculation such as $\frac{72}{9} + 3 \times 5$ we choose any convenient cell and start with an equal sign followed by the typed expression =72/9+3*5. This is shown in Figure 3.1. (If you omit the equal sign, Excel will not perform the calculation.) Note that the symbol * is used to indicate multiplication. Once the expression is correctly typed, we press **Enter** and the calculation is performed as is shown in Figure 3.2. Note that with the cell selected, we now read the answer from the cell and the expression from the formula box at the top of the spreadsheet.

Figure 3.1: Typing an expression into Excel

Figure 3.2: Completing the calculation

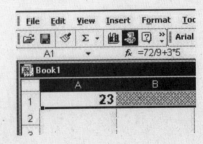

Exponentiation: To enter an expression like 2^3 we use the caret symbol ^. Typing =2^3 will produce the correct result as shown in Figure 3.3. Be sure to begin the expression with an equal sign.

Square roots: To get square roots, we use Excel's `sqrt` function. To calculate $\sqrt{7}$ we type `=sqrt(7)`. This is shown in Figure 3.4.

Figure 3.3: Entering exponents

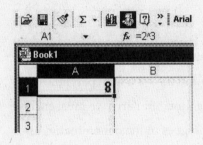

Figure 3.4: Calculating square roots

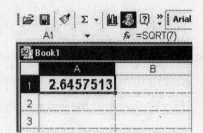

The number π: As with the square root, we get π by using an Excel function. We type `=pi()`. Excel will display a decimal approximation of π as shown in Figure 3.5. Note that the parentheses are essential here. If you leave them out, Excel will interpret your input as a word rather than a number.

The number e: The Excel function `exp(x)` is used to denote the exponential function e^x. Thus to get the number e, we use `=exp(1)`. This is shown in Figure 3.6. For practice, try calculating $3e^{0.08}$. You should get an answer of 3.24986. If you have trouble, remember that you must begin with an equal sign and that you need to insert the symbol `*` between the 3 and `exp` to indicate multiplication.

Figure 3.5: Entering π

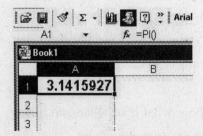

Figure 3.6: The number e

PROLOGUE: PARENTHESES

When you are performing arithmetic operations, two of the most important symbols on the keyboard are the parentheses . When parentheses are indicated in a calculation, you should use them, and sometimes it is necessary to add additional parentheses. To make a calculation such as 2.7(3.6+1.8) we type the expression just as shown, being careful to include the parentheses. (Also remember that the symbol * must be used to indicate multiplication.) This is shown in Figure 3.7. Some calculations such as

$$\frac{2.7 - 3.3}{6.1 + 4.7}$$

do not have parentheses when we write them on paper with a pencil, but in order to type them correctly we must supply parentheses. We must surround $2.7 - 3.3$ with parentheses so that Excel knows that the entire expression goes in the numerator. Similarly, we must enclose $6.1 + 4.7$ in parentheses to indicate that the entire expression goes in the denominator. Thus =(2.7-3.3)/(6.1+4.7) is the correctly typed expression. The result is shown in Figure 3.8. If the parentheses are not supplied as shown, Excel will misinterpret your input and produce the wrong answer.

Figure 3.7: *A calculation using parentheses*	**Figure 3.8:** *A calculation where additional parentheses must be supplied*

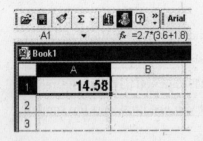

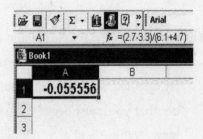

When you are in doubt about whether parentheses are necessary or not, it is good practice to use them. For example, if you calculate $\frac{3 \times 4}{7}$ using =3*4/7, you will get the right answer, 1.71. You will get the same answer if you use parentheses to emphasize that the entire expression 3×4 goes in the numerator: =(3*4)/76. If on the other hand you calculate $\frac{7}{3 \times 4}$, you must use parentheses to get the correct answer of 0.58. The correct entry is =7/(3*4). If you leave out the parentheses and simply type =7/3*4, Excel will think that only the 3 goes in the denominator, and you will get the wrong answer.

The use of parentheses leaves no room for doubt about what goes where, and their correct use is essential to the operation of any calculating device, including Excel.

PROLOGUE: CELL REFERENCE

The feature of Excel known as *cell referencing* is quite simple but adds greatly to the power of Excel. In fact, an understanding of cell referencing is fundamental to the operation of Excel. Notice that the columns of Excel are labeled by letters A, B, C, ... and that the rows are labeled by numbers 1, 2, 3, ... This gives a reference for each cell in the spreadsheet. For example, the reference **B3** refers to the third cell down in column **B**.

We can illustrate one important feature of cell referencing by calculating several values of the function $f(x) = x^2 + 1$. We will show two ways of doing this. Here is the first: Assuming that we start with a clean spreadsheet, get in cell **B1** and type =A1^2+1. This instruction tells Excel that it should calculate the value of $x^2 + 1$ using whatever is in cell **A1** in place of x. Note in Figure 3.9 that Excel interprets the blank in **A1** as 0 and evaluates $x^2 + 1$ at $x = 0$. In Figure 3.10 we have entered 2 into cell **A1**. Excel automatically recalculates the value in cell **B1** evaluating $x^2 + 1$ at $x = 2$. Try some other values in **A1** so that you have a clear understanding of how Excel handles cell referencing.

Figure 3.9: *Entering the formula using direct cell reference*

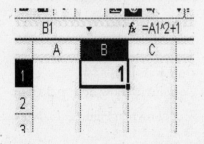

Figure 3.10: *Evaluating the formula at* $x = 2$

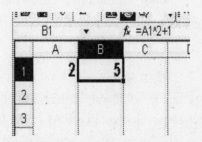

Employing cell referencing by directly typing in the cell column letter and row number has the advantage of conforming directly to Excel's method of operation. There is an alternative method of cell referencing which allows us to type in formulas exactly as they appear using the variable name rather than the direct cell reference. This is accomplished using Excel's ability to assign names to cells or groups of cells. Let's give the **A** column the name x. Highlight the **A** column and click on the name box which is just to the left of the formula box.

Replace A1 by x as shown in Figure 3.11. Just to remind us that column A now has the name x, we type x in cell A1.

Now we get in cell B2 and type =x^2+1. Note that since we have given the name x to column A, we can type the formula using x as the variable rather than A1. Since we are in cell B2, x refers to the value in A2. (If we were in cell B3, then x would refer to the value in A3.) Note that in Figure 3.12 we have typed 2 in cell A2 and the answer, $x^2 + 1$ evaluated at $x = 2$, appears in cell B2.

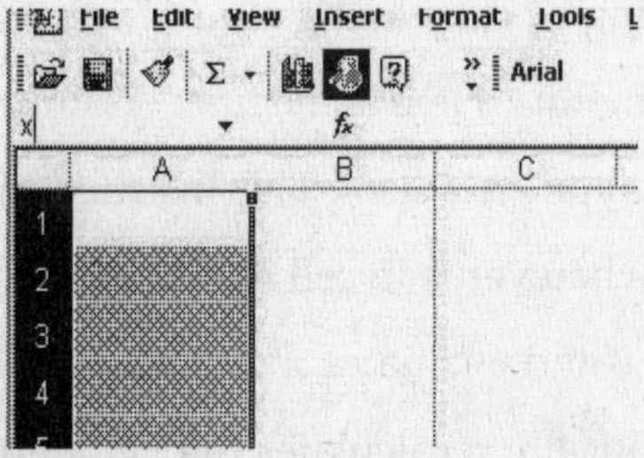

Figure 3.11: *Giving column A the name x*

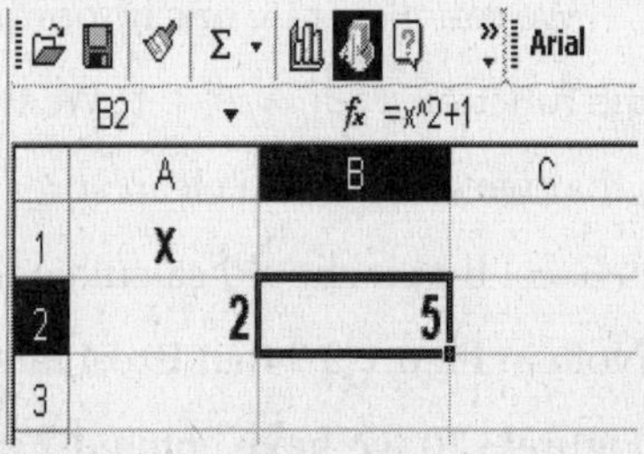

Figure 3.12: *Evaluating x^2+1 at $x = 2$*

PROLOGUE: CHAIN CALCULATIONS

Some calculations are most naturally done in stages. To show how to do this with Excel let's look at

$$\left(\frac{2+\pi}{7}\right)^{\left(3+\frac{7}{9}\right)}.$$

This can of course be done in a single calculation, but we want to show how to do it in two steps. First get in cell A1 and calculate the exponent $3 + \frac{7}{9}$ as shown in Figure 3.13. Now move to cell B1. We now want to calculate

$$\left(\frac{2+\pi}{7}\right)^{x},$$

where x is the value in cell A1. So we type =((2+pi())/7)^A1. Alternatively, if we have given the A column the name x we can type =((2+pi())/7)^x. The final result (showing the alternative approach) is in Figure 3.14.

Figure 3.13: *The first step in a chain calculation*

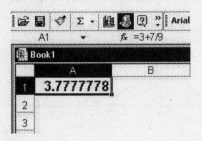

Figure 3.14: *Completing the chain calculation*

Chapter 2: THE FACTORIAL FUNCTION

The *factorial* function occurs most often in probabilistic and statistical applications of mathematics. It applies only to non-negative integers and is denoted by $n!$. It is defined as follows:

$$0! \; = \; 1$$

$$\text{For } n > 0, \; n! \; = \; n(n-1)(n-2)\ldots 3 \times 2 \times 1.$$

Thus for example $3! = 3 \times 2 \times 1 = 6$ and $5! = 5 \times 4 \times 3 \times 2 \times 1 = 120$. Excel can calculate the factorial function. To calculate 6! with Excel, we use Excel's **fact** function. We simply type =fact(6). Excel will display the answer 720. Try some other examples. If you have trouble, remember to begin with an equal sign and to use parentheses as indicated.

Tables and Graphs

Excel can make both tables and graphs, and it has a large collection of tools for further analysis.

CHAPTER 2: TABLES OF VALUES

In order to make a table of values in Excel we first need to make a *sequence*. Suppose for example that we wish to put the sequence of numbers 1, 2, 3, ... in column **A**. We could simply type in the numbers one at a time, but Excel offers a much simpler method. Type **1** in **A1** and **2** in **A2**. Next click on cell **A1**, hold the mouse button down and drag downward to select both **A1** and **A2** as shown in Figure 3.15. Now move the cursor to the bottom right corner of cell **A2**. When you do this, the thick cross will change to a thin cross hair. Hold down the left mouse button and drag downwards. Excel will look at the first two numbers you typed and try to determine which sequence you had in mind. It is pretty good at getting it right, as Figure 3.16 shows in this case.

Figure 3.15: *Two sequence cells selected*

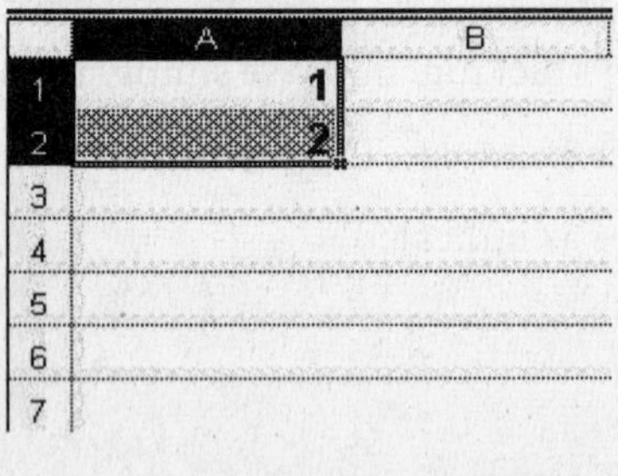

Figure 3.16: *Dragging the cross hair to extend the sequence*

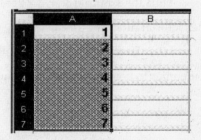

Now let's make a table of values for $f = 3x + 1$. In cell **B1** we type =3*A1+1 as shown in Figure 3.17. (If we had given the name x to column **A** we would use 3*x+1.) Don't forget to begin with the equal sign and use the symbol * for multiplication. Notice that in cell **B1** Excel has evaluated f at $x = 1$. We want to evaluate the entire column. To do this, select **B1** and move to the lower right corner until you get a cross hair. Click, hold, and drag downward to extend the sequence as shown in Figure 3.18.

There are any number of ways in which we may wish to alter the table. If, for example, we decide that we wish to extend the table further, we first use the mouse to select the last two cells of **A** and **B** (four cells in all) as shown in Figure 3.19. Move the mouse to the lower

Figure 3.17: *Entering the formula in* **B1**

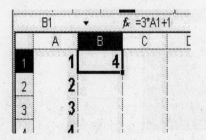

Figure 3.18: *Dragging the cross hair to complete the table*

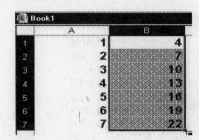

right corner of the selection and drag the cross hair downward. The table extends as you drag, as is shown in Figure 3.20.

Figure 3.19: *Selecting four cells at the bottom*

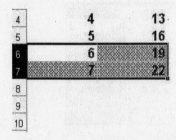

Figure 3.20: *Dragging the cross hair to extend the table*

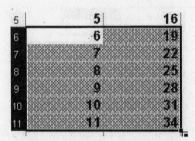

Suppose now that we want to see the table begin at 300 rather than at 1. Type `300` in **A1** and `301` in **A2**. Select **A1** and **A2**, and drag them downward just as we did to make our initial sequence. The values in the **B** column will adjust automatically. The completed table is in Figure 3.21.

It is just as easy to change the increment we use in the sequence. Let's start at 0 and move in steps of 0.2. We just enter **0** into cell **A1** and **0.2** into cell **A2**. Now click and drag to extend the sequence as before. The new table is in Figure 3.22.

Excel easily handles more than one function at a time, and this allows us to use tables to make comparisons. For example, let's show how to compare the values of $f = 3x + 1$ with those of $g = 4x - 2$. We begin with the table of values for $f = 3x + 1$ using x from 1 to 7 in steps of 1. In the **C1** cell, we type `=4*A1-2` as shown in Figure 3.23. (Alternatively, we could type `=4*x-2` if column **A** has been named x.) Now we use the mouse as before to drag downward from the **C1** column. The resulting table of values is in Figure 3.24. Note that f

Figure 3.21: Starting the table at 300

Figure 3.22: Changing the table increment

	A	B
1	300	901
2	301	904
3	302	907
4	303	910
5	304	913
6	305	916
7	306	919

	A	B
1	0	1
2	0.2	1.6
3	0.4	2.2
4	0.6	2.8
5	0.8	3.4
6	1	4
7	1.2	4.6

starts out larger than g. They have the same value when $x = 3$, and, after that g has the larger values.

Figure 3.23: Entering a second function

Figure 3.24: The completed comparison table

	A	B	C
1	1	4	2
2	2	7	
3	3	10	
4	4	13	
5	5	16	
6	6	19	
7	7	22	

	A	B	C
1	1	4	2
2	2	7	6
3	3	10	10
4	4	13	14
5	5	16	18
6	6	19	22
7	7	22	26

CHAPTER 2: GRAPHING

Excel can generate graphs of many types and present them in many ways. We will consider only two types of graphs (traditional function graphing here and discrete graphs in the next chapter). You are invited to explore other types of graphs that Excel can produce. Let's illustrate how to do this using $\dfrac{x^4}{50} - x$. Before beginning, some may wish to name the first column x. The first decision we need to make is what horizontal span to use for the graph. Often functions come from physical situations, and the physical situation might determine the span we want to use.

In this case, let's arbitrarily use the span from -5 to 5. We need to make a table of values running from -5 to 5. Now we have a second decision to make. What step size should we use in our table? A smaller step size gives a more accurate graph, but if we make the step size too small the table may be unnecessarily large. In this case, let's use a step size of 0.25. Thus we enter -5 into **A1** and -4.75 into **A2** (since $-5 + 0.25 = -4.75$). Now select the two cells and click and drag downward until we reach 5. This is partially shown in Figure 3.25. In the **B1** cell, we type `=A1^4/50-A1`. (If you named the first column x, you may type `=x^4/50-x`.) Click and drag downward to complete the table, which is partially shown in Figure 3.26.

Figure 3.25: *A sequence up to 5*

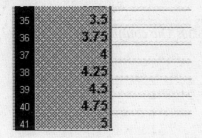

Figure 3.26: *Completing the table*

35	3.5	-0.49875
36	3.75	0.2050781
37	4	1.12
38	4.25	2.2750781
39	4.5	3.70125
40	4.75	5.4313281
41	5	7.5

Some will find it convenient before making the graph to give column B the name y. Next click on the **Chart Wizard** icon. You will be prompted for the type of graph you wish to make. In the left-hand column, select **Line** as shown in Figure 3.27. You will see several possible sub-types to the right. Select the very first, as we did in Figure 3.27. You are now prepared to go on to the **Next** step, where you will encounter the **Chart Source Data** screen. On the **Data Range** tab of that screen, delete whatever is in the **Data range** box. Now click on the worksheet data and select **B1** through **B41**. (If you gave the column B the name y, you may simply replace what is in the **Data range** box with y as shown in Figure 3.28.) Also be sure that **Series in** is set to **Columns**.

Figure 3.27: *Selecting the chart type*

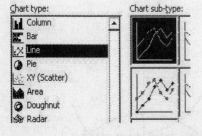

Figure 3.28: *The* **Data range** *tab of the* **Chart Source Data** *screen*

Data range: y

Series in: ○ Rows
             ⦿ Columns

Next click on the **Series** tab. Click in the **Category (X) axis labels** box and then use the mouse to select your sequence, A1 through A41. (This particular box will not accept x as the name of the sequence.) The result should appear as in Figure 3.29. At this point, you may if you wish simply click on Finish. Alternatively you may continue through the Chart Wizard to style your graph as you wish. Excel offers a seemingly unlimited collection of edits and modifications you may choose to make in order to improve the appearance of the graph. You are encouraged to explore these options. Our selection of options produced the graph in Figure 3.30. Making graphs with Excel may on first try seem a bit complicated. That is because Excel offers so many different types and styles of graphs. A bit of practice will make the procedure go quickly and easily.

Figure 3.29: *The* Series *tab of the* Chart Source Data *page*

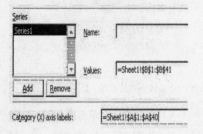

Figure 3.30: *The completed graph*

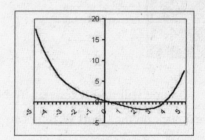

Let's see how to add another graph, say that of $5 - \dfrac{x^2}{3}$, to the picture we already have. You may wish to give column C the name z. Enter the formula in cell C1 as either =5-A1^2/3 or =5-x^2/3, whichever is appropriate. Now click and drag to make a table.

With the chart selected, go to the Chart menu at the top and select Add data. You will be presented with an Add Data page. Click in the Range box, and select C1 through C41. (If you named this column z, you may just type z in the Range box.) The properly configured box is in Figure 3.31. Finally click OK, and the graph shown in Figure 3.32 will appear.

Figure 3.31: *Selecting a new data range*

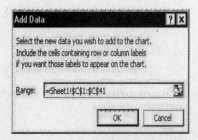

Figure 3.32: *Two graphs displayed*

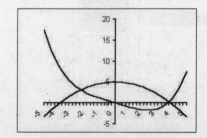

CHAPTER 2: SOLVING EQUATIONS

On TI calculators, graphs are totally interactive in that they allow one to analyze functions on the graphing screen. Excel does not offer these features, but most can be accomplished outside the graphing environment. It is not necessary in Excel to make a graph in order to solve an equation, but it is often helpful, and it is a much recommended practice.

The methods we use for solving equations using Excel apply to equations of the form $f(t) = 0$ and so correspond to the single-graph method in the text *Functions and Change*. (More generally, the methods apply to equations of the form $f(t) = c$ for a constant c.) As the text notes, all of the equations considered can be put in this form.

We will show two methods for solving equations using Excel. The first method uses the **Solver** add-in to solve equations. It is accessed from the **Tools** menu[1]. Let's see how to use **Solver** to solve $t^2 + t - 3 = 0$. We strongly recommend that you first make a graph as we have done in Figure 3.33. The graph shows us that there are two solutions of the equation: one near -2 and another near 1.5.

You may wish to give the cell **D1** the name t. To do this, select **D1**, and change the name in the **Name** box to t just as we did in naming columns. Next we want to enter the formula. Go to cell **E1** and enter =D1^2+D1-3. Alternatively, if you named **D1** t, you can type =t^2+t-3. Assuming there is nothing currently in cell **D1**, Excel will treat the entry as 0, and -3 will appear in cell **E1**, indicating that $t^2 + t - 3$ has the value -3 when $t = 0$. Now with cell **E1** selected, click on **Solver**. Excel will display the screen partially shown in Figure 3.34. Be sure the **Set Target Cell** box reads **E1**. (This will happen automatically if you selected **E1** before calling up **Solver**.) Be sure the **Value of** option is selected and the value in the box is set to 0. Finally put D1 (or t if you named that cell t) in the **By Changing Cells** box. The properly completed entries are in Figure 3.34. The entries can be deciphered to read "make the value of $t^2 + t - 3$ equal 0 by changing the value of t." Now click on **Solve**, and Excel will present the answer 1.302776 in **D1**.

We have found one of the solutions of the equation, but the graph shows us that there is another near -2. We will show two ways to get the other solution. The first is the easiest and is good practice in any case. It is always wise when using **Solver** to give Excel a little help

---

[1] If **Solver** does not appear on your **Tools** menu, then you should select **Add-Ins** and load the appropriate software from your Excel disk.

Figure 3.33: *The graph of* $t^2 + t - 3$

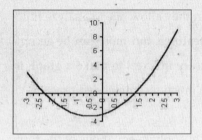

Figure 3.34: *Entering the* Solver parameters

by starting as near the point you want as you can. In this case, the solution we are looking for is near -2. We make Excel start its work there by first entering -2 into **D1**. Now select **E1** and proceed as before. This time Excel shows the second zero, -2.302775.

For most functions, just forcing **Solver** to start reasonably near the point you are looking for will get the job done. In some cases it may be necessary to resort to more serious means which we now present. As before, we select cell **E1** and choose **Solver** from the **Tools** menu. Again we select the **Value of** option and put 0 in the **Value of** box. Type t in the **By Changing Cells** box. We want to constrain the solution by $t \geq -2.5$ and $t \leq -2$. To do this go to the **Subject to the Constraints** section and click on **Add**. You will see the **Add Constraint** page shown in Figure 3.35. Enter t in the **Cell reference** box. There is a pull down menu in between the two boxes which lets us select $>=$. Now enter -2.5 in the **Constraint** box. The completed **Add Constraint** page is in Figure 3.35. Click on **Add** in order to enter the second constraint, $t \leq -2$. The properly configured **Add Constraint** box is in Figure 3.36. Click on **OK** and finally on **Solve**. Excel will present the solution -2.302775, as before, in cell **D1**.

Figure 3.35: *Entering the constraint* $t \geq -2.5$

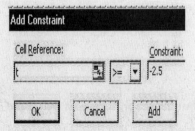

Figure 3.36: *Entering the constraint* $t \leq -2$

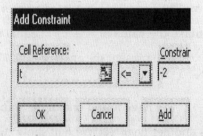

Let's see how to solve the same equation $t^2 + t - 3 = 0$ using the *Goal Seek* feature of Excel. Let's find the solution near $t = 1.5$ first. As in the case of **Solver** we enter the formula we want in cell **E1**. We can either type `=D1^2+D1-3`, or, if we have given **D1** the name t, we can use `=t^2+t-3`. As with **Solver**, it is good practice to give **Goal Seek** a good starting point. Thus we type `1.5` in cell **D1**. Next click on **E1**, and then from the **Tools** menu select **Goal Seek**. Excel will present you with the **Goal Seek** entry screen. Be sure **E1** is in the **Set cell** box. (If you clicked on **E1** before calling **Goal Seek** this will be done automatically.) Type `0` in the **To value** box. Finally in the **By changing cell** box type `t` if you have made the appropriate definition; otherwise, type `D1`. The proper entries are shown in Figure 3.37. If we now click on **OK**, we get the solution in cell **D1** as shown in Figure 3.38. To find the other zero, we type `-2` in cell **D1** before calling up **Goal Seek**. We emphasize that if we use either **Goal Seek** or **Solver** it is important first to graph the function so that we can give Excel a good place to start its work.

Figure 3.37: *The* **Goal Seek** *entry screen*

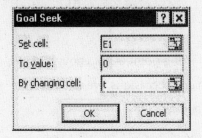

Figure 3.38: *One solution using* **Goal Seek**

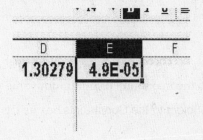

CHAPTER 2: OPTIMIZATION

Maxima and *minima*, or *peaks* and *valleys*, of a graph can be located using Excel in much the same way that we solve equations with **Solver**. To illustrate the method, we will look at the graph of $f = 2^t - t^2$. We have made the graph using a horizontal span of -1 to 4 in Figure 3.39. We see one maximum and one minimum. Let's locate the maximum first. Just as we did when solving an equation, either we first give the cell **D1** the name t and then type `=2^t-t^2` in **E1**, or, if we prefer not to use name definitions, we type `=2^D1-D1^2` in **E1**. The graph shows that the maximum occurs near $t = 0.5$, so let's enter that number into **D1** to give **Solver** a good starting point. Now select **E1** and click on **Solver**. As before the **Solver** constraint page is displayed. Select **Max** (short for Maximum) and enter the appropriate one of `t` or `D1` in the **By Changing Cells** box. The properly completed page is in Figure 3.40. Click on Solve, and in cells D1 and E1 Excel shows us that a maximum value of 1.164361 is reached at $t = 0.48509$.

To find the minimum value, we first note that it occurs near $t = 3$. Thus we type 3 in cell D1 to give Solver a good starting point. Now select E1 and click on Solver. The entries are the same as for the maximum except that we now select Min (short for Minimum). When we finish, Excel shows that a minimum value of -1.050602 is reached at $t = 3.212432$.

Figure 3.39: *The graph of* $2^t - t^2$

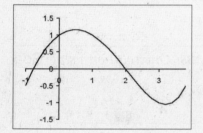

Figure 3.40: *The properly completed* Solver parameter *page*

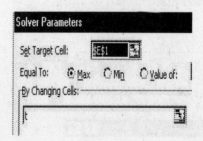

There may be cases when just giving Excel a good starting point is not enough to get to the maximum or minimum value we are looking for. On such occasions, we must use the Subject to the Constraints box just as we did in solving equations.

CHAPTER 5: PIECEWISE-DEFINED FUNCTIONS

Sometimes functions are defined by different formulas over different intervals. Such functions are known as *piecewise-defined* functions. For example, we define

$$f(x) = \begin{cases} x + 1, & \text{if } x < 1 \\ x - 1, & \text{if } x \geq 1. \end{cases}$$

This means that $f(x)$ is $x + 1$ if x is less than 1 but it is $x - 1$ when x is greater than or equal to 1. Thus $f(0) = 0 + 1 = 1$ since $0 < 1$. But $f(3) = 3 - 1 = 2$ since $3 \geq 1$. Excel will graph piecewise-defined functions correctly if we arrange the data right. We start by putting a sequence in column A as usual. Let's go from -1 to 3 in steps of 0.25. Since $-1 < 1$, the function formula there is $x - 1$. Thus we type =A1-1 in B1. This definition only applies for values less than 1. Thus we click on B1 and drag downward only to B8 (corresponding to $x = 0.75$). Leave the remainder of column B blank. Now get in cell C9 corresponding to

$x = 1$. From here on, the function is $x + 1$, and so we type =A9+1. We expand this down to A17, leaving the top half of column C blank. The correctly entered data is partially displayed in Figure 3.41. Now we click on the **Chart Wizard** icon, select **Line** as the type of graph and choose the first sub-type to the right, and go on to the **Next** step, where we encounter the **Chart Source Data** screen. On the **Data Range** tab of that screen, we delete whatever is in the **Data range** box, click on the worksheet data, and select the data from both columns **B** and **C**. That is, we include the complete rectangle with corners at **B1, B17, C17,** and **C1,** including the blank cells in the rectangle. For the **Category (X) axis labels** we choose **A1** through **A17**. Now complete the graph using appropriate options. Our graph is in Figure 3.42. Note that as might be expected, the graph shows a jump at $x = 1$.

Figure 3.41: *Entering data for a piecewise-defined function*

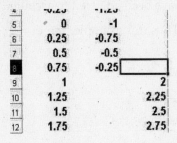

Figure 3.42: *The completed graph*

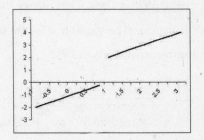

Discrete Data

Many times information about physical or social phenomena is obtained by gathering individual bits of data and then performing various kinds of analysis on the data. We will show several of Excel's abilities to handle data.

CHAPTER 3: PLOTTING DATA

We will show how to make data plots using the following table. It is taken from the *1996 Information Please Almanac*, and it shows median American family income I by year in terms of 1996 dollars. That means the dollar amounts shown have been adjusted to account for inflation. The variable d in the table is years since 1980. Thus, for example, the $d = 2$ column corresponds to 1982.

d = years since 1980	0	1	2	3	4	5
I = median income	21,023	22,388	23,433	24,580	26,433	27,735

We enter the d values in column A and the I values in column B. To plot the data, call up the **Chart Wizard**. Select **XY (Scatter)** as the chart type, and select the first chart sub type (the one without lines) as shown in Figure 3.43. Then go on to the **Next** step. Click in the **Data range** box and select the data in column B. (If you have given column B the name y, you may simply enter y in the **Data range** box.) Select the **Series** tab, click in the **X values** box, and select **A1** through **A6**. Complete the graph as before. Our graph is in Figure 3.44. This method works to plot any data set.

Figure 3.43: *Preparing for a data plot*

Figure 3.44: *The completed graph*

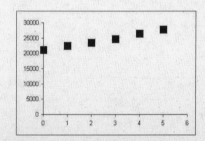

CHAPTER 3: LINEAR REGRESSION

Linear regression is a method of getting a linear function which approximates data that is almost linear. Excel will perform this and other types of regression automatically. We will use the plot from the previous section to show the method. With the chart selected, go to the **Chart** menu at the top and click on **Add Trendline**. (Alternatively you can left-click on the data points, then right-click, and you will get a menu containing **Add Trendline**.) On the **Type** tab, choose **Linear** as seen in Figure 3.45. On the **Options** tab select **Display equation on chart**. Now click **OK**, and the regression line and its equation $y = 1338.3x + 20919$ will be added to the plot as shown in Figure 3.46.

Figure 3.45: Plot of the data

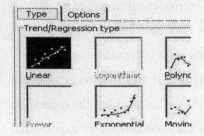

Figure 3.46: Regression line and equation added

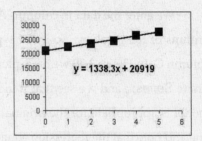

It is often convenient to copy and paste the regression formula into the spreadsheet so that you can do calculations with it. Highlight $= 1338.3x + 20919$ (leaving off the y) and **Copy**. Now click in **B7** and **Paste**. Note that you will need to edit the formula before it will work. You will need to insert the symbol * between 1338.3 and x to indicate multiplication, and, if you did not give column A the name x, you will need to replace x by A7. The properly edited formula is in Figure 3.47. In Figure 3.48 we have entered 2.7, and the regression equation evaluated there is automatically calculated in B7.

Figure 3.47: Pasting the regression equation into the spreadsheet

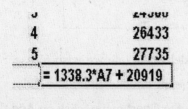

Figure 3.48: Regression equation evaluated at 2.7

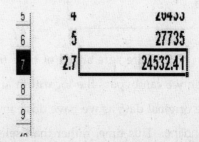

CHAPTER 3: EXPONENTIAL REGRESSION

We can use Excel to get an exponential model directly. Alternatively, we can use the logarithm to transform the data into linear data. This is often useful in determining if an exponential model is the appropriate one to use. Thus if data such as

input values	a	b	...
function values	z	w	...

is exponential data, then

input values	a	b	...
function values	$\ln z$	$\ln w$	...

is linear data.

We will illustrate with the following data table.

x	1	2	3	4	5
y	0.55	0.83	1.5	2.7	4.4

We enter the data in columns A and B. We want the entries of column C to be the logarithms of the y values. So in C1 type =ln(B1) and then click and drag to expand down column C. In Figure 3.49 we have plotted the logarithm of the data (after selecting the appropriate Series) , and we see that it indeed appears linear. This indicates that an exponential model is appropriate for the original data set. In Figure 3.50 we have added the regression line and note that the regression equation is $y = 0.5338x - 1.1823$. We can get the exponential model directly from this as follows:

$$e^{-1.1823}\left(e^{0.5338}\right)^x = 0.31 \times 1.71^x.$$

Figure 3.49: *Plot of the logarithm of the data*

Figure 3.50: *Regression line and equation added*

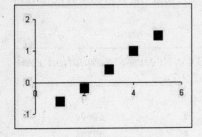

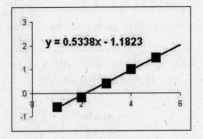

If we are sure ahead of time that the exponential model is the one we wish to use, then we can bypass the logarithm and do exponential regression directly as follows. Plot the original data as we have done in Figure 3.51. Now from the Chart menu, click on Add trendline. This time, rather than select Linear, we select Exponential. As before go to the

Options tab and select Display equation of chart. We see from Figure 3.52 that Excel displays the regression equation as

$$0.3066e^{0.5338x}.$$

This is the same (up to rounding) as the answer we got, because

$$e^{0.5338x} = \left(e^{0.5338}\right)^x = 1.71^x.$$

As with linear regression, you may wish to copy and paste the regression equation in to the spread sheet. Remember to edit the equation using the symbol * for multiplication and if necessary correcting the cell reference.

Figure 3.51: *Plot of the original data*   **Figure 3.52:** *Exponential model added*

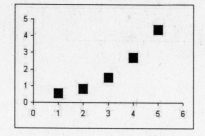
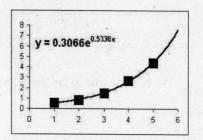

CHAPTER 5: POWER AND QUADRATIC REGRESSION

Power models can be constructed in much the same way as exponential models. We first show how to transform power data into linear data. The key difference is that we apply the logarithm to both input values and function values. If data such as

input values	a	b	$\ldots$
function values	z	w	$\ldots$

is power data, then

input values	$\ln a$	$\ln b$	$\ldots$
function values	$\ln z$	$\ln w$	$\ldots$

is linear data.

As with exponential data, this is how we test to see if a power model is appropriate. We illustrate using the following data table.

x	1	2	3	4	5
y	0.22	0.54	0.88	1.25	1.91

As before, we put the x data in column A and the y data in B. In column C get the logarithm of the x values (=ln(A1)) and in column D get the logarithm of the y values (=ln(B1)). Now we plot using column D as the **Data range** and column C as **X values**. The result is in Figure 3.53. The data does appear to be almost linear, and so it is reasonable to model the original data with a power function. In Figure 3.54, we have added the regression line, and we see that the regression equation is $y = 1.3075x - 1.5295$.

Figure 3.53: *Plot of the transformed data*

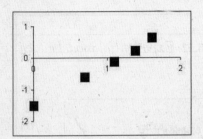

Figure 3.54: *Regression line added*

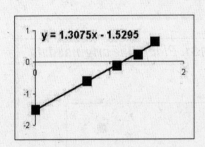

We can use this to get the power model as follows:

$$e^{-1.5295}x^{1.3075} = 0.22x^{1.31}.$$

As with exponential regression, we can bypass the transformation of the data and go directly to a power model. In Figure 3.55 we have plotted the original data. Next we select **Add trendline**, but this time we select **Power**. The result is in Figure 3.56.

Figure 3.55: *Plot of original data*

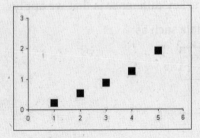

Figure 3.56: *Power model added*

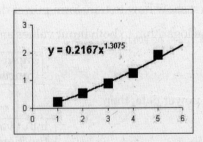

Quadratic (or higher-degree polynomial) regression is just as easy as other types of regression with Excel. Suppose that we want to fit the following data with a quadratic model.

x	-2	0	1	3	6
y	3.3	1.7	2.5	10.4	33.2

As usual, we enter the x data in column A and the y data in column B. The plot of the data is in Figure 3.57. Next choose **Add trendline**. This time we choose **Polynomial** regression and set the **order** to 2. Figure 3.58 shows that the quadratic model (rounded to two decimal places) for this data is $0.78x^2 + 0.61x + 1.43$.

Figure 3.57: Data plot

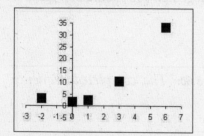

Figure 3.58: Quadratic model added

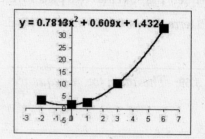

CHAPTER 5: SEQUENCE GRAPHS

Sometimes we are asked to graph data in a sequence where the next term is calculated from the one before. Often this is given as a starting point and an iterative formula. Let's look for example at the sequence we get starting at 1 and using the iterative formula $3x - 1$. Here is how we make the sequence.

The first term is 1.

Apply the formula to 1 to get the second term $3 \times 1 - 1 = 2$.

Apply the formula to 2 to get the third term $3 \times 2 - 1 = 5$.

Apply the formula to 5 to get the fourth term $3 \times 5 - 1 = 14$.

The process continues by applying the formula to the last term calculated to get the next. You should verify that the next term in the sequence is 41. We want to see how to get Excel to calculate the terms of such a sequence automatically and plot the points $(1, 1)$, $(2, 2)$, $(3, 5)$, $(4, 14)$, ...

We used the sequence above because the first few terms are easy to calculate, making it a simple illustration of what we are trying to do. But it is clear that, after a few more terms, the sequence will get quite large. For an example to illustrate the use of Excel we will look at a more interesting sequence. Let's start at 0 and use the formula $x^2 - 0.7$. It is not at all clear from looking at the formula how the sequence behaves. We will use Excel to see. First enter the sequence $1, 2, 3, \ldots 30$ in column A. In B1 enter the starting value, 0. In cell B2 enter the formula by typing =B1^2-0.7. Now click in B2 to get a cross hair and extend the sequence downward through B30. The properly entered data is partially displayed in Figure 3.59. We plot the data using a Scatter Plot as usual. The result is in Figure 3.60. It appears that the sequence is split into two pieces which seem to be coming together. You should plot the first 75 terms of the sequence to see if this trend continues.

Figure 3.59: *The data for a sequence plot*

	A	B
1	1	0
2	2	-0.7
3	3	-0.21
4	4	-0.6559
5	5	-0.26979519
6	6	-0.627210555
7	7	-0.306606919
8	8	-0.605992197

Figure 3.60: *The completed sequence plot*

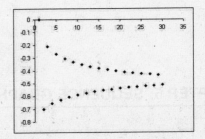

Index